Sprinkler Fitting

Level One

Trainee Guide
Third Edition

Boston Columbus Indianapolis New York San Francisco Upper Saddle River
Amsterdam Cape Town Dubai London Madrid Milan Munich Paris Montreal Toronto
Delhi Mexico City São Paulo Sydney Hong Kong Seoul Singapore Taipei Tokyo

NCCER

President: Don Whyte
Director of Product Development: Daniele Stacey
Sprinkler Fitting Project Manager: Patty Bird
Production Manager: Tim Davis
Quality Assurance Coordinator: Debie Ness
Editor: Chris Wilson
Desktop Publishing Coordinator: James McKay
Production Assistant: Megan Casey

NCCER would like to acknowledge the contract service providers for this curriculum:
Topaz Publications, Syracuse, New York and Paul Lagasse, Active Voice Writing & Editorial Services

ISBN 13: 978-0-13-380297-9
ISBN 10: 0-13-380297-3

Preface

To the Trainee

There are very few trades which are dedicated to the preservation of life and property. Sprinkler fitting is one of these trades, and carries with it a tremendous amount of responsibility. The work that you do as a sprinkler fitter may one day save lives and thousands of dollars in property. Therefore it is essential that you demand excellence and precision from yourself, and from your co-workers.

The sprinkler fitting industry began in England in 1809, when the first sprinkler system was patented by John Carey. His system used perforated pipes to discharge water on the coverage area. This method of fire stopping was costly, as water damaged areas that were not affected by fire. The sprinkler head was developed as a result. In America, sprinkler fitting began with the invention of the first practical automatic fire sprinkler by Henry Parmelee in 1874. The method of fire stopping by using automatic sprinklers soon gained popularity, and by 1896, over 25,000 fires had been controlled or extinguished by automatic sprinkler systems.

As the new industry grew, regional disparities in workmanship and materials made it necessary to establish industry best practices—a code for the trade. This led to the creation of the National Fire Protection Association (NFPA) in 1896. Today, the NFPA publishes numerous codes for the industry. Fire sprinkler codes and equipment have evolved over time. As you begin your training, pay close attention to codes and how they affect material choices.

Sprinkler fitters may install new systems, retrofit old buildings with updated sprinkler systems, or repair existing systems. Accordingly, the demand for sprinkler fitters grows with new construction and code changes. In addition, the U.S. Department of Labor's Bureau of Labor Statistics estimates that many of the workers currently employed in the industry will retire within the next ten years. All of this results in higher demand for trained workers, and higher demand generates higher wages. Stay on top of industry developments, code changes, and emerging technologies, and your earning potential will grow.

NCCER wishes you success as you embark on your first year of training in the sprinkler fitting trade, and we hope that you'll continue your training outside of this series. By taking advantage of training opportunities as they arise, you'll demonstrate initiative and a desire to learn—qualities that are present in the industry's best professionals.

We invite you to visit the NCCER website at www.nccer.org for information on the latest product releases and training, as well as online versions of the *Cornerstone* magazine and Pearson's product catalog.

Your feedback is welcome. You may email your comments to **curriculum@nccer.org** or send general comments and inquiries to **info@nccer.org**.

NCCER Standardized Curricula

NCCER is a not-for-profit 501(c)(3) education foundation established in 1996 by the world's largest and most progressive construction companies and national construction associations. It was founded to address the severe workforce shortage facing the industry and to develop a standardized training process and curricula. Today, NCCER is supported by hundreds of leading construction and maintenance companies, manufacturers, and national associations. The NCCER Standardized Curricula was developed by NCCER in partnership with Pearson, the world's largest educational publisher.

Some features of the NCCER Standardized Curricula are as follows:

- An industry-proven record of success
- Curricula developed by the industry for the industry
- National standardization providing portability of learned job skills and educational credits
- Compliance with the Office of Apprenticeship requirements for related classroom training *(CFR 29:29)*
- Well-illustrated, up-to-date, and practical information

NCCER also maintains a National Registry that provides transcripts, certificates, and wallet cards to individuals who have successfully completed a level of training within a craft in NCCER's Curricula. *Training programs must be delivered by an NCCER Accredited Training Sponsor in order to receive these credentials.*

NCCER Standardized Curricula

NCCER's training programs comprise more than 80 construction, maintenance, pipeline, and utility areas and include skills assessments, safety training, and management education.

Boilermaking
Cabinetmaking
Carpentry
Concrete Finishing
Construction Craft Laborer
Construction Technology
Core Curriculum:
  Introductory Craft Skills
Drywall
Electrical
Electronic Systems Technician
Heating, Ventilating, and
  Air Conditioning
Heavy Equipment Operations
Highway/Heavy Construction
Hydroblasting
Industrial Coating and Lining
  Application Specialist
Industrial Maintenance Electrical
  and Instrumentation Technician
Industrial Maintenance
  Mechanic
Instrumentation
Insulating
Ironworking
Masonry
Millwright
Mobile Crane Operations
Painting
Painting, Industrial
Pipefitting
Pipelayer
Plumbing
Reinforcing Ironwork
Rigging
Scaffolding
Sheet Metal
Signal Person
Site Layout
Sprinkler Fitting
Tower Crane Operator
Welding

Maritime

Maritime Industry Fundamentals
Maritime Pipefitting

Green/Sustainable Construction

Building Auditor
Fundamentals of Weatherization
Introduction to Weatherization
Sustainable Construction
  Supervisor
Weatherization Crew Chief
Weatherization Technician
Your Role in the Green
  Environment

Energy

Alternative Energy
Introduction to the Power Industry
Introduction to Solar Photovoltaics
Introduction to Wind Energy
Power Industry Fundamentals
Power Generation Maintenance
  Electrician
Power Generation I&C
  Maintenance Technician
Power Generation Maintenance
  Mechanic
Power Line Worker
Power Line Worker: Distribution
Power Line Worker: Substation
Power Line Worker: Transmission
Solar Photovoltaic Systems Installer
Wind Turbine Maintenance
  Technician

Pipeline

Control Center Operations, Liquid
Corrosion Control
Electrical and Instrumentation
Field Operations, Liquid
Field Operations, Gas
Maintenance
Mechanical

Safety

Field Safety
Safety Orientation
Safety Technology

Supplemental Titles

Applied Construction Math
Careers in Construction
Tools for Success

Management

Fundamentals of Crew Leadership
Project Management
Project Supervision

Spanish Titles

Acabado de concreto: nivel uno,
  nivel dos
Aislamiento: nivel uno, nivel dos
Albañilería: nivel uno
Andamios
Aparejamiento básico
Aparajamiento intermedio
Aparajamiento avanzado
Carpintería:
  Introducción a la carpintería,
  nivel uno; Formas para
  carpintería, nivel tres
Currículo básico: habilidades
  introductorias del oficio
Electricidad: nivel uno, nivel dos,
  nivel tres, nivel cuatro
Encargado de señales
Especialista en aplicación de
  revestimientos industriales: nivel
  uno, nivel dos
Herrería: nivel uno, nivel dos, nivel
  tres
Herrería) de refuerzo: nivel uno
Instalación de rociadores: nivel uno
Instalación de tuberías: nivel uno,
  nivel dos, nivel tres, nivel cuatro
Instrumentación: nivel uno, nivel
  dos, nivel tres, nivel cuatro
Orientación de seguridad
Mecánico industrial: nivel uno,
  nivel dos, nivel tres, nivel cuatro,
  nivel cinco
Paneles de yeso: nivel uno
Seguridad de campo
Soldadura: nivel uno, nivel dos,
  nivel tres

Portuguese Titles

Currículo essencial: Habilidades
  básicas para o trabalho
Instalação de encanamento
  industrial: nível um, nível dois,
  nível três, nível quatro

Acknowledgments

This curriculum was revised as a result of the farsightedness and leadership of the following sponsors:

Bamford Fire Sprinkler Company
Brenneco Fire Protection, Inc.
Cen-Cal Fire Systems
Fire and Life Safety America
Parsley Consulting
SimplexGrinnell
Southeast Fire Protection, Inc.
Strickland Fire Protection, Inc.
The Viking Corporation

This curriculum would not exist were it not for the dedication and unselfish energy of those volunteers who served on the Authoring Team. A sincere thanks is extended to the following:

Chris Kachura
John Denhardt
Eric Flora
Marc Haug
Joe Heinrich
Jack Medovich
Scott Martorano
Ken Wagoner
Bryon Weisz

NCCER Partners

American Fire Sprinkler Association
Associated Builders and Contractors, Inc.
Associated General Contractors of America
Association for Career and Technical Education
Association for Skilled and Technical Sciences
Carolinas AGC, Inc.
Carolinas Electrical Contractors Association
Center for the Improvement of Construction Management and Processes
Construction Industry Institute
Construction Users Roundtable
Construction Workforce Development Center
Design Build Institute of America
GSSC – Gulf States Shipbuilders Consortium
Manufacturing Institute
Mason Contractors Association of America
Merit Contractors Association of Canada
NACE International
National Association of Minority Contractors
National Association of Women in Construction
National Insulation Association
National Ready Mixed Concrete Association
National Technical Honor Society
National Utility Contractors Association
NAWIC Education Foundation
North American Technician Excellence
Painting & Decorating Contractors of America
Portland Cement Association
SkillsUSA®
Steel Erectors Association of America
U.S. Army Corps of Engineers
University of Florida, M. E. Rinker School of Building Construction
Women Construction Owners & Executives, USA

Contents

18101-13
Orientation to the Trade 1.i

Identifies sprinkler fitter career opportunities and looks at some typical work environments. Examines trade-specific safety hazards and identifies shop plans specific to the sprinkler fitting industry. Introduces workplace safety, material handling, and common tools. Illustrates the correct use of common tools. (5 Hours)

18102-13
Introduction to Components and Systems 2.i

Introduces testing laboratories and listing agencies. Provides an overview of the major types of sprinkler systems including wet pipe, dry pipe, preaction, and deluge systems. Defines sprinkler-head types, orifice size, and K-Factor. Underground and above-ground pipe and tubes are discussed, including hangers, bracing, and restraints. Summarizes valves, alarms, and fire department connections. (7.5 Hours)

18103-13
Steel Pipe...................... 3.i

Identifies steel piping materials along with tools used to cut and thread steel pipe. Describes methods for threading, cutting, and grooving pipe, including how to determine pipe length between fittings (takeouts). Discusses threaded, plain-end, and flanged fittings. Covers grooved pipe and fittings including installation techniques. (22.5 Hours)

18104-13
CPVC Pipe and Fittings.......................... 4.i

Describes handling and storing of CPVC pipe. Identifies CPVC safety concerns and cautions. Outlines methods and tools for cutting, chamfering, and cleaning CPVC pipe, including calculating takeouts. Joining techniques are described, particularly the solvent-cement (one-step) method. Rules for using plastic pipe hangers are explained. (10 Hours)

18105-13
Copper Tube Systems .. 5.i

Copper tubing and fittings are introduced along with cutting and bending tools. The soldering process is described along with techniques for measuring, cutting, reaming, and cleaning. Brazing is described as are brazing metals, fluxes, and brazing equipment. Support bracing for copper tube is discussed as are grooved couplings for copper pipe. (10 Hours)

18106-13
Underground Pipe......... 6.i

The types and properties of soil are identified, including sloping requirements. Guidelines for working in or near a trench are discussed. Methods are presented for digging trenches and for making them safe. Bedding and backfilling are described. Underground piping installations are detailed for various types of pipe. Thrust blocks and restraints are explained. In-building risers, hydrants, yard valves, and hydrant houses are discussed as are testing, inspection, flushing, and chlorinating. The underground test certificate is covered. (17.5 Hours)

Glossary G.1

Orientation to the Trade

18101-13

SPRINKLER FITTING LEVEL ONE

18106-13 Underground Pipe

18105-13 Copper Tube Systems

18104-13 CPVC Pipe and Fittings

18103-13 Steel Pipe

18102-13 Introduction to Components and Systems

18101-13 Orientation to the Trade

Core Curriculum: Introductory Craft Skills

This course map shows all of the modules in *Sprinkler Fitting Level One.* The suggested training order begins at the bottom and proceeds up. Skill levels increase as you advance on the course map. The local Training Program Sponsor may adjust the training order.

Objectives

When you have completed this module, you will be able to do the following:

1. Identify specific codes and standards that apply to the fire sprinkler industry.
2. Define the typical work environment of a sprinkler fitter.
3. Identify career opportunities in the fire sprinkler industry.
4. Describe the personal responsibilities of sprinkler fitters.
5. Recognize safety hazards that you may come across as a sprinkler fitter.
6. Describe procedures to best handle and store trade materials.
7. Recognize drawings typically seen by sprinkler fitters in the field.
8. Identify basic tools, materials, and fire sprinkler systems used in the sprinkler fitter trade.

Trade Terms

American Fire Sprinkler Association (AFSA)
Authority Having Jurisdiction (AHJ)
Block
Compressed gas
Deluge sprinkler system
Dry pipe sprinkler system
Flammable
Hazard
National Fire Protection Association (NFPA)
NCCER Standardized Curricula
NFPA 13, The Standard for the Installation of Sprinkler Systems
Office of Apprenticeship (OA)
On-the-job-learning (OJL)
Preaction sprinkler system
Rack
Regulations
Sprinkler
SprinklerNet
Wet pipe sprinkler system

Prerequisites

Before you begin this module, it is recommended that you successfully complete *Core Curriculum.*

NOTE

In some locations, sprinkler fitter trainees will be required to perform metric conversions. A metric conversion chart is provided in *Appendix A* at the back of this module.

Contents

Topics to be presented in this module include:

Contents (continued)

Figures and Tables

1.0.0 INTRODUCTION

Fire sprinklers are one of the most important methods for protecting life and property. They provide protection 24 hours a day, seven days a week, even when building occupants are asleep or absent from the premises.

History shows us that sprinklers are very effective. They have a success rate worldwide of over 99 percent, a major factor in the insurance industry's insistence on installation of sprinklers in high risk areas. But sprinkler systems provide adequate protection only if designed and installed properly. Therefore, sprinkler system installation is the primary focus of this training program.

Sprinkler systems are designed and installed based on specific codes and standards from organizations like the National Fire Protection Association (NFPA), with worker safety considerations governed by the Occupational Safety and Health Administration (OSHA).

This module provides a brief look at the beginnings of the industry, along with an introduction to the basic sprinkler system types. A brief look at codes and standards is presented, as well as trade opportunities and your responsibilities as an employee. Finally, this module provides an introduction to working relationships and employer-employee safety obligations.

1.1.0 The Beginnings of the Automatic Fire Sprinkler Industry

The first sprinkler system was invented in 1806 by an Englishman, John Carey. He came up with the idea for a heat-operated device for putting out fires that delivered water through perforated pipes. The first sprinkler was invented by Major Stewart Harrison of London, England. In 1864, Harrison also invented an automatic fire extinguishing system, but could not interest British businesses in buying it. He did not take out a patent. Perforated pipe systems proved unsuccessful in the long run due to excessive water damage, exhaustion of water supplies, and most of all due to the fact that they were manual rather than automatic.

The first practical application of sprinklers is credited to Americans Henry Parmelee and Frederick Grinnell. The modern sprinkler industry came about due to the pioneering automatic sprinkler patented by Henry Parmelee in August 1874. A second design was introduced that was the initial step toward commercialization. Three years later, Grinnell and Parmelee agreed to manufacture the Parmelee sprinklers on a royalty basis. After a great deal of research, in October 1881 Grinnell received four patents on the pendent sprinkler, known as the first sensitive sprinkler, which turned out to be the first standard-orifice sprinkler. Sprinklers underwent continuous development; up to January 1896 about 25,000 fires had been controlled and/or extinguished by Grinnell sprinklers.

So, from the earliest days of perforated pipe to today's complex and specialized components, fire sprinkler systems have undergone continuous change and improvement. Automatic fire sprinkler systems not only protect property, but have been proven to save lives as well.

2.0.0 FIRE SPRINKLER SYSTEMS

The principles of sprinkler operation are relatively simple. Pipes are run, generally overhead, from a water source to the areas needing protection. Sprinklers are fitted to the pipes at appropriate intervals. Thermal linkage in the sprinkler prevents water release by holding a seal closed until the sprinkler's activation temperature is reached, which is typically between 135°F – 286°F (57°C – 141°C). When the thermal linkage senses its activation temperature, it deforms, releases the seal, and sprays water over its area of coverage. Only the sprinkler or sprinklers that sense heat will open. All others remain closed, thereby minimizing the water damage to areas where there is no fire and thus limiting the amount of water needed.

Some design considerations are sprinkler spacing, anticipated water flow, the size of the building, the goods stored therein, and the type of space in which the sprinklers are located (enclosed roof spaces, floor space, etc.).

Systems typically have a control valve placed at the point where the water enters the sprinkler system. This valve is normally open and used for emergency shutoff and for system maintenance. When a sprinkler opens, water flows through the valve, routing water into a second pipe. Water flows through the control valve and past the alarm device to activate an alarm bell or horn.

Did You Know?

Beginning in 1895 complaints of improperly installed sprinklers and inconsistencies in the insurance rating industry became a concern to the fire insurance industry as well as sprinkler manufacturers and installers. These concerns led to the creation in 1896 of the NFPA by five insurance companies and Frederick Grinnell.

2.1.0 Overview of Sprinkler System Types

There are three fundamental types of sprinkler systems: wet pipe sprinkler systems, dry pipe sprinkler systems, and preaction sprinkler systems. A fourth system, the deluge sprinkler system, is a variation of the preaction type. Each of these systems is designed for specific applications, situations, and conditions.

2.2.0 Wet Pipe Sprinkler Systems

Wet pipe sprinkler systems, defined in the Automatic Sprinkler Systems Handbook and the National Fire Protection Association Standard *NFPA 13, The Standard for the Installation of Sprinkler Systems*, are the simplest and most reliable sprinkler systems *(Figure 1)*. Operation of one sprinkler activates the wet pipe system. Note that only those sprinklers activated by fire will open. All others remain closed.

In wet pipe systems, the pipes are always filled with water. When heat activates the thermal linkage in a sprinkler, water is discharged onto the fire. Wet pipe systems are simple, reliable, have fewer components, experience fewer problems, are the most common, and are preferred in most instances because they put water onto the fire quickly.

Wet pipe system installation and maintenance costs are lower because less installation and maintenance time is required for these dependable systems. Wet pipe systems can be modified easily; no additional work is required for the control and detection circuits required by other types of systems. There is less downtime following a

Did You Know?

- Fires are usually controlled or contained with only one or two sprinklers.
- In most systems, one head at a time goes off.
- Most fire sprinkler systems are designed to control or contain a fire, not necessarily to fully extinguish the fire.

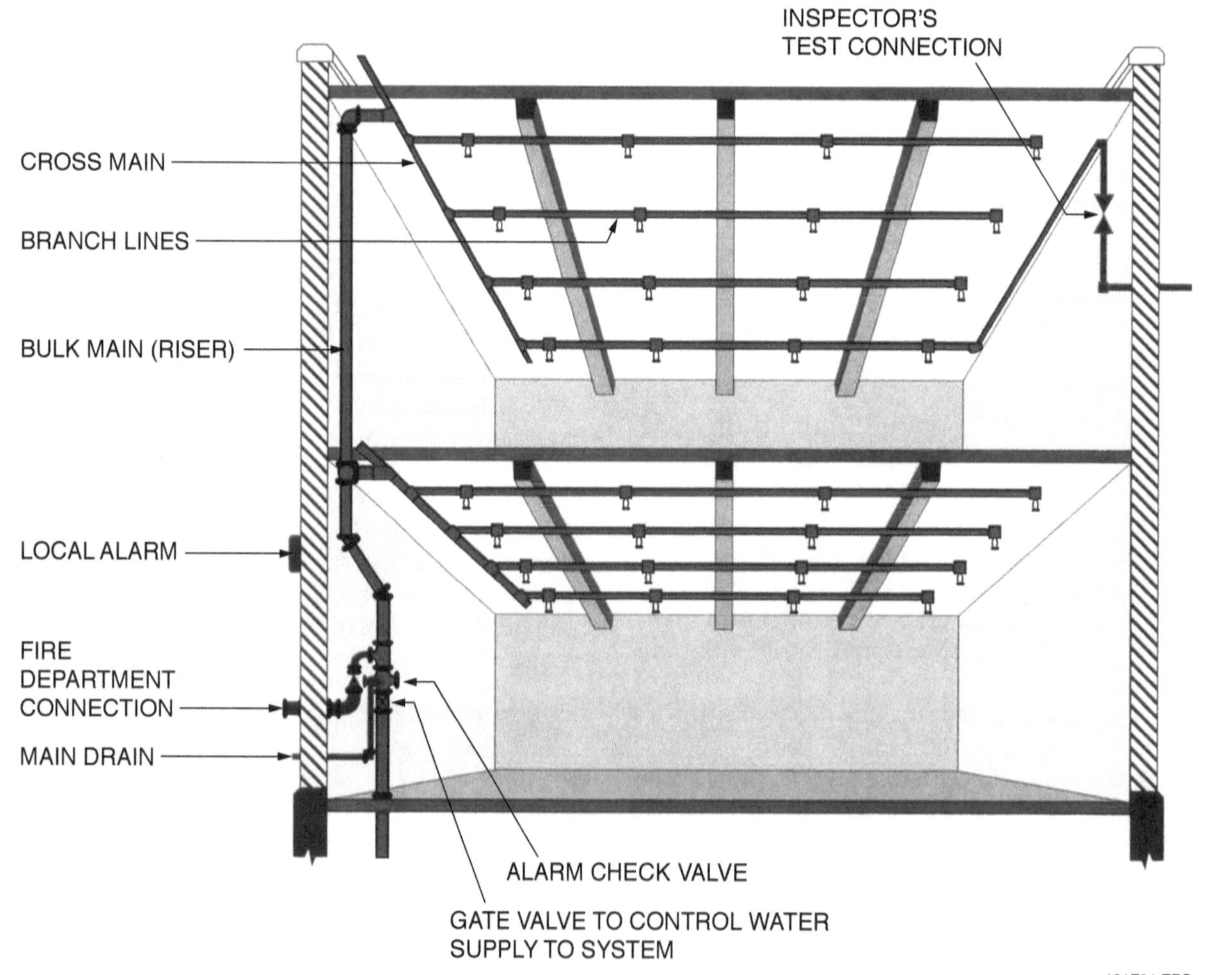

Figure 1 Typical wet pipe sprinkler system.

fire because usually all that is required is to replace the fused sprinklers and turn on the water. For all of these reasons, considering wet pipe systems first is preferable.

However, a wet pipe system is not a good choice for any location likely to freeze. Freezing can damage the piping and can render the system inoperable. A dry pipe or antifreeze system is a better choice for protection of areas with freezing temperatures.

Dry pipe and preaction systems use additional and more complex components than wet pipe systems. More components increase the need for inspecting, testing, and maintenance and increase the possibility that problems can occur. As a result, higher cost and more labor are incurred to keep the system working properly.

2.3.0 Dry Pipe Sprinkler Systems

Dry pipe sprinkler systems are filled with pressurized air or nitrogen, not water. The system has a dry pipe valve that is held closed by air pressure. When a fire fuses the sprinkler, the sprinkler releases the air pressure and the dry pipe valve opens. Water enters through the dry pipe valve and sprays the fire via the activated (open) sprinklers. The only significant advantage of dry pipe systems is fire protection in areas that might freeze.

Dry pipe sprinkler systems have several disadvantages compared to wet pipe systems: additional complexity, more costly installation and maintenance, a design resistant to change, greater chance of corrosion, and slower reaction time. These disadvantages are due to having more control and air pressure systems; limited system size; up to a 60-second delay before water is released on the fire; and the requirement that the system be completely drained after discharge.

2.4.0 Preaction Systems

A preaction system uses automatic sprinklers attached to a piping system. The piping system contains air and may or may not be under pressure. There is a supplemental detection system installed near the sprinklers.

2.5.0 Deluge Systems

A deluge system is a type of sprinkler system using open sprinklers rather than closed sprinklers. When a fire triggers the detection system, the deluge valve is released, producing immediate water flow through all sprinklers in a given area. Deluge systems are typically found in places like aircraft hangers and chemical plants where high volume and fast reaction are required.

2.6.0 Installation Drawings

Construction drawings are a very important part of the installation process. A sprinkler apprentice must understand and be able to read piping plans, as well as understand where to find information on the drawings.

A sprinkler installation begins with good drawings that have been prepared by a competent sprinkler layout technician or professional engineer. The drawings are also used for permit purposes and to provide information for fire departments. Following the installation drawings with little or no variation is very important. Proposed deviations must always be checked with the designer because such a change can affect the successful operation of the system.

3.0.0 Codes and Standards

Both codes and standards serve to protect life, health, and property and to ensure the quality of construction. A code specifies what has to be done. A standard tells how it should be done. Codes and standards have no legal status unless they are adopted by state law or local ordinance.

Codes and standards are usually generated by non-profit, third-party organizations on a consensus basis. Consensus basically means that all concerned parties agree. Consensus is a democratic process in which all interested parties have an opportunity to contribute their opinion. These opinions are harmonized into a consistent document and then accepted by a vote of the members of the sponsoring organization.

The codes and standards addressed in this module are primarily those that affect construction and the installation and maintenance of fire protection systems. A municipal code package typically consists of the following:

- Electrical Code
- Excavating and Grading Code
- Existing Buildings Code
- Fire Prevention Code
- Gas Code
- Housing Code
- Mechanical Code
- One- and Two-Family Dwelling Code
- Plumbing Code
- Swimming Pool Code
- Unsafe Building Abatement Code

In the sprinkler industry, the primary concern is with the effects of the building codes and the fire codes. However, there are provisions in the electrical codes, mechanical codes, and plumbing codes that may affect the work of a fitter indirectly.

Most communities adopt one of the nationally recognized building codes and related codes and standards. Most also have local ordinances that modify or add to these documents, which can have a significant effect on sprinkler work. Some jurisdictions do opt to write their own building code. Some communities use zoning ordinances to require sprinkler installations when they are not required by other documents.

The general rules are indicated in this module, but local rules take precedence. These can change from one side of the street to another, and if a building is built across municipal or state lines, part of the building may be covered by one set of rules and part by a different set of rules. It will be up to the sprinkler contractor to understand and operate under the local rules.

Knowing why things must be done a certain way, or where the standard or code can be found, is part of your training as a sprinkler fitter. The base reference for the sprinkler fitting industry is *NFPA 13 Standard for the Installation of Sprinkler Systems.*

Building codes and standards regulate all construction work performed in North America. Used together, codes and standards serve to protect life, health, and property and to ensure quality of construction.

3.1.0 Codes

Building codes and other codes normally address the minimum safety requirements that must be adhered to (*Figure 2*). They typically describe circumstances under which the use of a given type of system, equipment, or component is required, depending on the use and occupancy of a building or structure. Generally, codes are written so that they can easily be adopted into law.

Useful building codes are complex legal documents that have to evolve through experience from application. This makes the nationally recognized model building codes an attractive source for a useful code.

- ADDRESS MINIMUM SAFETY REQUIREMENTS
- DESCRIBE CIRCUMSTANCE(S) UNDER WHICH THE USE OF A GIVEN TYPE OF SYSTEM, EQUIPMENT, OR COMPONENT IS REQUIRED
- WRITTEN SO THAT THEY CAN BE EASILY ADOPTED INTO LAW

101F02.EPS

Figure 2 Codes.

Three of the following codes were formerly published and maintained nationally as accepted building codes. Even though they are no longer published, they still may be in force in some state and local jurisdictions.

- *International Building Code®*
- *NFPA 5000, Building and Construction Safety Code™*
- *National Building Code* (no longer published)
- *Uniform Building Code™* (no longer published)
- *Standard Building Code* (no longer published)

There are now only two active building codes: the *International Building Code®* and *NFPA 5000.* State and local governments may also have additional requirements that must be met. It is essential that everyone involved in the installation of sprinkler systems be familiar with the local and national codes in force in the locality where the systems will be installed.

3.1.1 Model Building Codes

The three model building code groups, BOCA (Building Officials and Code Administrators), ICBO (International Conference of Building Officials) and SBCCI (Southern Building Code Congress International) have combined their forces into the ICC (International Code Council) to establish a single series of model codes. The advantage of a single code is that it allows the United States to become more competitive in the global marketplace. It also helps reduce construction costs by creating a set of standards that will apply across the entire country.

The *International Building Code* (IBC) has been adopted in many areas. The ICC also has additional codes including the *International Plumbing Code* (IPC), the *International Mechanical Code* (IMC), the *International Fire Code* (IFC), and several others.

Codes are revised and updated on a regular basis, usually every two or three years. Always check to make sure that you are working with the most current codes that have been adopted in your area.

With the adoption of a single code, the sprinkler contractor operating in several states will be able to reduce design time on a project. The contractor will also be able to lessen the chance of missing an item, such as draft stop requirements, that now vary between the three existing codes.

3.1.2 NFPA 101 Code for Safety to Life from Fire in Buildings and Structures

The Life Safety Code is a mixture of building-code-style requirements and fire-code-style requirements. It addresses the construction problem strictly from a life safety standpoint, and there are specific requirements within it for fire sprinkler systems. At least 27 states and the federal government have adopted *NFPA 101* for use. There are conflicts between the Life Safety Code and the building codes that may require resolution where both codes are imposed.

3.1.3 Fire Codes

Each model building code has an associated fire code or fire prevention code. Usually, a community will adopt the fire code that is compatible with its building code. However, there are communities that have adopted a building code but have not adopted a fire code. Some communities write their own fire code separate and apart from the model codes.

Generally, a fire prevention code administers existing buildings, not new construction. It is primarily a maintenance code. The fire codes deal with operations, fire prevention, and maintenance of fire protection systems. Fire codes are usually administered and enforced by the local fire chief.

In some instances, there are provisions within the fire code that affect new construction. For example, the *Uniform Fire Code* incorporates the major fire protection requirements in the *Uniform Building Code*, giving the fire department the license to enforce the fire protection provisions of the Building Code.

The City of Dallas Fire Code has a provision that states "Every occupancy shall be sprinklered unless," and proceeds to identify those conditions under which a sprinkler system can be avoided under the fire code. However, this does not remove any of the building code requirements for sprinklering.

Conformance to the fire code is usually a requirement before a municipality will issue a Certificate of Occupancy (CO). By law, in many communities, the permanent utilities will not be connected without a CO.

3.1.4 The Purpose of Building Codes

The principal objective of a building code is to regulate the quality of building construction in a community. Community may mean village, city, township, parish, county, or state. Canada has a National Building Code. Provisions of building codes are administered by the local building officials, and in some instances, by state building officials. Building codes regulate new construction by establishing limits on the height and the floor area of buildings. The degree of combustibility of the construction materials used in that building are a major consideration. A building code also deals in separations between buildings. The minimum requirements for the fire rating of walls and partitions, finish details, etc. are established by usage categories.

Generally, building codes administer new construction and alterations. Approximately 80 percent of the content of building codes deals with fire protection. The most important characteristic of building codes, to the sprinkler industry, has to do with requirements for installation of fire sprinkler systems and standpipe systems. Certain occupancies must be sprinklered. Other occupancies can have their heights and/or areas increased by the installation of fire sprinkler systems that are not otherwise required. These mandatory sprinklering requirements and the available trade-offs are an important, expanding part of the model building codes.

3.2.0 Standards

Standards detail how the protection specified by the code is to be achieved (*Figure 3*). Performance standards specify functions and capabilities of hardware and conditions under which the equipment must operate. The primary standard for sprinkler fitters is *NFPA 13 Standard for the Installation of Sprinkler Systems.*

Standards should not be confused with technical specifications. For example, NFPA standards require that an audible fire alarm signal be "clearly heard throughout the building," which is usually defined as 15 dBA above ambient sound levels. It does not specify how many audible alarms to use or where to locate them to achieve the requirement.

3.2.1 Types of Standards

There are many types of standards (*Figure 4*). The more common types are defined as follows:

- *International standards* – Mainly used to enhance trade, these standards use a common language and establish equivalent measurements.

- STANDARDS DETAIL HOW THE PROTECTION SPECIFIED IN CODES IS TO BE ACHIEVED.
- PERFORMANCE STANDARDS SPECIFY FUNCTIONS AND CAPABILITIES OF HARDWARE AND CONDITIONS UNDER WHICH THE EQUIPMENT MUST OPERATE.
- PRESCRIPTIVE STANDARDS SPECIFY PRECISELY WHAT TO DO AND HOW TO DO IT USING APPROVED AND LISTED SYSTEMS AND DEVICES.

101F03.EPS

Figure 3 Standards.

TYPES OF STANDARDS

- INTERNATIONAL
- NATIONAL
- REGIONAL AND LOCAL
- SAFETY
- INSTALLATION
- INDUSTRY
- COMPANY
- MANUFACTURER'S INSTRUCTIONS

101F04.EPS

Figure 4 Types of standards.

- *National standards* – These standards usually define a method of manufacture or design. Two organizations that establish national standards are the American National Standards Institute (ANSI) and the National Electrical Manufacturers Association (NEMA).
- *Regional and local standards* – Both regional and local standards establish requirements that conform to state and local needs.
- *Safety standards* – These standards establish requirements related to safety issues.
- *Installation standards* – These standards establish requirements related to how the installation of systems and equipment is to be achieved. *NFPA 13* is an example of an installation standard.
- *Industry standards* – Standards developed over the years that form the foundation of how most companies in an industry deal with complexities relating to systems and devices are called industry standards. For example, Underwriters Laboratories (UL) has developed standards on most aspects of the telecommunications and electronic security industries, including equipment, installation, and maintenance procedures. To be listed by UL, a company must demonstrate that its products meet the standards for each category.
- *Company standards* – Most companies have developed specific procedures for handling many aspects of your job. These procedures are usually based on applicable codes and standards, as well as other good practice methods.
- *Manufacturer's instructions* – Instructions that come with the equipment are called the manufacturer's instructions, and they are part of the approval or listing. If the instructions are not followed, the approval or listing is void.

3.2.2 Adopted and Recognized Standards

Adopted standards are standards that have been adopted through statutory or administrative laws in a specific geographic area. Once a document is adopted by reference in a state statute or local ordinance, it has the same force and effect as if printed therein in its entirety.

Recognized standards are standards that are generally accepted as relevant in civil law, even in areas where no standards have been formally adopted. Where no applicable code or standard has been officially implemented, it is common for the courts to compare actual circumstances in a case with nationally recognized standards.

3.3.0 Determining Which Codes and Standards to Follow

There are many types of codes and standards that can apply to a particular system or job. The decision as to which codes and standards apply is made by the Authority Having Jurisdiction (AHJ). Each job has at least one AHJ. The AHJ is the person or agency that decides the acceptability of the systems and devices being installed in accordance with code requirements.

The AHJ can include one or all of the following: a building inspector; a fire code inspector; another federal, state, local, or regional authority; a designated insurance company representative; or the property owner. The AHJ determines which rules apply and ultimately resolves any conflicts between any conflicting codes or rules. The AHJ is the organization, office, or individual responsible for approving equipment, an installation, or a procedure.

3.4.0 NFPA Codes and Standards

The standards published by the National Fire Protection Association are almost universally accepted as guides for the design and installation of fire sprinkler systems. These are produced by the consensus process in the volunteer standards system used in the United States. However, these standards use international representation with significant participation from Canada.

Most communities codify these standards through adoption by local ordinance. This process makes these consensus standards law in that community. A knowledge of NFPA standards therefore becomes critical to the design and installation of fire protection systems. The standards define the results of empiricism in terms of spacing, location, and position of sprinklers, the materials that are acceptable, and the techniques that work in implementing and maintaining systems for a variety of circumstances.

NFPA is now requiring all standards committees to observe a three-year revision cycle. The revision years are staggered among the standards so that they do not all receive revision in the same calendar year. Active communities move each year to adopt these standards by ordinance. Not all communities stay current.

The model building codes cite specific editions of NFPA standards. When these building codes are adopted, specific editions of the NFPA standards are also adopted. If the local ordinances or fire code do not adopt different years of issue, by default, the issue that is law is determined by the building code. It thus can become a problem as to which edition of an NFPA standard should be used. Frequently, the standard of latest issue is used by the fire marshal when the actual, legally adopted issue, is somewhat older.

Small changes in a structure are usually permitted without upgrading the whole building to the current standard. However, if the dollar value of the changes becomes large, some building codes will require that the entire building be brought up to current standards. There are also retroactive laws that require all buildings within a jurisdiction to be brought up to a current standard.

Codes and standards are living documents. Most of these are on a three-year revision cycle, although there are annual meetings and annual changes that may be adopted by ordinance. This means that a building 20 years old may be under the standards of 20 years ago or it may be under current standards. The year of issue of the applicable NFPA standard must be determined to know the requirements for a specific project.

The NFPA standard system covers a wide range of subjects. These standards have a uniform format. Chapter 1 of each standard has a scope definition that defines the applicability of that standard. Use of the standard beyond its defined scope can produce undesirable liability if a problem develops. Those kinds of deviations should be covered by proper documentation (i.e. written permission of the appropriate authorities). The NFPA standards most significant to the fire sprinkler industry are cited in subsequent paragraphs. They are given in numerical order rather than in the order of importance.

Codes and standards published by NFPA that apply to fire sprinkler systems are briefly described in this section. These codes may be cited in project specifications; therefore, all installers must be familiar with their content.

NFPA 13 was initiated in 1896 as the first publication of NFPA. It has been periodically updated since that time. *NFPA 13* is the key standard for design and installation of fire sprinkler systems. Its stated purpose is to provide a "reasonable degree of protection for life and property." To provide this, it specifies minimum requirements.

When a sprinkler system is to be installed, this standard provides the engineering inputs so that technicians can design effective fire sprinkler systems and, by following this same guide, mechanics can install properly functioning fire sprinkler systems. All of the other fire sprinkler system standards in the United States depend upon this standard for the basics of pipe, fittings, valves, hangers, devices, and sprinkler locations.

Design is based on occupancies that include Light Hazard, Ordinary Hazard, Extra Hazard, and storage. The old design approach of pipe schedules is usable under very limited conditions. The more recent design criteria uses area/density curves. The design criteria includes sprinkler spacing, location of sprinklers relative to the ceiling or roof deck and position of the sprinklers relative to obstructions. These factors vary according to occupancy and construction type.

This standard goes into substantial detail to define various types of sprinkler systems and where and how they are to be used. It also defines acceptable pipe, fittings, valves, hangers, and installation methods and test methods to assure, as best as practicable, functional, reliable, economic fire sprinkler systems. Many of the hardware-related requirements of *NFPA 13* apply to the other NFPA standards and are referenced by these standards. *NFPA 13* is thus the most important and basic fire sprinkler system standard in the NFPA series. For interpretive guides that expand and/or deviate, see the section on *Interpretive Guides*.

3.4.1 Standard for the Installation of Sprinkler Systems (NFPA 13)

NFPA 13 covers the proper design and installation of typical sprinkler systems for all types of fire hazards. Coverage includes the following items:

- Design considerations
- Water supplies
- Equipment requirements
- System flow rates
- Sprinkler location and position
- General storage and rack storage of materials
- Specifically identified criteria for special occupancy hazards
- Minimum sizes for sprinklers used in storage applications
- Separation requirements between early suppression fast response and other types of sprinkler systems
- Rules for protecting sprinkler systems against seismic events.

3.4.2 Standard for the Installation of Sprinkler Systems in One- and Two-Family Dwellings and Manufactured Homes (NFPA 13D)

NFPA 13D covers the design and installation of automatic sprinkler systems for protection against fire hazards in one- and two-family dwellings and manufactured homes. *NFPA 13D* was originally issued in 1975. In 1980, the 1975 edition was rescinded and a completely rewritten standard was issued. *NFPA 13D* is on the same revision schedule as *NFPA 13* and is processed by the same committee.

3.4.3 Standard for the Installation of Sprinkler Systems in Residential Occupancies up to and Including Four Stories in Height (NFPA 13R)

NFPA 13R covers the design and installation of automatic sprinkler systems for protection against fire hazards in residential occupancies up to and including four stories in height. This standard was first issued in 1989. It, too, is processed by the *NFPA 13* committee and follows the same revision cycle as *NFPA 13*.

3.4.4 Standard for the Installation of Stationary Pumps for Fire Protection (NFPA 20)

NFPA 20 is concerned with the requirements for selecting and installing pumps supplying liquid for private fire protection. Included in this standard are liquid supplies; suction, discharge, and auxiliary equipment; power supplies; electric drive and control; diesel engine drive and control; steam turbine drive and control; and acceptance tests and operations.

Fire pumps are used when the normal automatic water supply cannot furnish either the pressure or the volume needed to counter a fire. Material on steam and rotary fire pumps was included in the first issue of *NFPA 13* in 1896. Subsequently, a separate standard was developed. *NFPA 20* deals with the selection and installation of pumps used for fire-fighting purposes.

Not covered are system liquid capacity and pressure requirements; periodic inspection, testing, and maintenance of fire pump systems; or the requirements for installing wiring of fire pump units.

3.4.5 Standard for Water Tanks for Private Fire Protection (NFPA 22)

NFPA 22 specifies the minimum requirements for design, construction, installation, and maintenance of tanks and other equipment providing water for private fire protection. This includes the following:

- Gravity, suction, pressure, and embankment-supported coated-fabric suction tanks
- Towers
- Foundations
- Pipe connections and fittings
- Valve enclosures
- Tank filling
- Protection against freezing

3.4.6 Standard for the Installation of Private Fire Service Mains and Their Appurtenances (NFPA 24)

NFPA 24 focuses on the minimum requirements for installing private fire service mains and their equipment. The standard covers the following seven areas:

- Automatic sprinkler systems
- Open sprinkler systems
- Water spray fixed systems
- Foam systems
- Private hydrants
- Monitor nozzles or standpipe systems relative to water supplies
- Hose houses

NFPA 24 applies to combined service mains that carry water for fire service.

3.4.7 Standard for the Inspection, Testing, and Maintenance of Water-Based Fire Protection Systems (NFPA 25)

NFPA 25 sets the minimum requirements for inspecting, testing, and maintaining water-based fire protection systems, both land-based and marine. The standard covers sprinkler, standpipe and hose, fixed water spray, and foam water systems. Also covered are the water supplies that are part of the systems, including private fire service mains and equipment, fire pump and water storage tanks, and valves that control system flow. Another area covered is impairment handling and reporting.

As a standard, *NFPA 25* covers all of the water-based fire extinguishing systems. It has been coordinated with the applicable design requirements of the related NFPA standards to provide a complete system inspection and maintenance standard. This standard will be accessory to fire codes as an inspection and maintenance requirement, if adopted. It is required to be given to the owner, usually at the end of the project.

NFPA 25 does not apply to systems installed improperly, and corrective actions are outside the scope of this standard. This standard does not apply to sprinkler systems designed and installed to the requirements of *NFPA 13D, Standard for the Installation of Sprinkler Systems in One- and Two-Family Dwellings and Manufactured Homes.*

3.5.0 National Building Codes

Building codes that are national in scope provide minimum standards to guard the life and safety of the public by regulating and controlling the design, construction, and quality of materials used in modern construction. They have also come to govern the use and occupancy, location of a type of building, and the on-going maintenance of all buildings and facilities. Once adopted by a local jurisdiction, these national building codes then become law. In order to meet more stringent requirements and/or local needs, it is common for localities to change or add new requirements to any adopted national code requirements. The provisions of the national building codes apply to the construction, alteration, movement, demolition, repair, structural maintenance, and use of any building or structure within the local jurisdiction.

The national building codes are the legal instruments that enforce public safety in construction of human habitation and assembly structures. They are used not only in the construction industry, but also by the insurance industry for compensation appraisals and claims adjustments, and by the legal industry for court litigation.

4.0.0 Sprinkler Fitter Careers

Once you complete your initial training and gain job experience, there are a number of career opportunities available to you in the sprinkler industry. If you have leadership qualities, you can become foreman of a sprinkler fitter installation crew. Some experienced sprinkler fitters become instructors at contractor or vocational training schools. Others can become quality control inspectors, estimators, or piping designers. With further education and training, a sprinkler fitter can become a sprinkler system layout technician.

Depending on your level of skill, experience, education, and interest, you may want to be a fire sprinkler contractor. Fire sprinkler contractors work with architects, general contractors, code officials, and sometimes building owners to provide fire protection systems for a structure.

Or, you may work for a contractor as a layout technician. Layout technicians must know specific fire sprinkler design requirements and how those requirements fit into the design of the overall structure. Perhaps sales or estimating would interest you. *Figure 5* shows general career paths for sprinkler fitters.

Sprinkler fitters are employed all over the United States and the world. As a sprinkler fitter, you will probably work for a sprinkler contractor doing various types of commercial sprinkler installations. Sprinkler systems are now required in most commercial structures, resulting in a continuing demand for sprinkler fitters. A journey-level sprinkler fitter will earn as much as, and sometimes more than, a college-educated person.

4.1.0 Sprinkler Fitter Work

Sprinkler fitters perform a wide variety of tasks relating to the installation, repair, and maintenance of all types of fire sprinkler systems. The work is primarily on automatic sprinkler systems. Fitters must also have the skills and knowledge to install and maintain a number of related systems such as standpipes, carbon dioxide systems, foam systems, and dry chemical systems.

As an apprentice, you most likely will be fabricating, installing, testing, maintaining, or repairing sprinkler systems in industrial facilities, department stores, office buildings, hotels, schools, hospitals, and residences. You will install wet and dry pipe sprinkler systems, preaction, and deluge systems, along with a variety of related systems.

Did You Know?

Building Codes

Up until 2000, there were three model building codes. The three code writing groups – Building Officials and Code Administrators (BOCA), International Conference of Building Officials (ICBO), and Southern Building Code Congress International (SBCCI) – combined into one organization called the International Code Council (ICC) with the purpose of writing one nationally accepted family of building and fire codes. The first edition of the *International Building Code®* was published in 2000 and the second edition in 2003. It is intended to continue on a three-year cycle.

In 2002, the NFPA published its own building code, *NFPA 5000.* There are now two nationally recognized codes competing for adoption by the 50 states.

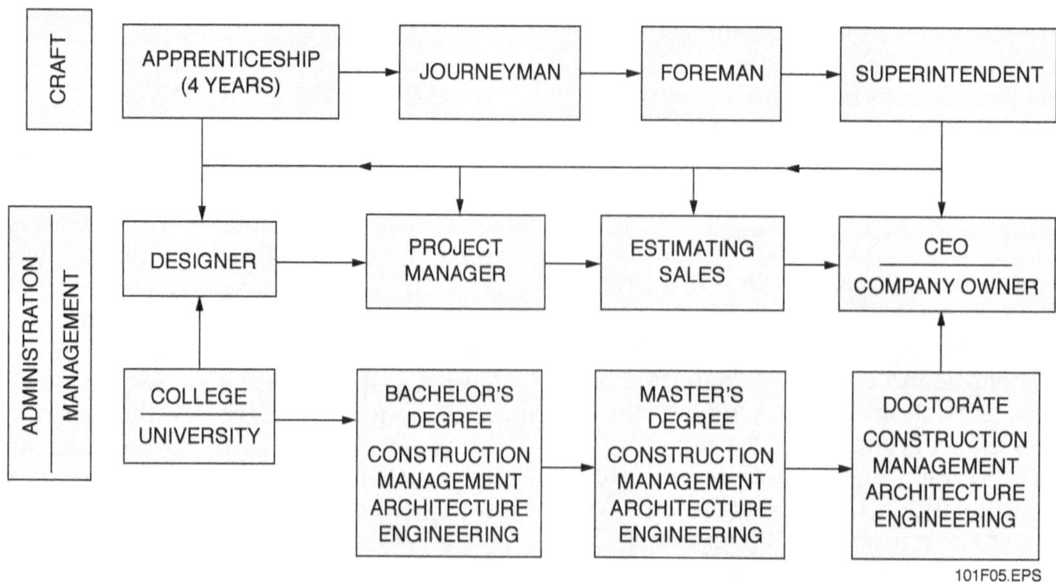

Figure 5 General career paths for sprinkler fitters.

As a journeyman sprinkler fitter, you must also be able to read and interpret engineering drawings and specifications, prepare cost estimates for clients, select the type and size of pipe required, and arrange piping to provide fire protection. You will install hangers to support the piping system, and solder, thread, and glue (solvent weld) pipe joints. You will test systems for leaks, inspect, maintain, and repair piping, fixtures, and controls including hydrants, pumps, and sprinkler connections.

4.2.0 General Requirements and Working Conditions

The work is done both indoors and outdoors. There is a substantial amount of climbing, kneeling, standing, and walking on scaffolding. Because of this activity, the sprinkler fitter must be able to move about with ease and be in good physical condition.

If you want to be a sprinkler fitter, you should not have a fear of heights and should have a good sense of balance. You must be safety conscious for yourself and for your co-workers. You need to be good with your hands and be willing to do hard physical labor when required.

You should be mechanically inclined, able to read blueprints, and be able to use power tools to cut and thread pipe both safely and efficiently.

4.3.0 Rewarding Career

There is a future in sprinkler fitting. During construction booms there is a strong demand for sprinkler fitters. But even when the economy weakens, there is always a demand for fitters to service, maintain, and repair existing systems, and to retrofit existing buildings that lack sprinkler systems.

There is always a strong demand for good service technicians. If you continue your education and can master the building and fire codes, you may become an inspector, superintendent, or manager.

Many sprinkler fitters go through a four-year apprenticeship leading to certification or licensing as a fire sprinkler fitter. As part of the apprenticeship program, you receive instruction and training via textbooks, classes, and lab work.

4.3.1 Coordinating with the Construction Industry

The construction industry recognizes the responsibility of working closely with schools in order to develop a positive industry image with young people and their parents. Changing technologies, improving materials, and better building techniques have all created new careers that can only be filled by a talented and diverse workforce.

The NCCER was created to address the critical workforce shortage facing the construction industry and to develop industry-driven standardized craft training programs with portable credentials.

4.4.0 NCCER

NCCER is a not-for-profit education foundation established by the nation's leading construction companies. NCCER was created to provide the industry with standardized construction education materials, the NCCER Standardized Curricula (the sprinkler fitting modules are part of this series), and a system for tracking and recognizing students' training accomplishments—NCCER's Automated National Registry (ANR). See *Appendix B* for samples of NCCER's apprentice training credentials, certificate, and transcript.

NCCER also offers accreditation, instructor certification, and skills assessments. NCCER is committed to developing and maintaining a training process that is internationally recognized, standardized, portable, and competency-based.

Working in partnership with industry and academia, NCCER has developed a system for program accreditation that is similar to those found in institutions of higher learning. NCCER's accreditation process ensures that students receive quality training based on uniform standards and criteria. These standards are outlined in NCCER's *Accreditation Guidelines* and must be adhered to by NCCER Accredited Training Sponsors.

More than 450 training sponsors and/or assessment centers across the U.S. and eight other countries are proud to be NCCER Accredited Training Sponsors and Assessment Centers. Millions of craft professionals and construction managers have received quality construction education through NCCER's network of Accredited Training Sponsors and the thousands of Training Units associated with the Sponsors. Every year the number of NCCER Accredited Training Sponsors increases significantly.

A craft instructor is a journeyman craft professional or career and technical educator trained and certified to teach NCCER's Standardized Curricula. This network of certified instructors ensures that NCCER training programs will meet the standards of instruction set by the industry. There are more than 4,300 master trainers and 45,000 craft instructors within the NCCER instructor network. More information is available at www.nccer.org.

4.5.0 The American Fire Sprinkler Association

The American Fire Sprinkler Association (AFSA) is the voice of the merit shop automatic fire sprinkler contractor. In support of the merit shop objective, the AFSA promotes the development of educational and training programs to maintain the quality and effectiveness of automatic fire sprinklers, encourages an expanded role for automatic fire sprinklers in protecting lives and property, disseminates information on labor, technology and business, and provides programs to enhance business practices for the merit shop contractor.

AFSA is a non-profit, international association representing merit shop fire sprinkler contractors, dedicated to the educational advancement of its members and promotion of the use of automatic fire sprinkler systems. AFSA was organized in 1981 to provide the merit shop fire sprinkler contractor with training, consulting, communication, representation, and many more services, all of which have expanded over the course of its existence.

Membership is open to contractors, manufacturers, suppliers, designers and Authorities Having Jurisdiction. Currently, AFSA represents nearly 1,200 companies and organizations in the United States and throughout the world. AFSA believes that the installation of fire sprinklers could save thousands of lives and billions of dollars that are lost to fire each year. Increasingly, public officials are realizing this.

AFSA acts as a liaison with other national associations involved in fire safety. Working together with government agencies, such as the Federal Emergency Management Agency (FEMA), the NFPA, and the Home Fire Sprinkler Coalition, AFSA helps to increase awareness concerning the dangers of fire.

4.5.1 Training

To assist fire sprinkler contractors in maintaining a quality workforce, AFSA co-sponsored with NCCER the development of this standardized training program specifically for apprentices in the fire sprinkler trade.

In addition, AFSA offers the following specialized schools:

- A two-week Beginning Fire Sprinkler System Planning School offering a comprehensive practical approach to preparing a fire sprinkler system drawing
- A one-week System Layout School for Residential 1 and 2 Family Dwellings covering basics of residential system layout including requirements of *NFPA 13D* and the International Residential Code
- Fire Sprinkler eCampus, a website offering online courses for topics including sprinkler system layout, hydraulic calculations, seismic bracing, AutoCAD, and more
- Frequent webinars on various subjects such as NFPA standards updates, special systems, inspections, system layout, installations, etc.
- Training tools for project management, estimating and bidding, NICET study aids, foremanship, and on-the-job safety
- The Center for Life Safety Education, an educational affiliate of AFSA, offers specialized training for fire department personnel and building officials on the latest technology in fire sprinklers

4.5.2 SprinklerNet

The Fire Sprinkler Network is sponsored by the AFSA. It is a resource for current information related to automatic fire sprinklers. See Sprinkler-Net at www.sprinklernet.org.

5.0.0 Responsibilities of the Employee

In order to be successful, the professional must be able to use current trade materials, tools, and equipment to finish the task quickly and efficiently. A sprinkler fitter must be adept at adjusting methods to meet each situation. The successful sprinkler fitter must continuously train to remain knowledgeable about technical advancements and to gain the skills to use them. A professional never takes chances with regard to personal safety or the safety of others.

5.1.0 Professionalism

The word *professionalism* is a broad term that describes the desired overall behavior and attitude expected in the workplace. Professionalism is too often absent from the construction site and the various trades. Most people would argue that professionalism must start at the top in order to be successful. It is true that management support of professionalism is important to its success in the workplace, but it is more important that individuals recognize their own responsibility for professionalism.

Professionalism includes honesty, productivity, safety, civility, cooperation, teamwork, clear and concise communication, being on time and prepared for work, and regard for one's impact on one's co-workers. It can be demonstrated in a variety of ways every minute you are in the workplace. Most important is that you do not tolerate the unprofessional behavior of co-workers. This is not to say that you shun the unprofessional worker; instead, you work to demonstrate the benefits of professional behavior.

Professionalism is a benefit to both the employer and the employee. It is a personal responsibility. Our industry is what each individual chooses to make of it; choose professionalism and the industry image will follow.

5.2.0 Honesty

Honesty and personal integrity are important traits of the successful professional. Professionals pride themselves in performing a job well, and in being punctual and dependable. Each job is completed in a professional way, never by cutting corners or reducing materials. A valued professional maintains work attitudes and ethics that protect property such as tools and materials belonging to employers, customers, and other trades from damage or theft at the shop or job site.

Honesty and success go hand-in-hand. It is not simply a choice between right and wrong, but a choice between success and failure. Dishonesty will always catch up with you. Whether you are stealing materials, tools, or equipment from the job site or simply lying about your work, it will not take long for your employer to find out. Of course, you can always go and find another employer, but this option will ultimately run out on you.

If you plan to be successful and enjoy continuous employment, consistency of earnings, and being sought after as opposed to seeking employment, then start out with the basic understanding of honesty in the workplace and you will reap the benefits. Honesty means more, however, than just not taking things that do not belong to you. It means giving a fair day's work for a fair day's pay. It means carrying out your side of a bargain. It means that your words convey true meanings and actual happenings. Our thoughts as well as our actions should be honest. Employers place a high value on an employee who is strictly honest.

5.3.0 Loyalty

Employees expect employers to look out for their interests, to provide them with steady employment, and to promote them to better jobs as openings occur. Employers feel that they, too, have a right to expect their employees to be loyal to them—to keep their interests in mind, to speak well of them to others, to keep any minor troubles strictly within the company, and to keep absolutely confidential all matters that pertain to the business. Both employers and employees should keep in mind that loyalty is not something to be demanded; rather, it is something to be earned.

5.4.0 Willingness to Learn

Every company has its own way of doing things. Employers expect their workers to be willing to learn these ways. Adapting to change and being willing to learn new methods and procedures as quickly as possible are key. Sometimes, a change in safety regulations or the purchase of new equipment makes it necessary for even experienced employees to learn new methods and operations. Employees often resent having to accept improvements because of the retraining that is involved. However, employers will no doubt think they have a right to expect employees to put forth the necessary effort. Methods must be kept up to date in order to meet competition and show a profit. It is this profit that enables the owner to continue in business and provide jobs and advancement for the employees.

5.5.0 Willingness to Take Responsibility

Most employers expect their employees to see what needs to be done, then go ahead and do it. It is very tiresome to have to ask again and again that a certain job be done. It is obvious that having been asked once, an employee should assume the responsibility from then on. Once the responsibility has been delegated, the employee should continue to perform the duties without further direction. Every employee has the responsibility for working safely.

5.6.0 Willingness to Cooperate

To cooperate means to work together. In our modern business world, cooperation is the key to getting things done. Learn to work as a member of a team with your employer, supervisor, and fellow workers in a common effort to get the work done efficiently, safely, and on time.

5.7.0 Rules and Regulations

People can work well together only if there is some understanding about what work is to be done, when and how it will be done, and who will do it. Rules and regulations are a necessity in any work situation and should be so considered by all employees.

5.8.0 Tardiness and Absenteeism

Tardiness means being late for work, and absenteeism means being off the job for one reason or another. Consistent tardiness and frequent absences are an indication of poor work habits, unprofessional conduct, and a lack of commitment.

We are all creatures of habit. What we do once we tend to do again unless the results are too unpleasant. The habit of always being late may have begun back in our school days when we found it hard to get up in the morning. This habit can get us into trouble at school, and it can continue to get us into trouble when we are through with school and go to work.

Our work life is governed by the clock. We are required to be at work at a definite time, and so is everyone else. Failure to get to work on time results in confusion, lost time, and resentment on the part of those who do come on time. In addition, it may lead to penalties up to and including dismissal. Although it may be true that a few minutes out of a day are not very important, you must remember that a principle is involved. It is your obligation to be at work at the time indicated. We agree to the terms of work when we accept the job. Perhaps you can see this more clearly from the point of view of the boss. Supervisors cannot keep track of people if they come in at any time they please. It is not fair to others to ignore tardiness. Failure to be on time may hold up the work of fellow workers. Better planning of your morning routine will often keep you from being delayed and so prevent a late arrival. In fact, arriving a little early indicates your interest and enthusiasm for your work, which is appreciated by employers. The habit of being late is another one of those things that stand in the way of promotion.

It is sometimes necessary to take time off from work. No one should be expected to work when sick or when there is serious trouble at home. However, it is possible to get into the habit of letting unimportant and unnecessary matters keep us from the job. This results in lost production and hardship on those who try to carry on the work. Again, there is a principle involved. The person who hires us has a right to expect us to be on the job unless there is some very good reason for staying away. Certainly, we should not let some trivial reason keep us home. We should not stay up nights until we are too tired to go to work the next day. If we are ill, we should use the time at home to do all we can to recover quickly. This, after all, is no more than most of us would expect of a person we had hired to work for us, and on whom we depend to do a certain job.

If it is necessary to stay home, then at least phone the office early in the morning so that the boss can find another worker for the day. Time and again, employees have remained home without sending any word to the employer. This is the worst possible way to handle the matter. It leaves those at work uncertain about what to expect. They have no way of knowing whether you have merely been held up and will be in later, or whether immediate steps should be taken to assign your work to someone else. Courtesy alone demands that you let the boss know if you cannot come to work.

The most frequent causes of absenteeism are illness, death in the family, accidents, personal business, and dissatisfaction with the job. Some of the causes are legitimate and unavoidable, while others can be controlled. One can usually plan to carry on most personal business affairs after working hours. Frequent absences will reflect unfavorably on a worker when promotions are being considered.

Employers sometimes resort to docking pay, demotion, and even dismissal in an effort to control tardiness and absenteeism. No employer likes to impose restrictions of this kind. However, in fairness to those workers who do come on time and who do not stay away from the job, an employer is sometimes forced to discipline those who will not follow the rules.

5.9.0 Setting Goals

Goal setting is very important to achieving success. Having short-term and long-term goals is critical to improving your value on the job site. An employee's wages are dependent on how the employee excels at work. Setting goals and working toward them are the keys to improving your performance at the job site and gaining improved wages and job satisfaction.

The goals that people set for themselves must include pride, professionalism, reputation, and recognition. If employees do not take pride in their work, it will affect their reputation and their company's reputation.

5.10.0 Employee Reviews and Evaluations

Some employers have annual or semi-annual employee reviews to evaluate job performance. The employer evaluates the quality of performance relative to wages. When actual job performance is greater than or equal to expected job performance, a raise and/or promotion is a reward for a job well done. Attitude on the job will also be considered. Developing and maintaining a positive attitude now will pay off in the future. *Figure 6* shows a typical employee review form.

6.0.0 Human Relations

Most people underestimate the importance of working well with others. There is a tendency to pass off human relations as nothing more than common sense. What exactly is involved in human relations? One response would be to say that part of human relations is being friendly, pleasant, courteous, cooperative, adaptable, and sociable.

6.1.0 Making Human Relations Work

As important as the previously noted characteristics are for personal success, they are not enough. Human relations is much more than just getting people to like you. It is also knowing how to handle difficult situations as they arise.

Human relations is knowing how to work with supervisors who are often demanding and sometimes unfair. It involves understanding your own personality traits and those of the people you work with. Building sound working relationships in various situations is important. If working relationships have deteriorated for one reason or another, restoring them is essential. Human relations is learning how to handle frustrations without hurting others.

6.2.0 Human Relations and Productivity

Effective human relations is directly related to productivity. Productivity is the key to business

On Site

Ethical Principles for Members of the Construction Trades

Honesty – Be honest and truthful in all dealings. Conduct business according to the highest professional standards. Faithfully fulfill all contracts and commitments. Do not deliberately mislead or deceive others.

Integrity – Demonstrate personal integrity and the courage of your convictions by doing what is right even where there is pressure to do otherwise. Do not sacrifice your principles because it seems easier.

Loyalty – Be worthy of trust. Demonstrate fidelity and loyalty to companies, employers and sponsors, co-workers, and trade institutions and organizations.

Fairness – Be fair and just in all dealings. Do not take undue advantage of another's mistakes or difficulties. Fair people are open-minded and committed to justice, equal treatment of individuals, and tolerance for and acceptance of diversity.

Respect for others – Be courteous and treat all people with equal respect and dignity.

Obedience – Abide by laws, rules, and regulations relating to all personal and business activities.

Commitment to excellence – Pursue excellence in performing your duties, be well informed and prepared, and constantly try to increase your proficiency by gaining new skills and knowledge.

Leadership – By your own conduct, seek to be a positive role model for others.

Employee ____________________ Fitter ____________ Apprentice ____________

Employer ____________________ Evaluation Date ____________

Factors	UNSATISFACTORY	MARGINAL	GOOD	VERY GOOD	EXCELLENT
	1 2 3 4 5	6 7 8 9 10	11 12 13 14 15	16 17 18 19 20	21 22 23 24 25
1. QUALITY OF WORK	Is very slip-shod with frequent, avoidable errors	Makes frequent errors, work often careless, leaves loose ends.	Is generally satisfactory, sometimes uncorrected errors slip through.	Is almost always accurate and neat. Corrects own errors.	Work is always of superior quality.
2. COOPERATIVENESS	Seems unable to cooperate with others. Is argumentative over every innovation.	Often fails to cooperate/frequently disagreeable, difficult to get along with.	Gets along well with others most of the time, is sometimes obstructive.	Often cooperates well/willing to try new methods.	Works effectively with co-workers and supervisors, falls in readily with new ideas.
3. JOB KNOWLEDGE	Does not know enough about the job.	More knowledge of job would be desirable.	Knows job well enough to get along.	Thorough knowledge of work.	Has excellent mastery of work.
4. INITIATIVE	Routine worker, usually waits to be told what to do, needs constant supervision.	Often at a loss in other than routine situations, requires frequent checking.	Does regular work without waiting for direction. Requires some supervision.	Resourceful; needs minimal supervision, alert to ways to improve work.	Always gets on with job on his own, highly ingenious.
5. INDUSTRIOUSNESS	Socializes on the job whenever possible and/or excessive phone use.	Frequently neglects work, needs to improve work attitude.	Usually sticks to the job but with occasional wandering.	Conscientious most of the time.	Can always be relied on to get things done.
6. QUANTITY OF WORK	Very slow.	Below average in output.	Turns out required volume, seldom more.	Above average producer.	Unusual output, exceptionally fast, does more than required.
7. ABILITY TO LEARN	Unable to grasp job without constant re-instruction.	Learns slowly.	Learns moderately fast and checks occasionally with supervisor.	Learns fast and remembers well.	Unusually quick and complete, full grasp.
8. PAPERWORK	Fails to turn in completed paperwork.	Paperwork is turned in, but is often incomplete.	Usually turns in paperwork on time. Sometimes needs prompting.	Paperwork is turned in, but sometimes lacks signatures.	Paperwork is always filled out neatly and accurately, and signed.
9. ATTENDANCE	Excessively absent or tardy.	Frequently absent or tardy.	Occasionally absent or tardy.	Rarely absent or tardy, then only for good cause.	Never absent or tardy.

ATTENDANCE RECORD

For Rating Period: Times Tardy: ________ Excused Absence: ________ Unexcused Absence: ________

Length of Time Employed: Years: ________ Months: ________ Days: ________

Remarks: ____

____

Signature of Employee ____________________ Date ____________

Signature of Superintendent or Foreman ____________________ Title ____________ Date ____________

101F06.EPS

Figure 6 Typical employee review form.

success. Every employee is expected to produce at a certain level. Employers quickly lose interest in an employee who has a great attitude but is able to produce very little. There are work schedules to be met and jobs that must be completed.

All employees, both new and experienced, are measured by the amount of quality work they can safely turn out. The employer expects every employee to do his or her share of the workload.

However, doing one's share in itself is not enough. If you are to be productive, you must do your share (or more than your share) without antagonizing your fellow workers. You must perform your duties in a manner that encourages others to follow your example. It makes little difference how ambitious you are or how capably you perform. You cannot become the kind of employee you want to be or the type of worker management wants you to be without learning how to work with your peers.

Employees must do everything they can to build strong, professional working relationships with fellow employees, supervisors, and clients.

6.3.0 Attitude

A positive attitude is essential to a successful career. First, being positive means being energetic, highly motivated, attentive, and alert. A positive attitude is essential to safety on the job. Second, a positive employee contributes to the productivity of others. Both negative and positive attitudes are transmitted to others on the job. A persistent negative attitude can spoil the positive attitudes of others. It is very difficult to maintain a high level of productivity while working next to a person with a negative attitude. Third, people favor a person who is positive. Being positive makes a person's job more interesting and exciting. Finally, the kind of attitude transmitted to management has a great deal to do with an employee's future success in the company. Supervisors can determine a subordinate's attitude by their approach to the job, reactions to directives, and the way they handle problems.

6.4.0 Maintaining a Positive Attitude

A positive attitude is far more than a smile, which is only one example of an inner positive attitude. As a matter of fact, some people transmit a positive attitude even though they seldom smile. They do this by the way they treat others, the way they look at their responsibilities, and the approach they take when faced with problems. Here are a few suggestions that will help you to maintain a positive attitude:

- Remember that your attitude follows you wherever you go. If someone makes a greater effort to be a more positive person in their social and personal lives, it will automatically help them on the job. The reverse is also true. One effort will complement the other.
- Negative comments are seldom welcomed by fellow workers on the job. Neither are they welcome on the social scene. The solution: talk about positive things and be complimentary. Constant complainers do not build healthy and fulfilling relationships.
- Look for the good things in people on the job, especially your supervisor. Nobody is perfect, and almost everyone has a few worthwhile qualities. If you dwell on people's good features, it will be easier to work with them.
- Look for the good things where you work. What are the factors that make it a good place to work? Is it the hours, the physical environment, the people, the actual work being done? Or is it the atmosphere? Keep in mind that you cannot expect to like everything. No work assignment is perfect, but if you concentrate on the good things, the negative factors will seem less important and bothersome.
- Look for the good things in the company. Just as there are no perfect assignments, there are no perfect companies. Nevertheless, almost all organizations have good features. Is the company progressive? What about promotional opportunities? Are there chances for self-improvement? What about the wage and benefit package? Is there a good training program? You cannot expect to have everything you would like, but there should be enough to keep you positive. In fact, if you decide to stick with a company for a long period of time, it is wise to look at the good features and think about

On Site

Tips for a Positive Attitude

Here is a short checklist of things that you should keep in mind to help you develop and maintain a positive attitude:

- Remember that your attitude follows you wherever you go.
- Helpful suggestions and compliments are much more effective than negative ones.
- Look for the positive characteristics of your teammates and supervisors.

them. If you think positively, you will act the same way.
- You may not be able to change the negative attitude of another employee, but you can protect your own attitude from becoming negative.

7.0.0 Employer and Employee Safety Obligations

An obligation is like a promise or a contract. In exchange for the benefits of your employment and your own well-being, you agree to work safely. In other words, you are obligated to work safely. You are also obligated to make sure anyone you happen to supervise or work with is working safely. Your employer is obligated to maintain a safe workplace for all employees. Safety is everyone's responsibility.

Some employers have safety committees. If you work for such an employer, you are then obligated to that committee to maintain a safe working environment. This means two things:

- Follow the safety committee's rules for proper working procedures and practices.
- Report any unsafe equipment and conditions directly to the committee or your supervisor.

On the job, if you see something that is not safe, report it! Do not ignore it. It will not correct itself. You have an obligation to report it.

In the long run, even if you do not think an unsafe condition affects you, it does. Do not mess around; report what is not safe. Do not think your employer will be angry because your productivity suffers while the condition is corrected. On the contrary, your employer will be more likely to criticize you for not reporting a problem.

Your employer knows that the short time lost in making conditions safe again is nothing compared with shutting down the whole job because of a major disaster. If that happens, you are out of work anyway. So do not ignore an unsafe condition. In fact, the Occupational Safety and Health Administration (OSHA) regulations require you to report hazardous conditions.

This applies to every part of the construction industry. Whether you work for a large contractor or a small subcontractor, you are obligated to report unsafe conditions. The easiest way to do this is to tell your supervisor. If that person ignores the unsafe condition, report it to the next highest supervisor. If it is the owner who is being unsafe, let that person know your concerns. If nothing is done about it, report it to OSHA. If you are worried about your job being on the line, think about it in terms of your life, or someone else's, being at risk.

The U.S. Congress passed the *Occupational Safety and Health Act* in 1970. This act also created OSHA. It is part of the U.S. Department of Labor. The job of OSHA is to set occupational safety and health standards for all places of employment, enforce these standards, ensure that employers provide and maintain a safe workplace for all employees, and provide research and educational programs to support safe working practices.

OSHA was adopted with the stated purpose "to assure as far as possible every working man and woman in the nation safe and healthful working conditions and to preserve our human resources." OSHA requires each employer to provide a safe and hazard-free working environment. OSHA also requires that employees comply with OSHA rules and regulations that relate to their conduct on the job. To gain compliance, OSHA can perform spot inspections of job sites, impose fines for violations, and even stop any more work from proceeding until the job site is safe.

According to OSHA standards, you are entitled to on-the-job safety training. As a new employee, you are entitled to the following:

- Being shown how to do your job safely
- Being provided with the required personal protective equipment
- Being warned about specific hazards
- Being supervised for safety while performing the work

The enforcement of this act of Congress is provided by the federal and state safety inspectors, who have the legal authority to make employers pay fines for safety violations. The law allows states to have their own safety regulations and agencies to enforce them, but they must first be approved by the U.S. Secretary of Labor. For states that do not develop such regulations and agencies, federal OSHA standards must be obeyed.

These standards are listed in *OSHA Safety and Health Standards for the Construction Industry* (29 CFR, Part 1926), sometimes called *OSHA Standards 1926*. Other safety standards that apply to construction are published in *OSHA Safety and Health Standards for General Industry* (29 CFR, Parts 1900 to 1910).

The most important general requirements that OSHA places on employers in the construction industry are as follows:

- The employer must perform frequent and regular job site inspections of equipment.
- The employer must instruct all employees to recognize and avoid unsafe conditions, and to know the regulations that pertain to the job so they may control or eliminate any hazards.
- No one may use any tools, equipment, machines, or materials that do not comply with *OSHA Standards 1926*.

- The employer must ensure that only qualified individuals operate tools, equipment, and machines.

7.1.0 Carrying Methods

A major part of a sprinkler fitter's job involves lifting, carrying, and lowering materials to be used on the job. Follow these guidelines when lifting and carrying materials:

- Lift the load with your legs instead of your back.
- Do not carry material so large or bulky that it obstructs vision.
- Make sure all walkways are free of obstacles.
- Once a secure grip is obtained, lift the load properly keeping the load close to your body.
- If more than one person is carrying the load, all team members should carry it on the same side shoulder, walk in step, and move slowly.
- When lowering the load, use your legs and not your back.
- Make sure that your fingers and toes are not in the way before setting the load down.
- Since many back and strain injuries occur early in the workday, spend a few minutes stretching your muscles before any heavy lifting.

7.2.0 Storing Materials

OSHA provides strict regulations concerning the way materials are to be stored at the job site. Aisles and pathways must be kept clear for the safe movement of materials. Materials not needed for immediate operations should not be stored on scaffolds or runways. This creates a safety hazard and slows the work process. All storage areas used for flammable, explosive, or toxic material must be ventilated properly.

OSHA regulations state that all materials stored in tiers must be stacked, racked, blocked, interlocked, or otherwise secured to prevent sliding, falling, or collapse. The safest way to store pipe and other cylindrical materials is on racks. When cylindrical materials are stored in stacks, they must be blocked to prevent spreading and tilting. The same size, type, and length of material should be stacked together. Neat stacks allow more efficient use of the materials.

When material is stored inside a building that is under construction, the material must not be placed within 6 feet of any hoistway or inside floor opening. Material must not be stored within 10 feet of an exterior wall that does not extend above the top of the material. These regulations are designed to prevent injuries caused by falling material.

Special safety precautions apply to compressed gas cylinders stored and used on the job site. Follow these requirements when storing and using compressed gas cylinders:

- Compressed gas cylinders must be stored in an upright position.
- The contents label must be located on the cylinder where it can easily be seen.
- Valves must be closed except when they are in use.
- Valve protection caps must be in place when the cylinders are not in use.
- When using acetylene, a key must be left on the bottle to quickly turn the gas off if necessary.
- Compressed gas cylinders should never be stored near a welding operation because they can be ignited by sparks, hot slag, or flames. If it is not possible to store the cylinders a safe distance away, a fire-resistant shield must be used.
- Do not store empty cylinders with full or partially full cylinders. Place the empty cylinders in a separate area and mark the cylinder MT. Only the gas supplier is permitted to mix gases in a cylinder.
- The owner of the cylinder is the only person permitted to refill a cylinder.
- Propane tanks must be stored in an orderly manner and must be at least 25 feet from floor openings.
- Signs must be posted in storage areas to warn of flammable and explosive materials.

8.0.0 Tools

After lifting, the second most common cause of on-the-job accidents is the improper use of hand tools. Use hand tools only for the purpose for which they were designed. Tools can be damaged and can create a potential safety hazard if used improperly. Hand tools also last longer if they are used and maintained properly. This section explains the purpose and proper use of the more common hand tools that a sprinkler fitter will use on the job site.

8.1.0 Pipe Wrenches

Many types and sizes of pipe wrenches are used in the sprinkler fitting trade. Pipe wrenches are heavy-duty wrenches made of steel or aluminum. They are used to grip and turn pipe or to grip and hold pipe stationary. It is critical to select the right size pipe wrench for a specific job. A pipe wrench that is too small will not hold the pipe firmly and a handle that is too short will not provide enough leverage. A pipe wrench that is too large can strip the threads, break the pipe or fitting, or cause excessive marring or scratching of the pipe. Sprinkler fitters use several types of pipe wrenches, including the following:

- Straight pipe wrenches
- Offset pipe wrenches
- Chain wrenches
- Strap wrenches

8.1.1 Straight Pipe Wrenches

Straight pipe wrenches have two jaws. The stationary jaw is known as the heel and the moveable jaw is known as the hook jaw. Both jaws have teeth cut into them to give the jaws a firm grip on the pipe. The teeth on the hook jaw face inward and the teeth on the heel face outward. *Figure 7* shows the components of a straight pipe wrench.

The hook jaw of a straight pipe wrench is adjusted by turning the nut located below the housing. A spring inside the housing allows the hook jaw to float in the housing. This causes the hook jaw to open slightly as you apply the wrench to the pipe. Straight pipe wrenches are sized by the length measured from the inside top of the hook jaw to the end of the handle when the jaws of the wrench are fully open. Straight pipe wrenches range in size from 6 to 60 inches. The size does not refer directly to the size pipe it fits, but each size wrench has recommended pipe capacities. *Table 1* shows pipe wrench sizes and capacities.

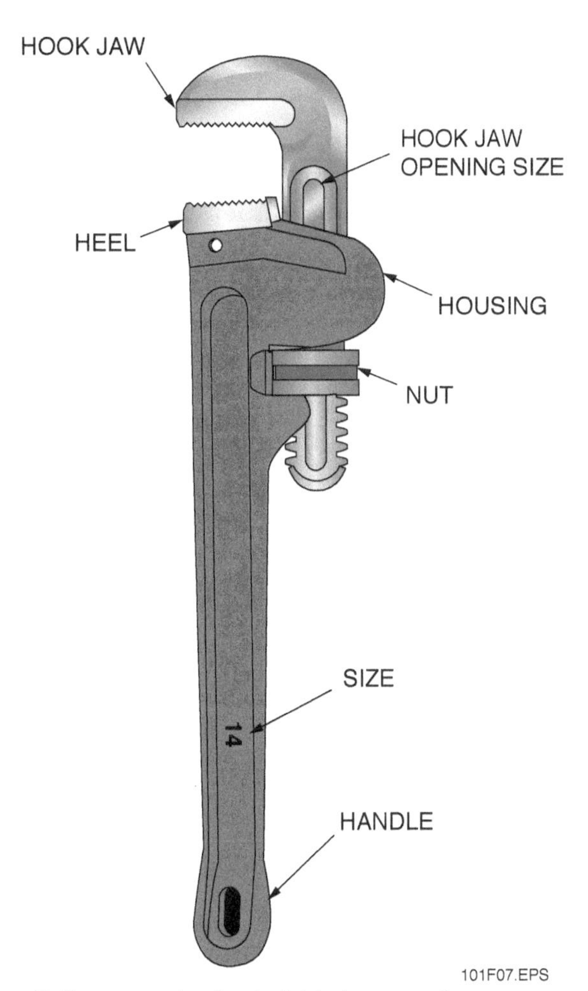

Figure 7 Components of a straight pipe wrench.

To use a pipe wrench, grip the pipe with the wrench so that the jaw opening faces you and you can tighten the pipe or fitting by pulling the handle towards your body. If there is not enough space to position the wrench this way, face the jaw opening away from you and push on the handle to tighten or remove the pipe. Be sure to brace yourself when pushing on the wrench so that you do not fall if the wrench slips off the pipe.

Follow these steps to tighten a fitting onto the threaded end of a pipe:

Step 1 Select a pipe wrench suitable for the size of pipe being used.

Step 2 Inspect the selected wrench for excessively worn jaw teeth and corners, and for any noticeable damage to the handle. Repair or replace the wrench if necessary.

Step 3 Secure the pipe in a chain vise, allowing the threaded end to extend from the vise about 8 inches.

Step 4 Apply pipe thread compound or Teflon® tape and tighten the fitting onto the end of the pipe by hand.

Step 5 Position your body so that the fitting is on the left side.

Step 6 Turn the nut on the pipe wrench until the jaws are opened slightly wider than the diameter of the fitting.

Table 1 Pipe Wrench Sizes and Capacities

Wrench Size (Handle Length)	Capacity	Recommended for Pipe Sizes
6	1⁄8–1⁄2	1⁄8
8	1⁄8–3⁄4	1⁄4–3⁄8
10	1⁄8–1	1⁄2–3⁄4
14	1⁄4–1 1⁄2	1
18	1⁄4–2	1 1⁄4–1 1⁄2
24	1⁄2–2 1⁄2	2–2 1⁄2
36	1⁄2–3 1⁄2	3
48	1–5	4–5

NOTE: All dimensions in inches. 101T01.EPS

NOTE

Many straight pipe wrenches have pipe sizes marked on the hook jaw just above the housing. If your wrench has these marks, line up the correct mark with the top of the housing.

Step 7 Hold the pipe wrench by the handle so that the open jaws are toward you.

Step 8 Slide the jaws onto the shoulder of the fitting.

NOTE

Be sure the jaws fit snugly onto the shoulder of the fitting. The shoulder should be centered inside the jaws, and there should be a gap between the fitting and the back of the hook jaw.

Step 9 Pull the handle toward your body to tighten the fitting.

Step 10 Push the handle in the opposite direction to release the jaws from the fitting.

Step 11 Regrip the fitting with the wrench and pull the wrench toward your body again.

Step 12 Repeat Steps 10 and 11 until the fitting is tight on the pipe.

CAUTION

Do not overtighten the fitting on the pipe as the threads may strip.

Step 13 Remove the wrench from the pipe fitting.

Step 14 Clean the wrench with a wire brush, lubricate it, and inspect it for any damage.

Step 15 Close the jaws and store the wrench.

You may have to install or remove a fitting from a piece of pipe where a vise is not available or the fitting is part of a piping system. In this case, use two pipe wrenches. Follow these steps to tighten a fitting without using a vise:

Step 1 Position yourself with the fitting on the right side.

Step 2 Place one pipe wrench on the pipe with the jaw opening facing you.

Step 3 Place the other pipe wrench on the fitting with the jaw opening facing away from you. *Figure 8* shows the use of two pipe wrenches.

Step 4 Push on the handle of the wrench that is on the pipe while at the same time pulling on the handle of the wrench that is on the fitting.

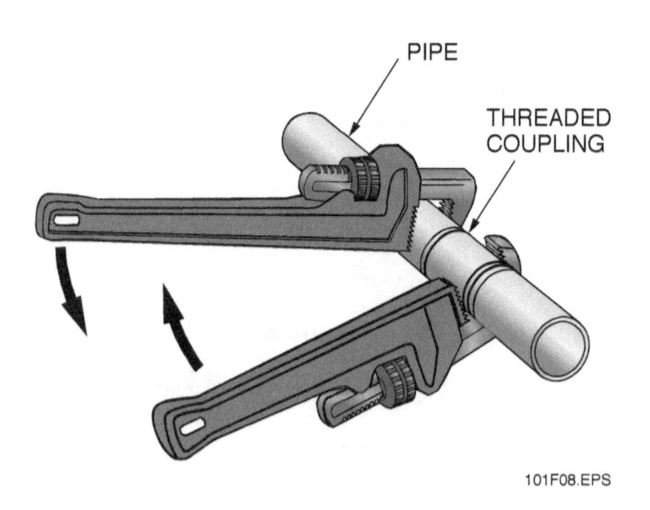

Figure 8 Tightening pipe.

Step 5 Regrip the pipe and the fitting when you have turned them as far as you can.

Step 6 Repeat Steps 2 through 5 until the fitting is tight.

Step 7 Remove the wrenches from the pipe and fitting.

Step 8 Clean the wrenches with a wire brush, lubricate them, and inspect for any damage.

Step 9 Close the jaws and store the wrenches.

8.1.2 Offset Pipe Wrenches

The offset pipe wrench operates the same way as the straight pipe wrench. It is made the same way and has similar components. The only difference is that the hook jaw and the heel are offset from the handle at either 45 degrees or 90 degrees. This wrench is useful when working in tight spaces such as in areas where pipes are located close to a ceiling or wall, or in other areas where you cannot grip and turn a pipe with a straight pipe wrench. *Figure 9* shows offset pipe wrenches. Follow the same steps for using straight pipe wrenches when using an offset pipe wrench.

8.2.0 Use and Care of Pipe Wrenches

Follow these guidelines to use and care for pipe wrenches:

- Use pipe wrenches only to turn pipe and fittings. Do not use a pipe wrench to bend, raise, or lift a pipe.
- Do not use the pipe wrench as a hammer.
- Do not drop or throw a pipe wrench. You may break or crack the wrench, and you may cause injury to yourself or others.

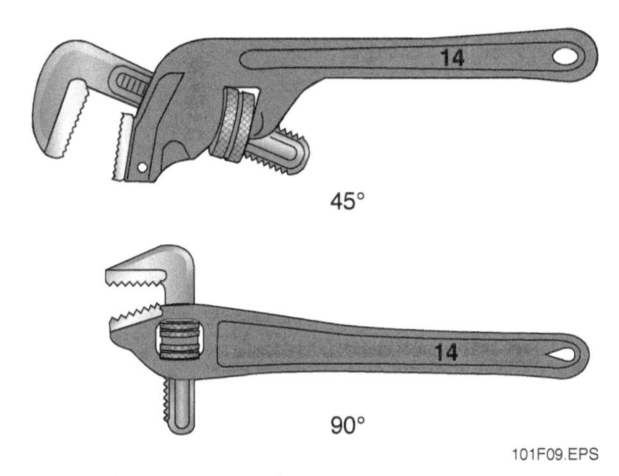

Figure 9 Offset pipe wrenches.

- Check the teeth of the wrench often and keep them clean and sharp. This will keep the wrench from slipping on the pipe.
- Apply penetrating oil to the wrench threads to prevent them from sticking.
- Never use a cheater or extension bar on a pipe wrench handle to increase leverage. This may cause the wrench components or the pipe to bend or break.
- Use the proper size wrench for the pipe.
- Ensure that the jaws are completely on the pipe before turning.
- Close and open the jaws by turning the threaded wheel.
- Pull on the wrench instead of pushing it. If pushing is necessary, use the base of your palm. This avoids skinning the knuckles if the pipe moves suddenly.
- Apply the torque at right angles to the pipe. Side force can spring the jaws and can break the grip unexpectedly.
- Double check to ensure that the applied force will close the jaws instead of opening them.
- Use a steady application of force and do not jerk on the handle or use a hammer on it.
- Clean all wrenches after each use and lubricate the moving parts.
- Repair or replace the pipe wrench if the jaws are sprung.
- Never use a pipe wrench on nuts or bolts. The jaws of a pipe wrench will strip the edges off a nut or bolt and make it difficult to move the nut or bolt with a standard wrench.

8.3.0 Chain Wrenches

Chain wrenches are used in the same manner as pipe wrenches. A chain wrench is a pipe wrench with a fixed, serrated jaw. A length of steel chain, attached to the pipe wrench, wraps around a pipe and connects to the other side of the jaw. This wrench holds the pipe firmly and distributes the bite of the jaw evenly. A ratchet action allows movement to turn the pipe either way and to tighten or loosen the pipe without removing the chain from the pipe. *Figure 10* shows a chain wrench.

The chain wrench is especially useful in close quarters. Follow these steps to use a chain wrench:

Step 1 Place the jaw of the wrench against the pipe.

Step 2 Wrap the chain around the pipe and lock the chain into the wrench frame.

NOTE

Be sure that the chain is tight around the pipe to ensure a good grip on the pipe.

Step 3 Push the handle to tighten the pipe; pull the handle to loosen the pipe.

Step 4 Slide the chain wrench around the pipe to regrip the pipe before turning again.

Step 5 Repeat Step 3 until the pipe is tightened or loosened enough to remove by hand.

Step 6 Remove the wrench from the pipe.

Step 7 Clean the wrench with a wire brush, lubricate it, and inspect for any damage.

Step 8 Store the wrench.

8.4.0 Strap Wrenches

A strap wrench is similar to a chain wrench except that a heavy-duty cloth or leather strap replaces the chain. A strap wrench is used to remove or install polished or plated valves and fittings that could be cosmetically damaged by these types of wrenches. The strap serves as the jaw of the wrench and is adjustable to provide a tight, secure grip. *Figure 11* shows a strap wrench. Follow the same steps for using chain wrenches when using strap wrenches.

Standard wrenches and vises cannot be used on finished surfaces such as brass or chrome.

Figure 10 Chain wrench.

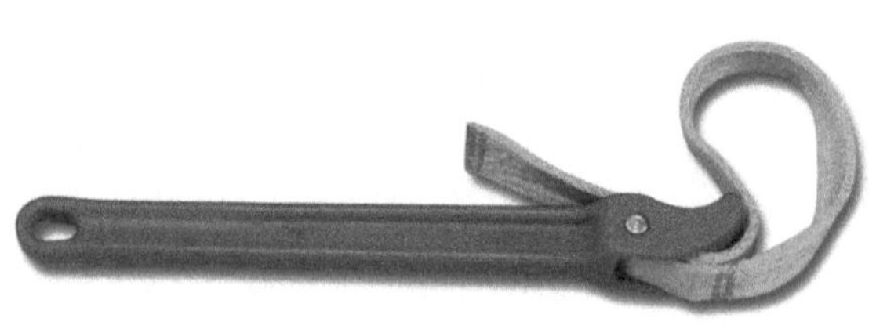

101F11.EPS

Figure 11 Strap wrench.

8.5.0 Sprinkler Wrenches

Installation of sprinklers requires a wrench manufactured by the sprinkler manufacturer for that particular make and model of sprinkler. This wrench is designed to fit snuggly on the sprinkler and not slip off and damage the sprinkler. Many sprinkler failures are the result of using improper wrenches. Often the damage is so small that it is not clearly visible. Only after catastrophic water damage and laboratory analysis is the damage revealed. *Figure 12* shows some typical sprinkler wrenches.

8.6.0 Torque Wrenches

Torque wrenches (see *Figure 13*) measure resistance to turning. You need them when you are installing fasteners that must be tightened in sequence without distorting the workpiece. You will use a torque wrench only when a torque setting is specified for a particular bolt.

Torque specifications are usually stated in inch-pounds for small fasteners or foot-pounds for large fasteners.

8.6.1 How to Use a Torque Wrench

Take these steps to use a torque wrench properly:

Step 1 Look on the tool to find out how many inch-pounds or foot-pounds you need to torque to. Set the controls on the wrench to the desired torque level (wrench models vary).

Step 2 Find out the torque sequence (which fastener comes first, second, and so on). If you use the wrong sequence, you could damage what you are fastening.

Step 3 Place the torque wrench on the object to be fastened, such as a bolt. Hold the head of the wrench with one hand to support the bolt and to make sure it is properly aligned.

Step 4 Watch the torque indicator to listen for the click (depending on the model of the wrench) as you tighten the bolt.

8.6.2 How to Calculate Torque When Using an Adaptor

When you use an adaptor or extension, the torque wrench becomes longer. The extra length and the applied torque will be greater than the torque indicator. Use this formula to determine the correct torque:

$$\text{Preset torque} = \frac{\text{length of torque wrench} \times \text{desired torque}}{\text{length of torque wrench} + \text{length of extension}}$$

To determine the length of the torque wrench, measure the distance from the center of the square drive of the wrench to the center of the handle. To determine the length of the extension, measure the distance from the center of the square drive of the extension to the center of the bolt or nut. Be sure to measure only the length that is parallel to the handle.

8.6.3 Safety and Maintenance

A torque wrench can cause property damage and injury if used incorrectly. Follow these guidelines when using a torque wrench:

- Always follow the manufacturer's recommendations for safety, maintenance, and calibration.
- Always store the wrench in its case.
- Never use the wrench as a ratchet or as anything other than its intended purpose.

8.6.4 Torque Wrench Calibration

The calibration of torque wrenches must be checked periodically to guarantee accuracy. The following terms, sometimes used in manufacturers' service literature, must be understood when using a torque wrench:

- *Breakaway torque* – The torque required to loosen a fastener. This is generally lower than the torque to which it has been tightened. For a given size fastener, there is a direct relationship between tightening torque and breakaway torque. This relationship is determined by an actual test. Once known, the tightening torque can be checked by loosening and checking breakaway torque.
- *Set or seizure* – In the last stages of rotation in reaching a final torque, seizing or set of the fastener may occur. When this happens, there is usually a noticeable popping sound and vibration. To break the set, back off and then again apply the tightening torque. Accurate torque settings cannot be made if the fastener is seized.

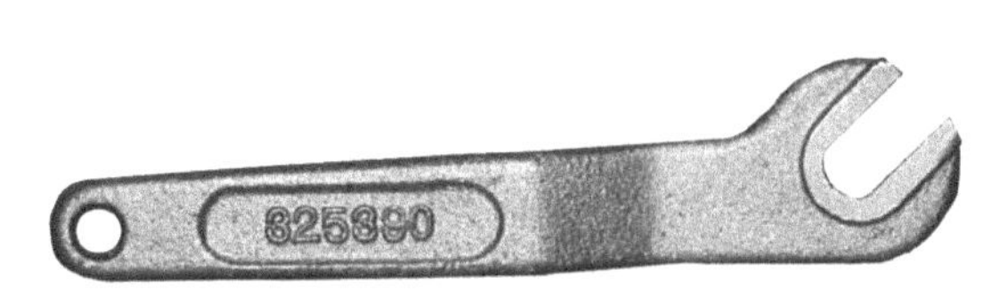

Figure 12 Typical sprinkler head wrenches.

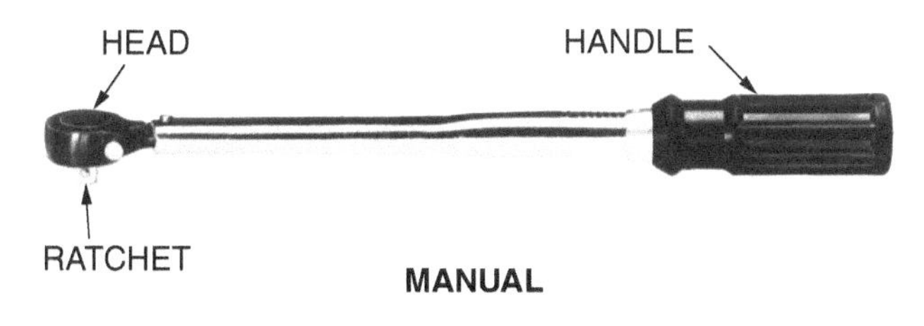

Image courtesy of Futek Advanced Sensor Technology.
DIGITAL

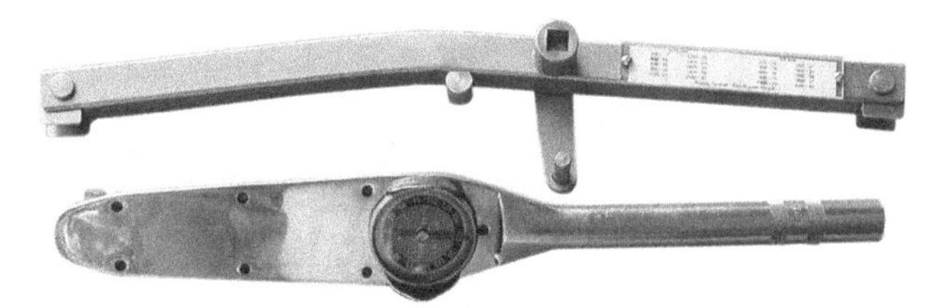

Figure 13 Torque wrenches.

- *Run-down resistance* – The torque required to overcome the resistance of associated hardware, such as locknuts and lockwashers, when tightening a fastener. To obtain the proper torque value where tight threads on locknuts produce a run-down resistance, add the resistance to the required torque value. Run-down resistance must be measured on the last rotation or as close to the makeup point as possible.

CAUTION

Tighten fasteners only small amounts at a time, following the proper tightening sequence for the bolt pattern and fastener type before reaching the final torque. Failure to follow the correct tightening sequence or overtightening can result in damage to the fasteners, any sealing gaskets, or the object being fastened.

8.7.0 Pliers

There are many types of pliers that are useful to sprinkler fitters. The most commonly used pliers are tongue-and-groove pliers and slip-joint pliers.

Tongue-and-groove pliers, sometimes called slip-groove pliers or channel-locks, are versatile utility pliers that work well on objects of all shapes. They range in size from 4½ to 16 inches with a jaw capacity of ⅞ to 4¼ inches. Pliers with long handles provide more leverage. The tongue-and-groove pliers are available with straight jaws or curved jaws. The jaws are joined with a heavy-duty nut and bolt and are self locking. They are especially useful for adjusting hanger rods. *Figure 14* shows straight jaw tongue-and-groove pliers.

Because of their shape, curved jaws are best for use with pipe and are preferable for the sprinkler fitter. *Figure 15* shows curved jaw tongue-and-groove pliers.

Slip-joint pliers are all-purpose tools that are most useful for gripping and turning round objects, such as hanger rods. Slip-joint pliers are available in thin nose and bent nose shapes. The bent nose type has a nose that is bent at a 30-degree angle, which keeps the hand clear of a flat surface while using the tool. The pliers range in length from 5 to 10 inches with jaw opening capacities from ¾ to 1½ inches. *Figure 16* shows slip-joint pliers.

101F14.EPS

Figure 14 Straight jaw tongue-and-groove pliers.

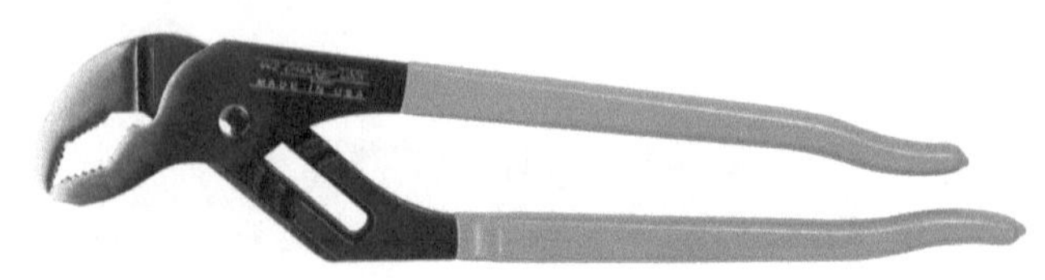

101F15.EPS

Figure 15 Curved jaw tongue-and-groove pliers.

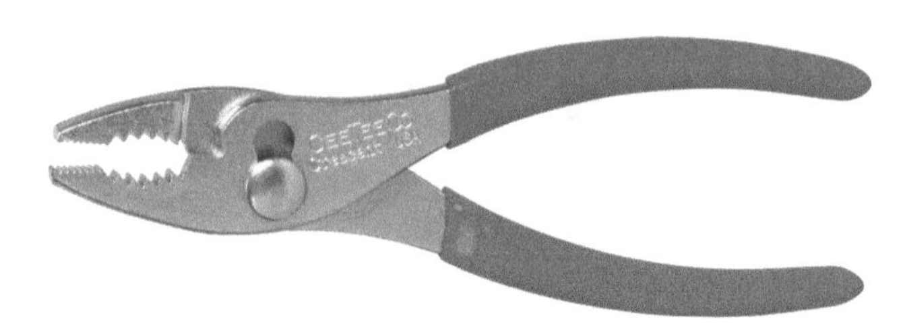

101F16.EPS

Figure 16 Slip-joint pliers.

8.8.0 Levels

A level is a device used to determine the trueness of a line. All levels have liquid-filled vials containing a bubble and two lines. Level occurs when the bubble is centered between the two lines. The longer the level, the more accurate the measurement.

The most important level for the fitter is a torpedo level. A torpedo level has three vials. A fitter can check pipes for vertical trueness, horizontal level, and true 45-degree angles with the torpedo level. The torpedo level is usually 9 inches long and is useful in tight spaces. Torpedo levels are available with magnetized bottoms to allow hand freedom while leveling steel pipe. *Figure 17* shows a torpedo level.

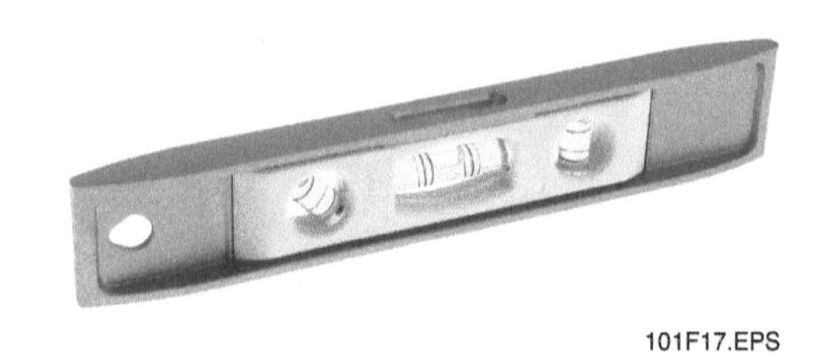

101F17.EPS

Figure 17 Torpedo level.

8.9.0 Hacksaws

Hacksaws are utility saws that are used to cut metal that is too heavy for bolt cutters or tin snips. The saw blades can be changed out to accommodate any particular sawing need. Replacement blades should be carried in case a blade breaks. The teeth of the blade should be curved forward, away from the handle, and the blade should be tight. *Figure 18* shows hacksaws.

Hacksaws cut on the push stroke. Apply pressure on the push stroke and lift on the return stroke. It is best to saw with long, slow, steady strokes.

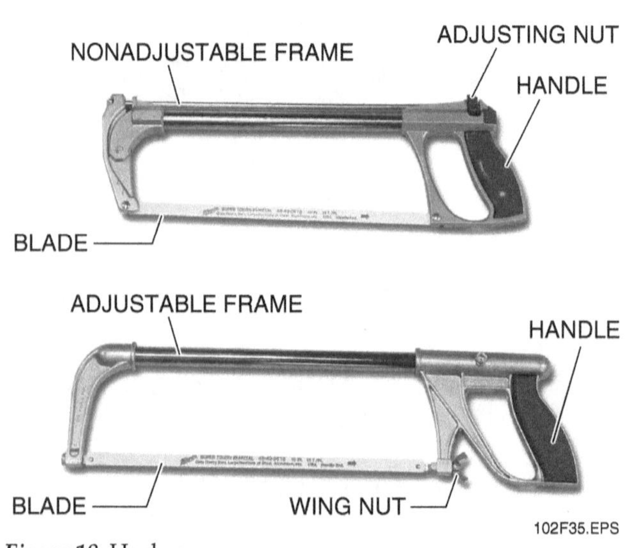

102F35.EPS

Figure 18 Hacksaws.

On Site

Laser Levels

Laser levels are widely used today to determine level and plumb. These levels do not use liquid-filled vials, but instead use low-power lasers that project either single or multiple continuous beams. Lasers are generally battery operated and self leveling. They are most commonly used in large open rooms where measurement to a common reference line is necessary.

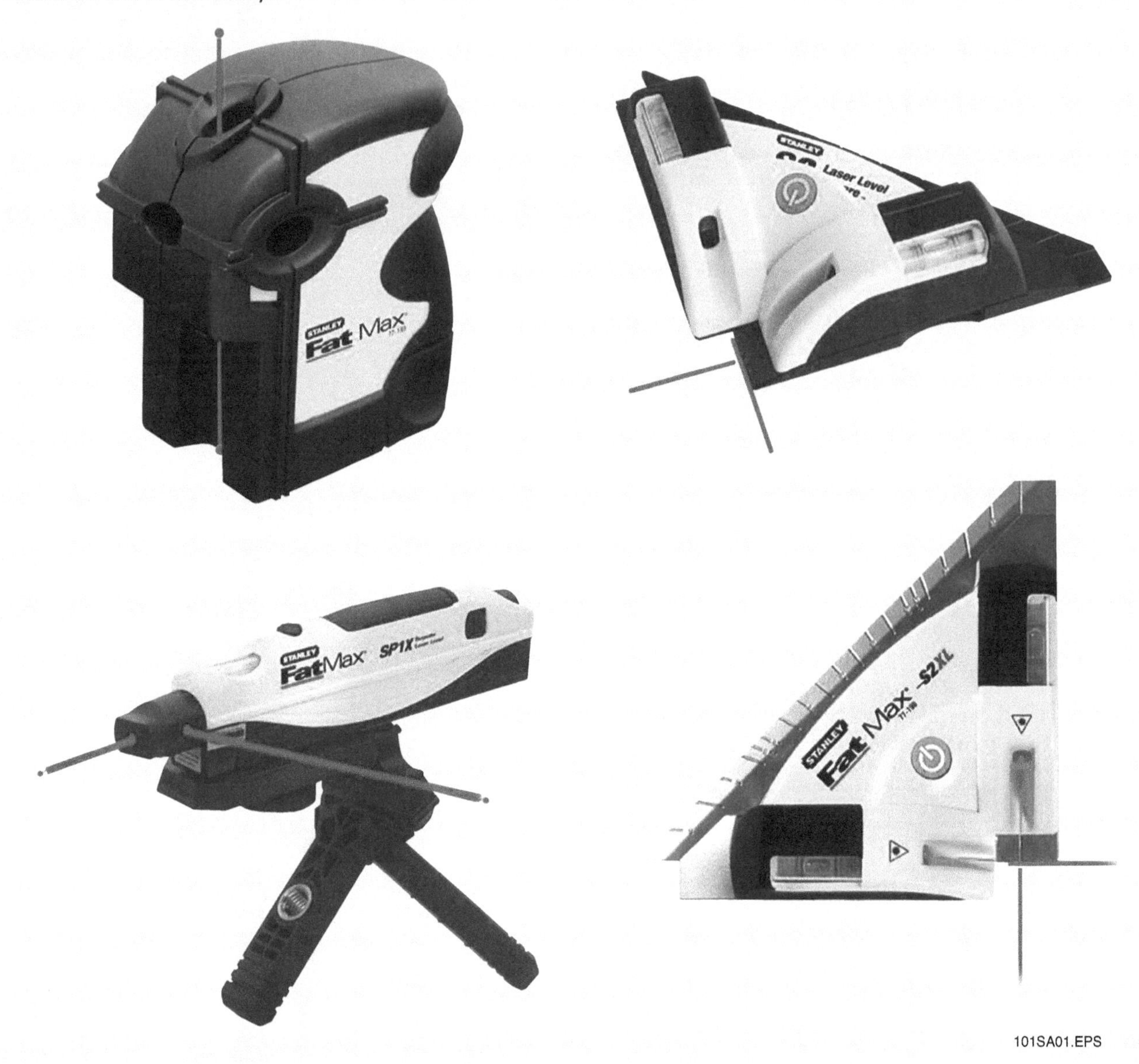

101SA01.EPS

8.10.0 Tool Boxes

A good quality tool box is needed to ensure that small hand tools are properly stored and protected. The tool box should have compartments so tools can be separated by size and type. When tools are well organized, time will not be wasted looking for one tool. A good lock for the tool box is needed to prevent tools from being stolen.

8.11.0 Ladders

The most commonly used ladder is the A-frame ladder (*Figure 19*). As with all ladders, they come in many different sizes and materials. Wood and fiberglass are the two most common materials used.

Extension trestle ladders, commonly referred to as extension or strut ladders, are three-piece ladders that allow for work at heights higher than the usual A-frame ladder. They are usually constructed of wood or fiberglass.

The extension trestle ladder (*Figure 20*) has a primary adjustable vertical ladder in the center with legs or trestles extending on either side. The safety rules for extension trestle ladders are the same as those for other ladders. Fall protection is required.

Observe the following basic safety rules when using a ladder:

- Never use a ladder horizontally as a platform.
- Never use a painted ladder or paint one. Paint can hide defects that may lead to an accident.
- Never stand on the top two steps of a stepladder or the third rung of an extension type ladder.
- Step in the middle of the rung or step so that your weight is evenly distributed.
- Always face the ladder when ascending or descending.
- Use only one rung or step at a time.
- Never reach beyond an easy arm's length.
- Ensure that the ladder is resting against a solid surface and that each leg is firmly positioned on a solid base.
- Ensure that a step ladder is always fully extended with the braces in the locked position.
- Extension ladders must be secured and extended three feet above the landing.

On Site

Portable Band Saws

Portable band saws are common where large amounts of metal cutting is necessary. These tools are electrically driven and have a metal cutting blade that is circular and moves in one direction around two wheels.

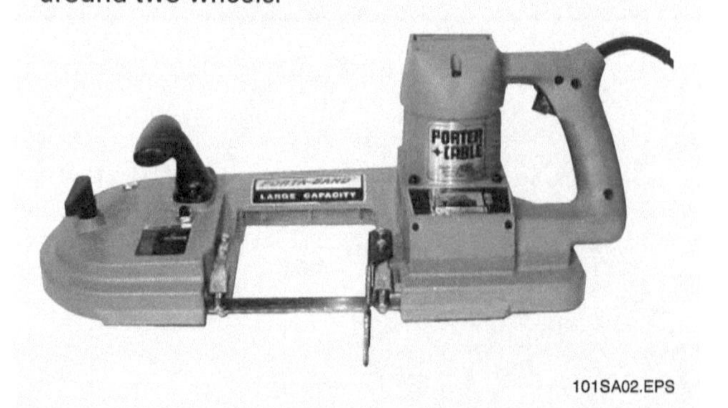

101SA02.EPS

On Site

Movable Leg Ladders

Movable leg, or multi-position ladders, are common in the industry. These ladders are collapsible and easily carried and stored compared to A-frame or extension ladders. Movable leg ladders can be adjusted for use on uneven surfaces such as stairs, or extended out to reach high areas only accessible with an extension trestle ladder. They are actually many different ladders in one.

101SA03.EPS

101F19.EPS

Figure 19 A-frame ladder.

9.0.0 Your Training Program

The Department of Labor's (DOL) **Office of Apprenticeship (OA)** sets the minimum standards for training programs across the country. These programs rely on mandatory classroom or correspondence courses of instruction and on-the-job learning (OJL). They require 144 hours of classroom or related instruction per year and 2,000 hours of OJL per year.

In a typical OA sprinkler fitter apprentice program, trainees spend 576 hours in classroom or related instruction and 8,000 hours in OJL before receiving journey certificates.

To address the training needs of the professional communities, NCCER developed a four-year sprinkler fitter training program. NCCER uses the minimum Department of Labor standards as a foundation for comprehensive curricula that provide trainees with in-depth classroom and OJL experience.

This NCCER Standardized Curriculum provides trainees with industry-driven training and education. It adopts a purely competency-based teaching philosophy. This means that trainees must demonstrate to the instructor that they possess the understanding and the skills necessary to perform the hands-on tasks that are covered in each module before they can advance to the next stage of the curriculum.

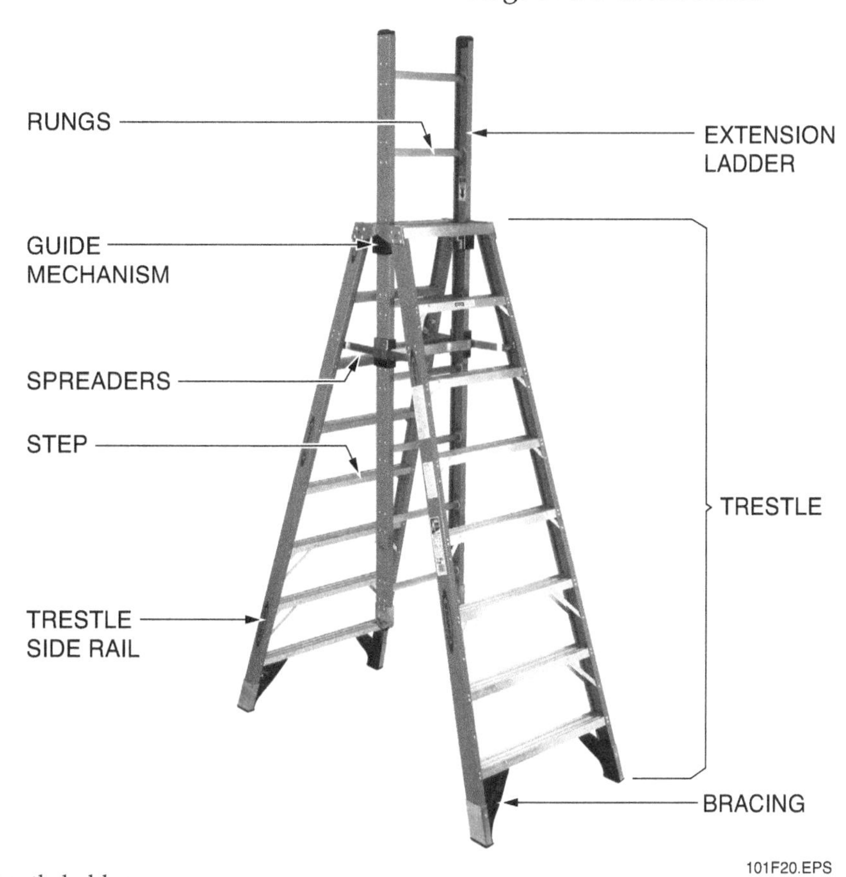

Figure 20 Extension trestle ladder.

When the instructor is satisfied that a trainee has successfully demonstrated the required knowledge and skills for a particular module, that information is sent to NCCER and kept in the National Registry. The National Registry can then confirm training and skills for workers as they move from state to state, company to company, or even within a company (see *Appendix B*).

The American Fire Sprinkler Association (AFSA) offers a correspondence course using the same NCCER curriculum. The AFSA course is highly flexible, combining correspondence course study, DVD audio/video supplemental training materials, support for supplemental classroom training, and on-the-job training. Test materials are provided to allow student testing on a periodic basis. Tests are sent to the AFSA headquarters for grading, and the company is notified of each student's test score. Options for paper-based or online testing are offered. The AFSA Sprinkler Fitter Correspondence Course has been approved for use in state or federally approved apprenticeship programs in most states. AFSA sample apprenticeship credentials are shown in *Appendix C*.

Whether you enroll in an NCCER program or another OA-approved program, ensure that you work for an employer or sponsor who supports a nationally standardized training program that includes credentials to confirm your skill development.

9.1.0 Apprenticeship Program

Apprentice training goes back thousands of years, and its basic principles have not changed in that time. First, it is a means for individuals entering the craft to learn from those who have mastered the craft. Second, it focuses on learning by doing; real skills versus theory. Although some theory is presented in this coursework, it is always presented in a way that helps the trainee understand the purpose behind the skill that is to be learned.

9.1.1 Apprenticeship Standards

All apprenticeship standards prescribe certain work-related or on-the-job training. This on-the-job training is broken down into specific tasks in which the apprentice receives hands-on training during the period of the apprenticeship. In addition, a specified number of hours is required in each task. The total number of on-the-job training hours for the sprinkler fitter apprenticeship program is traditionally 8,000, which amounts to about four years of training. In a competency-based program, it may be possible to shorten this time by testing out of specific tasks through a series of performance exams.

In a traditional program, the required on-the-job training may be acquired in increments of 2,000 hours per year. Layoff or illness may affect the duration.

The apprentice must log all work time and turn it in to the apprenticeship committee so that accurate time control can be maintained. After each 1,000 hours of related work, the apprentice may receive a pay increase as prescribed by the apprenticeship standards.

The course materials and related instruction and work-related training will not always run concurrently due to such reasons as layoffs, type of work needed to be done in the field, etc. Furthermore, apprentices with special job experience or coursework may obtain credit toward their classroom requirements. This reduces the total time required in the classroom while maintaining the total 8,000-hour on-the-job training requirement. These special cases will depend on the type of program and the regulations and standards under which it operates.

Informal on-the-job training provided by employers is usually less thorough than that provided through a formal apprenticeship program. The degree of training and supervision in this type of program often depends on the size of the employing firm. A small contractor may provide training in only one area, while a large company may be able to provide training in several areas.

For those entering an apprenticeship program, a high school or technical school education is desirable, as are courses in shop, mechanical drawing, and general mathematics. Manual dexterity, good physical condition, and quick reflexes are important. The ability to solve problems quickly and accurately and to work closely with others is essential. You must have a high concern for safety.

The prospective apprentice must submit certain information to the apprenticeship committee. This may include the following:

- Aptitude test (General Aptitude Test Battery or GATB Form Test) results (usually administered by the local Employment Security Commission)
- Proof of educational background (candidate should have school transcripts sent to the committee)
- Letters of reference from past employers and friends

- Results of a physical examination
- Proof of age
- If the candidate is a veteran, a copy of Form DD214
- A record of technical training received that relates to the construction industry and/or a record of any pre-apprenticeship training
- High school diploma or General Equivalency Diploma (GED)

The apprentice must do the following:

- Wear proper safety equipment on the job
- Purchase and maintain tools of the trade as needed and required by the contractor
- Submit a monthly on-the-job training report to the committee
- Report to the committee if a change in employment status occurs
- Attend classroom-related instruction and adhere to all classroom regulations such as attendance requirements

On Site

Child Labor Laws

Federal law establishes the minimum standards for workers under the age of 18. Some municipal jurisdictions may enforce stricter regulations. Employers are required to abide by the laws that apply to them.

The child labor provisions of the *Fair Labor Standards Act* forbid employers from using illegal child labor, and also forbid companies from doing business with any other business that does. The DOL investigates alleged abuses of the law. In such cases, employers have to provide proof of age for their employees.

In addition to the child labor provisions, employers in the construction trades are required to follow the DOL's *Child Labor Bulletin No. 101, Child Labor Requirements in Nonagricultural Occupations Under the Fair Labor Standards Act. Bulletin No. 101* does the following:

- Explains the coverage of the child labor provisions
- Identifies minimum age standards
- Lists the exemptions from the child laborprovisions
- Sets out employment standards for 14- and 15-year-old workers
- Defines the work that can be performed in hazardous occupations
- Provides penalties for violations of the child labor provisions
- Recommends the use of age certificates for employees

SUMMARY

In this module, you were introduced to four basic fire sprinkler systems: wet pipe, dry pipe, preaction, and deluge. Those fundamental systems form a starting point for your learning process and for the initial steps in your career. You saw that all sprinkler fitter work, including safety, is governed by codes and standards, particularly *NFPA 13* and the OSHA requirements.

There are a relatively large number of career choices in the sprinkler fitter industry and you learned that to advance and to be well respected in your field you must assume certain responsibilities such as honesty and integrity and many equally important personal traits.

It is not just learning the technical side and knowing codes and standards. You must also get along with and communicate effectively with your fellow workers. Human relations is a key aspect of sprinkler fitter work. Another key element is being aware of job safety.

You also saw some of the tools you will be using on the job. Tools are a crucial aspect of working as a sprinkler fitter. Know how to safely and efficiently use your tools.

You got a brief overview of your training and apprenticeship program. You should be aware that the construction industry in general, and the sprinkler fitter industry in particular, have invested a great deal of time and money to provide the training and career opportunities that are available to you.

There is always a strong demand for good service technicians. If you continue your education and can learn the building and fire codes, there are many opportunities to advance in the fire protection industry.

Review Questions

1. The most common type of fire sprinkler system is _____.
   a. wet pipe
   b. dry pipe
   c. proactive
   d. deluge

2. The sprinkler system you would consider for a below-freezing environment is _____.
   a. wet pipe
   b. dry pipe
   c. proactive
   d. deluge

3. The key advantage of a deluge sprinkler system is _____.
   a. it can be used anywhere
   b. that it is designed for below-freezing applications
   c. high volume and fast reaction
   d. it is much cheaper to install and maintain

4. The primary standard reference for sprinkler fitters is _____.
   a. *NFPA 13*
   b. *NFPA 20*
   c. *NFPA 24*
   d. *NFPA 5000*

5. Building codes and standards regulate all construction work in North America. Therefore, codes and standards are essentially the same thing.
   a. True
   b. False

For Questions 6 through 9, match the standard with the correct definition.

6. _____ *NFPA 13D*

7. _____ *NFPA 13R*

8. _____ *NFPA 20*

9. _____ *NFPA 24*

   a. Covers the selection and installation of stationary pumps supplying liquid for private fire protection.
   b. Covers the design and installation of automatic sprinkler systems in one- and two-family dwellings and manufactured homes.
   c. Sets the minimum requirements for installing private fire service mains and their equipment.
   d. Covers the design and installation of automatic sprinkler systems in residential occupancies up to four stories in height.
   e. Specifies the minimum requirements for design and installation of tanks and other equipment providing water for private fire protection.

10. The wrench most useful in removing or installing polished or plated devices such as valves is _____.
    a. a pipe wrench
    b. a chain wrench
    c. a strap wrench
    d. curved jaw pliers

Trade Terms Quiz

Fill in the blank with the correct trade term that you learned from your study of this module.

1. The online source for current information related to automatic fire sprinkler systems is _________.
2. The fire sprinkler system that has a supplemental detection system installed near the sprinklers is a(n) _________.
3. The type of fire sprinkler system that uses open sprinklers rather than closed sprinklers is a(n) _________.
4. A(n) _________ is a device used to secure pipe stored in tiers.
5. The _________ sets the minimum standards for training programs across the United States.
6. The simplest, most common type of sprinkler system is a(n) _________.
7. A(n) _________ is a heat-sensing device used in sprinkler systems to discharge water onto a fire.
8. _________ refers to job-related training acquired while working.
9. A possible source of danger or potential injury is called a(n) _________.
10. The standardized construction education materials produced by NCCER are called _________.
11. The person or agency that decides the acceptability of the systems and devices being installed is referred to as the _________.
12. _________ is a non-profit, international association representing merit shop fire sprinkler contractors.
13. The rules and specifications provided to the sprinkler industry are called _________.
14. Material that is easily ignited and burns rapidly is called _________.
15. _________ is stored under pressure in cylinders.
16. A(n) _________ is a device used to support pipe in tiers.
17. The industry reference for installing fire sprinkler systems is _________.
18. A sprinkler system with pipes filled with pressurized air or nitrogen is a(n) _________.
19. _________ publishes codes and standards with the goal of preventing loss of life and property.

Trade Terms

American Fire Sprinkler Association (AFSA)
Authority Having Jurisdiction (AHJ)
Block
Compressed gas
Deluge fire sprinkler system
Dry pipe sprinkler system
Flammable
Hazard
National Fire Protection Association (NFPA)
NCCER Standardized Curricula
NFPA 13, Standard for the Installation of Sprinkler Systems
Office of Apprenticeship (OA)
On-the-job-learning (OJL)
Preaction sprinkler system
Rack
Regulations
Sprinkler
SprinklerNet
Wet pipe sprinkler system

Appendix A

Conversion Table

METRIC CONVERSION CHART

INCHES		METRIC	INCHES		METRIC	INCHES		METRIC
Fractional	Decimal	mm	Fractional	Decimal	mm	Fractional	Decimal	mm
.	0.0039	0.1000	.	0.5512	14.0000	.	1.8898	48.0000
.	0.0079	0.2000	9/16	0.5625	14.2875	.	1.9291	49.0000
.	0.0118	0.3000	.	0.5709	14.5000	.	1.9685	50.0000
1/64	0.0156	0.3969	37/64	0.5781	14.6844	2	2.0000	50.8000
.	0.0157	0.4000	.	0.5906	15.0000	.	2.0079	51.0000
.	0.0197	0.5000	19/32	0.5938	15.0813	.	2.0472	52.0000
.	0.0236	0.6000	39/64	0.6094	15.4781	.	2.0866	53.0000
.	0.0276	0.7000	.	0.6102	15.5000	.	2.1260	54.0000
1/32	0.0313	0.7938	5/8	0.6250	15.8750	.	2.1654	55.0000
.	0.0315	0.8000	.	0.6299	16.0000	.	2.2047	56.0000
.	0.0354	0.9000	41/64	0.6406	16.2719	.	2.2441	57.0000
.	0.0394	1.0000	.	0.6496	16.5000	2 1/4	2.2500	57.1500
.	0.0433	1.1000	21/32	0.6563	16.6688	.	2.2835	58.0000
3/64	0.0469	1.1906	.	0.6693	17.0000	.	2.3228	59.0000
.	0.0472	1.2000	43/64	0.6719	17.0656	.	2.3622	60.0000
.	0.0512	1.3000	11/16	0.6875	17.4625	.	2.4016	61.0000
.	0.0551	1.4000	.	0.6890	17.5000	.	2.4409	62.0000
.	0.0591	1.5000	45/64	0.7031	17.8594	.	2.4803	63.0000
1/16	0.0625	1.5875	.	0.7087	18.0000	2 1/2	2.5000	63.5000
.	0.0630	1.6000	23/32	0.7188	18.2563	.	2.5197	64.0000
.	0.0669	1.7000	.	0.7283	18.5000	.	2.5591	65.0000
.	0.0709	1.8000	47/64	0.7344	18.6531	.	2.5984	66.0000
.	0.0748	1.9000	.	0.7480	19.0000	.	2.6378	67.0000
5/64	0.0781	1.9844	3/4	0.7500	19.0500	.	2.6772	68.0000
.	0.0787	2.0000	49/64	0.7656	19.4469	.	2.7165	69.0000
.	0.0827	2.1000	.	0.7677	19.5000	2 3/4	2.7500	69.8500
.	0.0866	2.2000	25/32	0.7813	19.8438	.	2.7559	70.0000
.	0.0906	2.3000	.	0.7874	20.0000	.	2.7953	71.0000
3/32	0.0938	2.3813	51/64	0.7969	20.2406	.	2.8346	72.0000
.	0.0945	2.4000	.	0.8071	20.5000	.	2.8740	73.0000
.	0.0984	2.5000	13/16	0.8125	20.6375	.	2.9134	74.0000
7/64	0.1094	2.7781	.	0.8268	21.0000	.	2.9528	75.0000
.	0.1181	3.0000	53/64	0.8281	21.0344	.	2.9921	76.0000
1/8	0.1250	3.1750	27/32	0.8438	21.4313	3	3.0000	76.2000
.	0.1378	3.5000	.	0.8465	21.5000	.	3.0315	77.0000
9/64	0.1406	3.5719	55/64	0.8594	21.8281	.	3.0709	78.0000
5/32	0.1563	3.9688	.	0.8661	22.0000	.	3.1102	79.0000
.	0.1575	4.0000	7/8	0.8750	22.2250	.	3.1496	80.0000
11/64	0.1719	4.3656	.	.8858	22.5000	.	3.1890	81.0000
.	0.1772	4.5000	57/64	.89063	22.6219	.	3.2283	82.0000
3/16	0.1875	4.7625	.	.9055	23.0000	.	3.2677	83.0000
.	0.1969	5.0000	29/32	.90625	23.0188	.	3.3071	84.0000
13/64	0.2031	5.1594	59/64	.92188	23.4156	.	3.3465	85.0000
.	0.2165	5.5000	.	.9252	23.5000	.	3.3858	86.0000
7/32	0.2188	5.5563	15/16	.93750	23.8125	.	3.4252	87.0000
15/64	0.2344	5.9531	.	.9449	24.0000	.	3.4646	88.0000
.	0.2362	6.0000	61/64	.95313	24.2094	3 1/2	3.5000	88.9000
1/4	0.2500	6.3500	.	.9646	24.5000	.	3.5039	89.0000
.	0.2559	6.5000	31/32	.96875	24.6063	.	3.5433	90.0000
17/64	0.2656	6.7469	.	.9843	25.0000	.	3.5827	91.0000
.	0.2756	7.0000	63/64	.98438	25.0031	.	3.6220	92.0000
9/32	0.2813	7.1438	1	1.000	25.40	.	3.6614	93.0000
.	0.2953	7.5000	.	1.0039	25.5000	.	3.7008	94.0000
19/64	0.2969	7.5406	.	1.0236	26.0000	.	3.7402	95.0000
5/16	0.3125	7.9375	.	1.0433	26.5000	.	3.7795	96.0000
.	0.3150	8.0000	.	1.0630	27.0000	.	3.8189	97.0000
21/64	0.3281	8.3344	.	1.0827	27.5000	.	3.8583	98.0000
.	0.3346	8.5000	.	1.1024	28.0000	.	3.8976	99.0000
11/32	0.3438	8.7313	.	1.1220	28.5000	.	3.9370	100.0000
.	0.3543	9.0000	.	1.1417	29.0000	4	4.0000	101.6000
23/64	0.3594	9.1281	.	1.1614	29.5000	.	4.3307	110.0000
.	0.3740	9.5000	.	1.1811	30.0000	4 1/2	4.5000	114.3000
3/8	0.3750	9.5250	.	1.2205	31.0000	.	4.7244	120.0000
25/64	0.3906	9.9219	1 1/4	1.2500	31.7500	5	5.0000	127.0000
.	0.3937	10.0000	.	1.2598	32.0000	.	5.1181	130.0000
13/32	0.4063	10.3188	.	1.2992	33.0000	.	5.5118	140.0000
.	0.4134	10.5000	.	1.3386	34.0000	.	5.9055	150.0000
27/64	0.4219	10.7156	.	1.3780	35.0000	6	6.0000	152.4000
.	0.4331	11.0000	.	1.4173	36.0000	.	6.2992	160.0000
7/16	0.4375	11.1125	.	1.4567	37.0000	.	6.6929	170.0000
.	0.4528	11.5000	.	1.4961	38.0000	.	7.0866	180.0000
29/64	0.4531	11.5094	1 1/2	1.5000	38.1000	.	7.4803	190.0000
15/32	0.4688	11.9063	.	1.5354	39.0000	.	7.8740	200.0000
.	0.4724	12.0000	.	1.5748	40.0000	8	8.0000	203.2000
31/64	0.4844	12.3031	.	1.6142	41.0000	.	9.8425	250.0000
.	0.4921	12.5000	.	1.6535	42.0000	10	10.0000	254.0000
1/2	0.5000	12.7000	.	1.6929	43.0000	20	20.0000	508.0000
.	0.5118	13.0000	.	1.7323	44.0000	30	30.0000	762.0000
33/64	0.5156	13.0969	1 3/4	1.7500	44.4500	40	40.0000	1016.000
17/32	0.5313	13.4938	.	1.7717	45.0000	60	60.0000	1524.000
.	0.5315	13.5000	.	1.8110	46.0000	80	80.0000	2032.000
35/64	0.5469	13.8906	.	1.8504	47.0000	100	100.0000	2540.000

TO CONVERT TO MILLIMETERS; MULTIPLY INCHES X 25.4
TO CONVERT TO INCHES; MULTIPLY MILLIMETERS X 0.03937*
*FOR SLIGHTLY GREATER ACCURACY WHEN CONVERTING TO INCHES; DIVIDE MILLIMETERS BY 25.4

201A01.EPS

SAMPLES OF NCCER TRAINING CREDENTIALS

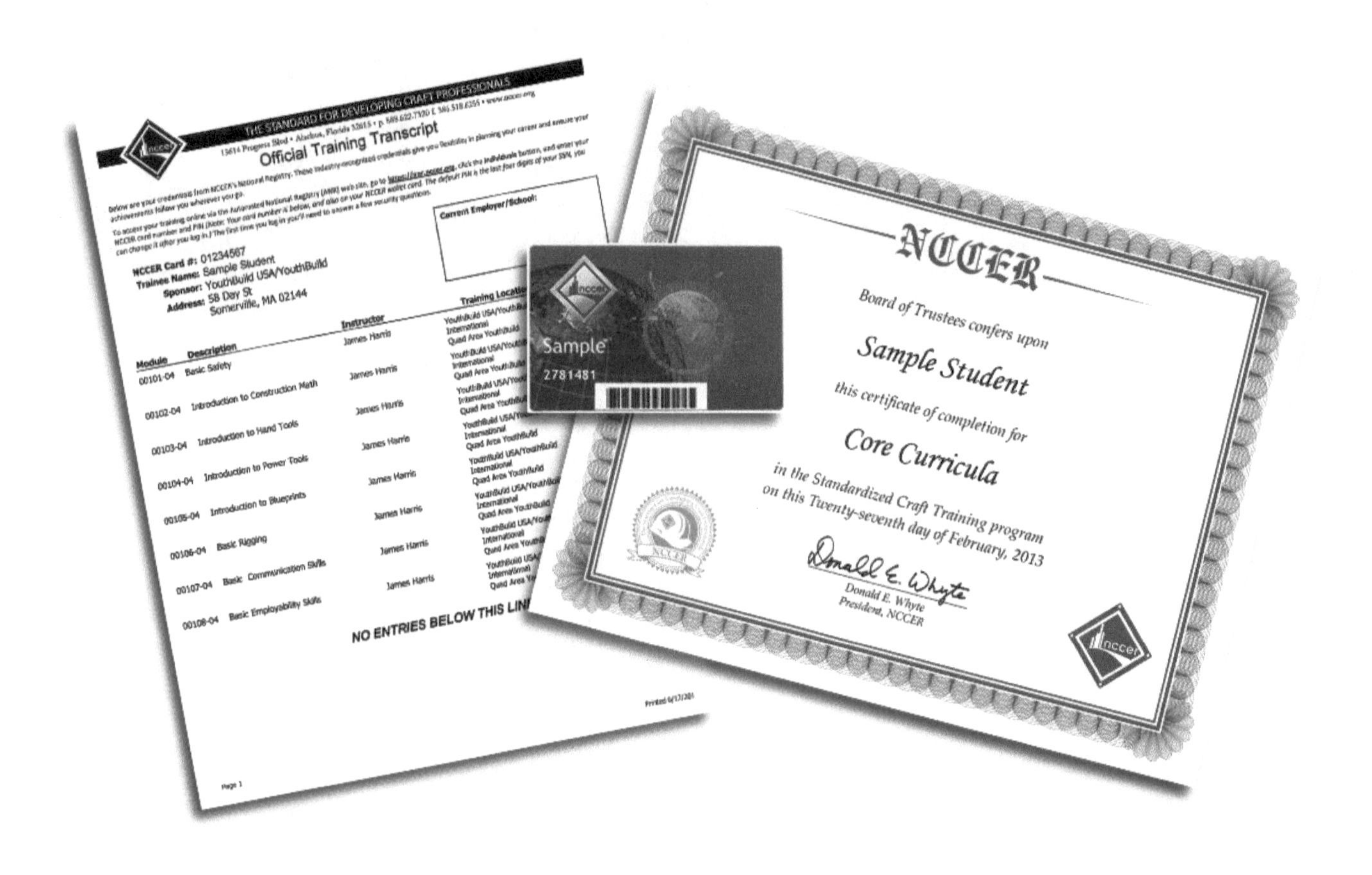

SAMPLES OF AFSA APPRENTICE TRAINING CREDENTIALS

AMERICAN FIRE SPRINKLER ASSOCIATION

AFSA

John Doe

HAS BEEN AWARDED A CERTIFICATE OF ACHIEVEMENT
IN RECOGNITION OF SUCCESSFUL COMPLETION OF

American Fire Sprinkler Association's
Contren® Learning Series
Four-Book Training Program in Automatic Fire Sprinkler Protection
For Apprentice Fire Sprinkler Fitters

Providing a Minimum of 576 Hours of Training

February 15, 2006
Date

President

American Fire Sprinkler Association
Apprenticeship Identification Card

Issued to: John Doe

The above named individual is enrolled in a course of study in an Apprenticeship Program for Automatic Fire Sprinkler Systems using the curriculum registered as a part of standards approved by the Bureau of Apprenticeship and Training, U.S. Department of Labor.

N-92021
Registration No.

Steve A. Muncy, CAE
President

Apprentice Credits

Book Completed	Date	CEU's Credited
Level I	6/30/10	16.0

12345
Student I.D. #

6/30/10
Date of Enrollment

AFSA Director of Education

12750 Merit Drive, Suite 350 • Dallas, Texas 75251 • (214) 349-5965

101A05.EPS

Cornerstone of Craftsmanship

Josh Stephens

Affordable Fire Protection, Inc. , Sales
Norcross, Georgia

During his apprenticeship, Josh Stephens won state and regional competitions. He took home a silver medal at the Associated Building Contractors (ABC) Craft Championships and the American Fire Sprinkler Association (AFSA) National Competition. He maintained an "A" average throughout his training. Not too bad for a guy who dropped out of college. After completing his apprenticeship, he has gone on to earn additional certifications and is working toward several more. He is motivated to complete his certification process, so he can open his own shop. Josh's story proves that with the right motivation, you can achieve anything.

How did you choose a career in the sprinkler fitting field?
I was working in another construction job, and the secretary told me that her husband was looking for someone over at Affordable Fire Protection. I took the new job because I thought there was more room in this field for advancement. This is more of a career than my previous job in carpet cleaning and fire restoration.

What types of training have you been through?
I had a few semesters of college, but I dropped out and went to work. Affordable offered me the opportunity to take the three-year apprenticeship program through the Construction Education Foundation of Georgia (CEFGA). During my apprenticeship, I was able to compete in the ABC and the AFSA national competitions. I won a silver medal at the national ABC Competition one year and a silver medal at the AFSA Competition the following year.

I like my company because they really encourage training. We have a lot of in-house training. I am also enrolled in the National Institute for Certification in Engineering Technologies (NICET). I am Level-II certified in Inspection and Testing of Water-Based Systems and am working on my Level III certification. When I achieve Level III, I will be able to perform inspections.

I am also working on my NICET certification in Automatic Sprinkler System Layout. I hope to own my own company one day. In this state, you must be certified in layout and design before you can own your own company.

What kinds of work have you done in your career?
I started five-and-one-half years ago as a helper carrying pipe. I was an installer for five years and worked my way up to foreman. After the national competitions, I was offered a job in sales, where I am now.

What do you like about your job?
Currently, I am enjoying sales because it gets me out of the elements. I get to meet new people every day and talk to them about life safety systems. I enjoy this industry. There is a lot of freedom in this business. You don't always have someone standing over you. I enjoy building systems. There is a lot of gratification when you can step back and say, "I built that." There is also a lot of satisfaction in knowing that you could save someone's life.

What factors have contributed most to your success?
Working for a good company has really helped me. CEFGA is also a great organization. They provided training from 5 to 9 p.m. one day a week. I was willing to put in the time and work hard because I knew that there was advancement in the future. These certification exams are really tough. It's like taking the SAT, an all-day exam on a Saturday. But it is worth it because I am working toward my goals.

What advice would you give to those new to the sprinkler fitter field?
Work hard and show that you are willing to grow in this industry, and you will be taken care of in turn.

Trade Terms Introduced in This Module

American Fire Sprinkler Association (AFSA): A non-profit, international association representing merit shop fire sprinkler contractors, dedicated to the educational advancement of its members and promotion of the use of automatic fire sprinkler systems.

Authority Having Jurisdiction (AHJ): The person or agency that decides the acceptability of the systems and devices being installed in accordance with code requirements.

Block: Device used to secure pipe stored in tiers.

Compressed gas: Gas stored under pressure in cylinders.

Deluge sprinkler system: A type of sprinkler system that uses open sprinklers rather than closed sprinklers. When a fire triggers the detection system, the deluge valve is released, producing immediate water flow through all sprinklers in a given area.

Dry pipe sprinkler system: Dry pipe systems are filled with pressurized air or nitrogen. Water flows into the system only when activated by a fire.

Flammable: Material that is easily ignited and burns rapidly.

Hazard: A possible source of danger or potential injury.

National Fire Protection Association (NFPA): Organization that publishes codes and standards with the goal of preventing loss of life and property.

NCCER Standardized Curricula: Standardized construction education materials produced by NCCER.

NFPA 13, The Standard for the Installation of Sprinkler Systems: The organization (NFPA) that provides the basic reference to the requirements (*NFPA 13*) for installing fire sprinkler systems.

Office of Apprenticeship (OA): The U.S. Department of Labor office that sets the minimum standards for training programs across the country.

On-the-job-learning (OJL): Job-related learning acquired while working.

Preaction sprinkler system: Uses automatic sprinklers attached to a piping system. The piping system contains air and may or may not be under pressure. There is a supplemental detection system installed near the sprinklers.

Rack: Device used to support pipe in tiers.

Regulations: Rules and specifications for the sprinkler industry provided by various organizations.

Sprinkler: A heat-sensing device used in sprinkler systems to discharge water onto a fire.

SprinklerNet: An online source for current information related to automatic fire sprinklers. www.sprinklernet.org.

Wet pipe sprinkler system: The simplest, most common type of sprinkler system. Wet pipe systems are filled with water at all times, compared to dry pipe systems which contain no water until the system is activated by a fire.

Additional Resources

This module presents thorough resources for task training. The following resource material is suggested for further study.

Multimedia Apprenticeship Training Supplement for Fire Sprinkler Fitters (CD set: Level 1 through Level 4). American Fire Sprinkler Association. www.sprinklernet.org.

NFPA 13, Standard for the Installation of Sprinkler Systems, Latest Edition. Quincy, MA: National Fire Protection Association.

NFPA 13D, Standard for the Installation of Sprinkler Systems in One- and Two-Family Dwellings and Manufactured Homes, Latest Edition. Quincy, MA: National Fire Protection Association.

NFPA 13R, Standard for the Installation of Sprinkler Systems in Residential Occupancies up to and Including Four Stories in Height, Latest Edition. Quincy, MA: National Fire Protection Association.

Figure Credits

© Wendy Kaveney Photography, 2010. Used under license from Shutterstock.com, Module opener

Tyco Fire and Building Products, 101F01

Western Fire Protection, Inc., 101F05, 101F06

Ridge Tool Co./Ridgid®, 101F10

Reed Manufacturing Company, 101F11

Globe Fire Sprinkler Corporation, 101F12

Klein Tools, Inc., 101F13 (manual torque wrench)

Futek Advanced Sensor Technology, 101F13 (digital torque wrench)

Holloway Engineering, 101F13 (tensionmeter)

Channellock, Inc., 101F14, 101F15

Courtesy Cooper Hand Tools, 101F16

The Stanley Works, 101F17, 101SA01

Topaz Publications, Inc., 101F18

Porter-Cable Corporation, 101SA02

Werner Company, 101F19, 101F20, 101SA03

American Fire Sprinkler Association, Appendix C

MODULE 18101-13 — ANSWERS TO REVIEW QUESTIONS

Answer		Section
1.	a	2.2.0
2.	b	2.3.0
3.	c	2.5.0
4.	a	3.0.0
5.	b	3.1.0; 3.2.0
6.	b	3.4.2
7.	d	3.4.3
8.	a	3.4.4
9.	c	3.4.6
10.	c	8.4.0

MODULE 18101-13 — ANSWERS TO TRADE TERMS QUIZ

1. SprinklerNet
2. Preaction sprinkler system
3. Deluge sprinkler system
4. Block
5. OA
6. Wet pipe sprinkler system
7. Sprinkler head
8. OJL
9. Hazard
10. NCCER Standardized Curricula
11. Authority Having Jurisdiction
12. AFSA
13. Regulations
14. Flammable
15. Compressed gas
16. Rack
17. *NFPA 13*
18. Dry pipe sprinkler system
19. NFPA

NCCER CURRICULA — USER UPDATE

NCCER makes every effort to keep its textbooks up-to-date and free of technical errors. We appreciate your help in this process. If you find an error, a typographical mistake, or an inaccuracy in NCCER's curricula, please fill out this form (or a photocopy), or complete the online form at **www.nccer.org/olf**. Be sure to include the exact module ID number, page number, a detailed description, and your recommended correction. Your input will be brought to the attention of the Authoring Team. Thank you for your assistance.

Instructors – If you have an idea for improving this textbook, or have found that additional materials were necessary to teach this module effectively, please let us know so that we may present your suggestions to the Authoring Team.

NCCER Product Development and Revision

13614 Progress Blvd., Alachua, FL 32615

Email: curriculum@nccer.org
Online: www.nccer.org/olf

❑ Trainee Guide ❑ AIG ❑ Exam ❑ PowerPoints Other ______________________________

Craft / Level: Copyright Date:

Module ID Number / Title:

Section Number(s):

Description:

Recommended Correction:

Your Name:

Address:

Email: Phone:

Introduction to Components and Systems

18102-13

SPRINKLER FITTING LEVEL ONE

18106-13 Underground Pipe

18105-13 Copper Tube Systems

18104-13 CPVC Pipe and Fittings

18103-13 Steel Pipe

18102-13 Introduction to Components and Systems

18101-13 Orientation to the Trade

Core Curriculum: Introductory Craft Skills

This course map shows all of the modules in *Sprinkler Fitting Level One*. The suggested training order begins at the bottom and proceeds up. Skill levels increase as you advance on the course map. The local Training Program Sponsor may adjust the training order.

Objectives

When you have completed this module, you will be able to do the following:

1. Define the term *Listed* and explain how the term relates to sprinkler systems.
2. Explain the purpose of a Listing agency.
3. Describe the characteristics of common sprinklers.
4. State the important characteristics of aboveground pipe, including wall thickness and joining methods.
5. Define C-factor and list the advantages of a higher C-factor.
6. Describe the types of pipe hangers and sway bracing.
7. Identify the characteristics of control valves, check valves, water flow alarms, and fire department connections.

Trade Terms

C-factor
Check valve
Control valve
Drain valve
Drop/suspended ceiling
Electrolysis
FM Global
Galvanized
Inside diameter (ID)
K-factor
Labeled
Listed
Outside diameter (OD)
Sway bracing
Underwriters Laboratories® (UL)
Water hammer

Prerequisites

Before you begin this module, it is recommended that you successfully complete *Core Curriculum* and *Sprinkler Fitting Level One*, Module 18101-13.

NOTE

Unless otherwise specified, references in parentheses following figure and table numbers refer to *NFPA 13*.

Contents

Topics to be presented in this module include:

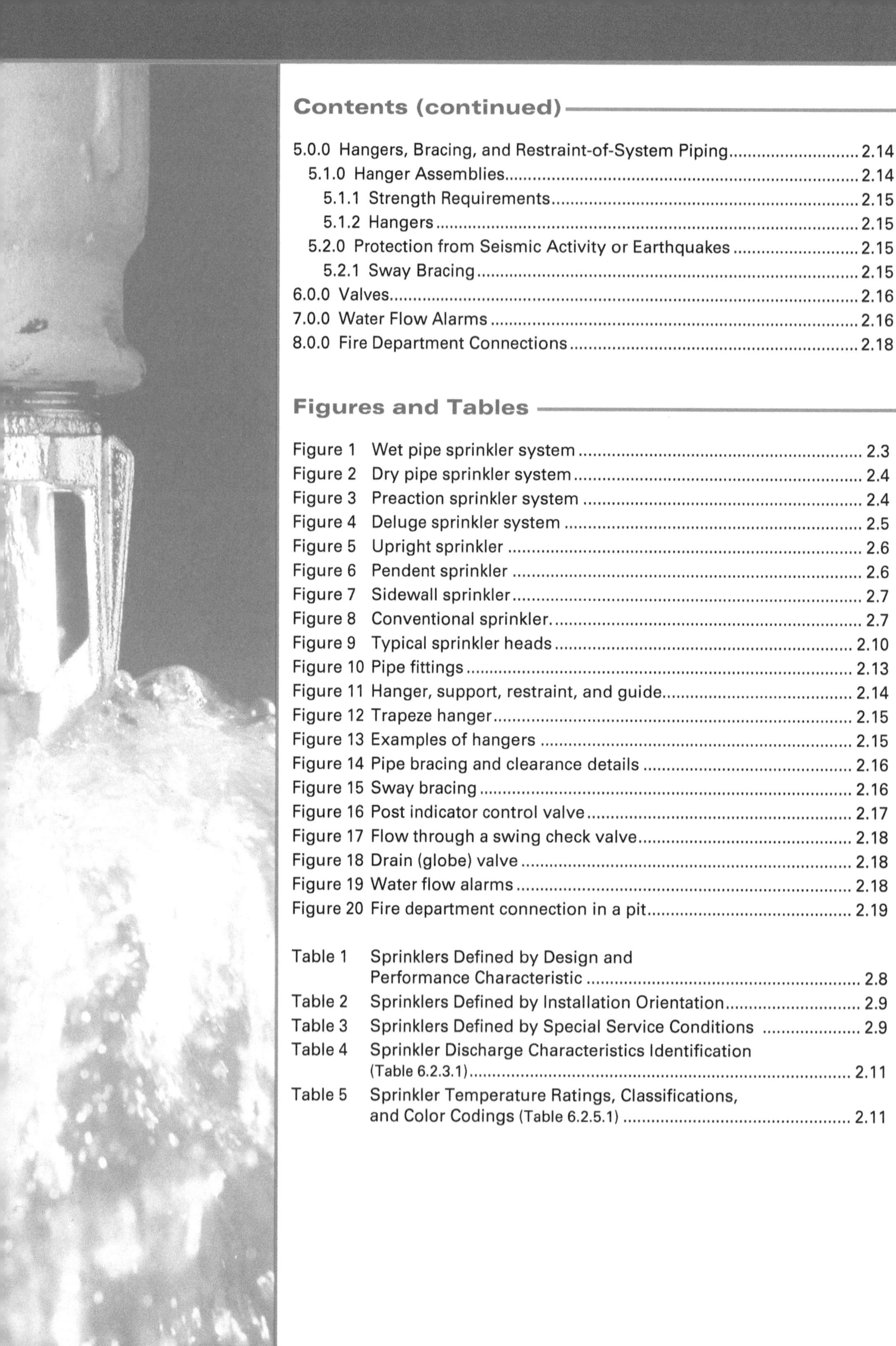

Contents (continued)

Figures and Tables

1.0.0 INTRODUCTION

Before you can understand how a fire sprinkler system works, you must learn the language of the industry and understand the components that make up such a system.

Many other trades, including plumbing, pipefitting, and HVAC workers install and join steel, plastic, and copper piping, valves, and fittings as part of their day-to-day activities. An added element in the sprinkler fitter trade is that the systems are being designed and installed to protect life and property. Therefore, sprinkler systems are subject to more rigorous codes and must meet more stringent installation, testing, and maintenance requirements.

The basic written reference for the fire sprinkler industry is *NFPA 13*. This agency, working with experts in various industries, develops and publishes codes related to fire safety, electrical work, medical care facilities, and other areas.

1.1.0 Testing Laboratories

Testing laboratories are an integral part of the development of codes and standards. The NFPA and other organizations rely on testing laboratories to conduct research into fire protection equipment and its safety. These laboratories perform extensive testing of new products to make sure that they are built to appropriate standards for fire safety. They receive statistics and reports from agencies all over the United States concerning fires and their causes. Upon seeing developing trends concerning the association of certain equipment and dangerous situations or circumstances, the equipment is specifically targeted for research. Reports from these laboratories are considered in the revisions of codes.

Did You Know?

The National Fire Protection Association (NFPA) reports that from 2003 to 2007, property damage from hotel fires was 53 percent less when the structures were protected with sprinklers, compared to structures without sprinklers. The buildings with sprinklers had a $9,000 average loss while the buildings with no sprinklers had losses averaging $19,000.

1.2.0 Listing Agencies

The main components of a sprinkler system, such as sprinkler heads, valves, and fire pumps, must be Listed for use in a given fire sprinkler application. The term *Listed* means that a specific part has been tested by a recognized listing agency and found suitable for a specified use in a fire sprinkler system. Two common listing agencies are Underwriters Laboratories® (UL) and FM Global. Listed items may only be installed for use in fire suppression applications and must be installed in accordance with the manufacturer's instructions.

1.2.1 Underwriters Laboratories®

Underwriters Laboratories® is a nonprofit organization whose principle business is to evaluate products to determine their compliance with defined standards for safety. UL evaluates electrical and mechanical products, building materials, construction systems, fire protection equipment, and marine products. UL's standards for safety establish the basis for testing products and levels of acceptability. Once a product has successfully passed testing, it may be Listed and have the UL label affixed to it. Through international agreements, UL is authorized to test and certify

On Site

Underwriters Laboratories®

The Chicago World's Fair was opened in 1893 and thanks to Edison's introduction of the electric light bulb, included a light display called the Palace of Electricity. But all was not perfect—wires soon sputtered and crackled, and, ironically, the Palace of Electricity caught fire. The fair's insurance company called in a troubleshooting engineer, who, after careful inspection, found faulty and brittle insulation, worn out and deteriorated wiring, bare wires, and overloaded circuits.

This engineer, William Henry Merrill, called for standards in the electrical industry, and then set up a testing laboratory above a Chicago firehouse to do just that. Underwriters Electrical Bureau, an independent testing organization, was born. UL is now an internationally recognized authority on product safety testing and safety certification and standards development.

specific types of products for use in both U.S. and Canadian markets.

UL's fire protection division provides insurance companies, Authorities Having Jurisdiction (AHJs), retailers, product manufacturers, and building owners with certification and customized testing services in many areas including:

- Fire modeling
- Ignition, smoke, and fire effluent (by-product) testing
- Large-scale fire research

Products tested at UL get Listed in a UL product directory. The product directory is a resource used to determine the correct application of a product or system. A product directory contains the names of companies qualified to use a UL mark on products that comply with UL requirements. Architects, AHJs, purchasing agents, and others use UL's product directories to find information about products bearing the UL mark.

As a fundamental element of UL's safety business, the fire protection division offers enhanced services including customer solutions and bundles services. The fire protection division can also develop custom certification and testing programs with technical expertise.

The fire protection division's suppression team supports manufacturers, AHJs, building owners, and insurance companies by offering certification and customized testing services in the following areas:

- Automatic sprinklers
- Extinguishing system units
- Fire extinguishers
- Fire pumps and engines
- Residential sprinklers

On Site

UL Testing

For more than 100 years, UL has been testing and certifying automatic sprinklers. The tests and certifications are based on the *Standards for Automatic Sprinklers for Fire-Protection Services (UL 199)* and *Early Suppression—Fast Response Sprinklers (UL 1767)*. UL is also capable of testing sprinklers sampled from field installations as specified in *NFPA 25, Standard for Testing and Maintenance of Water Extinguishing Systems*. This type of testing is a service to building owners, AHJs, and insurance companies. UL provides both large-scale testing (fires) and small-scale testing (water drop size, velocity, and heat release rate).

- Sprinkler piping systems
- Valves for fire suppression systems
- Water mist nozzles

1.2.2 FM Global

Because project specifications, contracts, and building codes often require fire protection systems to comply with FM Global standards, you should be familiar with the FM Global approval process. FM Global is a nonprofit scientific research and testing organization managed by FM Global, the world's largest commercial and industrial property insurance organization.

FM Global distinguishes between approval and acceptance. An approval is a confirmation and a subsequent listing by FM Global that a product has been examined according to FM Global's applicable requirements and found suitable for use in all instances, subject to any limitations stated in the approval. An acceptance confirms that materials installed at a specific location are suitable for their intended uses. In a sense, approval applies to multiple locations, and acceptance is considered on a case-by-case basis. FM Global's acceptances typically are limited to properties FM Global insures.

FM Global offers worldwide certification and testing services of industrial and commercial loss prevention products. FM Global approval certification assures customers that a product or service has been objectively tested and conforms to national and international standards. Some services provided by FM Global are as follows:

- *Product certification* – FM Global offers certification services to manufacturers of fire protection equipment, electrical equipment, hazardous location equipment, fire detection, signaling and other electrical equipment, materials, and roofing products. Products that earn approval are listed in the approval guide.
- *Specification testing* – Narrower in scope than the approvals program, FM Global specification testing allows manufacturers to test a single performance characteristic of a system or assembly.
- *Global certification* – FM Global has mutual agreements with testing and certification labs around the world.

FM Global is an international leader in third-party certification and approval of commercial and industrial products. FM Global provides certification for various types of products, including the following:

- Fire detection, signaling, and other electrical equipment
- Fire protection equipment
- Functional safety assessment and certification

1.3.0 Definitions

- *Approved* – Acceptable to the Authority Having Jurisdiction. (Also called *accepted*.)
- *Authority Having Jurisdiction (AHJ)* – An organization, office, or individual responsible for enforcing the requirements of a code or standard, or for approving equipment, materials, an installation, or a procedure.
- *Equipment Listed for the application* – When used, this term requires that equipment used in a system be properly Listed for the specific application. It is a common misunderstanding for people to assume that when an item is Listed, it is Listed for all applications. In the listing process, a device becomes Listed under one or more standards, and should therefore meet the performance requirements of the standard(s) if properly installed, tested, and maintained.
- *Equivalent* – The "or equivalent" clause in most standards allows the use of systems, methods, or devices of equal or better quality, as long as sufficient documentation is provided to the AHJ to demonstrate the equivalency.
- *Labeled* – Equipment or materials to which has been attached a label, symbol, or other identifying mark of an organization that is acceptable to the AHJ. Labeling requires periodic inspection of production of labeled equipment or materials and the manufacturer indicates compliance with appropriate standards or performance in a specified manner.
- *Listed* – Equipment, materials, or services included in a list published by an organization that is acceptable to the AHJ and concerned with evaluation of products or services, that maintains periodic inspection of production of listed equipment or materials or periodic evaluation of services, and whose listing states that either the equipment, material, or service meets appropriate designated standards or has been tested and found suitable for a specified purpose.
- *Shall* – Indicates a mandatory requirement.
- *Should* – Indicates a recommendation or that which is advised but not required.
- *Standard* – A document, the main text of which contains only mandatory provisions using the word "shall" to indicate requirements and which is in a form generally suitable for mandatory reference by another standard or code or for adoption into law. Nonmandatory provisions shall be located in an appendix or annex, footnote, or fine-print note and are not considered a part of the requirements of a standard.

2.0.0 Sprinkler Systems

The following sections describe the operation of four sprinkler systems and briefly describes their major components and accessories.

2.1.0 System Types and General Operation

There are three fundamental types of sprinkler systems: wet pipe (by far the most common system in use today), dry pipe, and preaction. A fourth system, deluge, is a variation of the preaction type. Each of these systems is designed for specific applications, situations, and conditions.

2.1.1 Wet Pipe Sprinkler System

In wet pipe systems (*Figure 1*), the pipes are always filled with water. When heat activates the thermal linkage in a sprinkler, water is discharged onto the fire. When a sprinkler (1) opens, water flow in the system lifts the alarm valve clapper (2), which allows water to flow to the open sprinkler and through the alarm port (3).

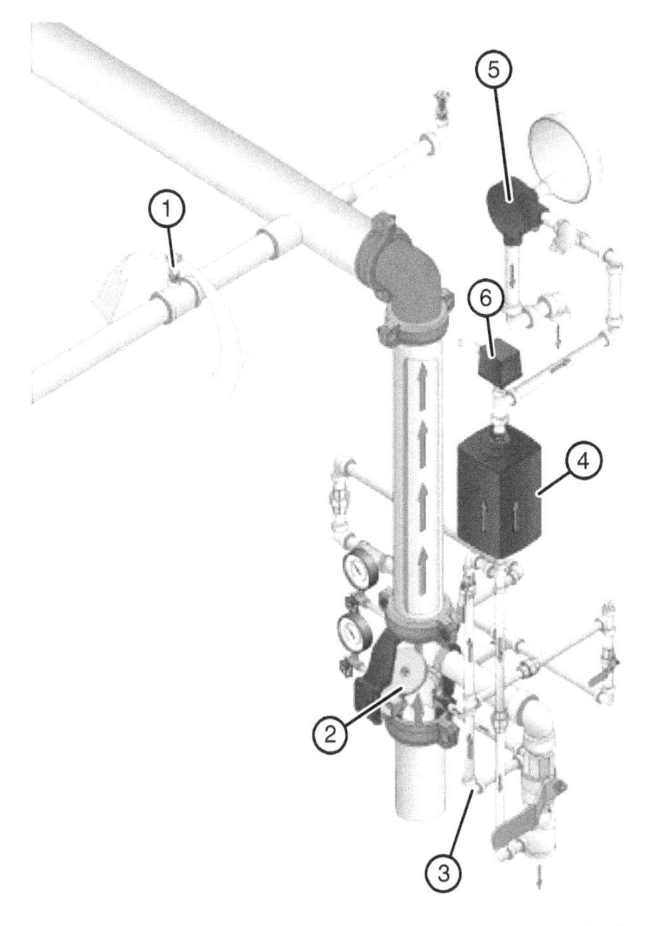

Figure 1 Wet pipe sprinkler system.

Flow from the alarm port enters the retard chamber (4). Water flow from the retard chamber enters the water motor alarm (5) and/or the optional pressure switch (6) causing an electric alarm bell to sound.

2.1.2 Dry Pipe Sprinkler System

Dry pipe sprinkler systems (*Figure 2*) are filled with pressurized air or nitrogen, not water. If an area is subject to freezing temperatures, a wet pipe system should not be used. Dry pipe systems are better for applications subject to freezing temperatures.

When a sprinkler (1) opens, the air in a dry pipe system escapes and is replaced by water. Once that happens, the system functions just like a wet pipe system. After the sprinkler opens, a reduction of air pressure in the system causes the dry valve clapper (2) to open. The open clapper floods the system with water. An accelerator (3) can be added to increase the opening speed of the clapper. The accelerator is protected against flooding by an integral anti-flood device. A pressure switch (4) is activated, and an alarm sounds, due to water flow from the intermediate chamber of the dry valve.

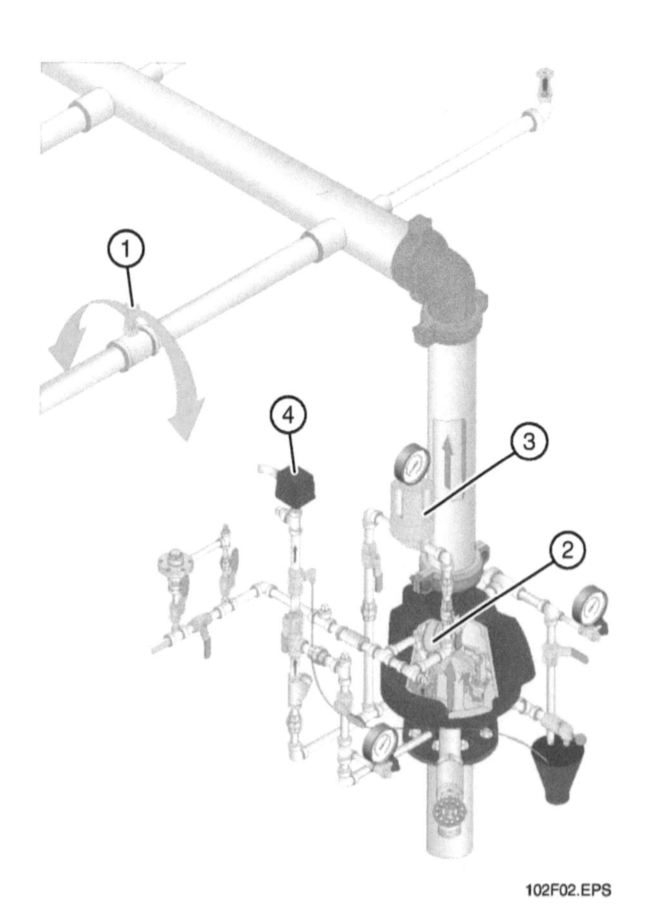

Figure 2 Dry pipe sprinkler system.

2.1.3 Preaction Sprinkler System

A preaction system uses automatic sprinklers attached to a piping system. The piping system contains air and may or may not be under pressure. There is a supplemental detection system installed near the sprinklers.

The preaction system operates somewhat differently than a wet or dry pipe system. The alarm is given before the sprinkler opens to discharge water.

Preaction systems are designed to protect properties in which the danger of serious water damage from premature operation of sprinklers or broken pipes is unusually severe, or those properties that, because of climate, must be equipped with a dry system. In a preaction system, the action of a heat-responsive device releases a preaction valve, which causes water to fill the piping system and an advance alarm to sound before the activation of sprinklers. The alarm often gives an opportunity to extinguish the fire by manual means before a sprinkler activates.

System operation (*Figure 3*) begins when a detector (1) is activated by fire and sends a signal to the release control panel (2). The control panel sends a release signal to a solenoid valve (3). The released solenoid valve allows the valve priming chamber (4) to vent faster than water is supplied through the restricted orifice (5). The venting allows the deluge valve to open, causing water to enter the system. However, until a sprinkler (6) opens, no water is discharged.

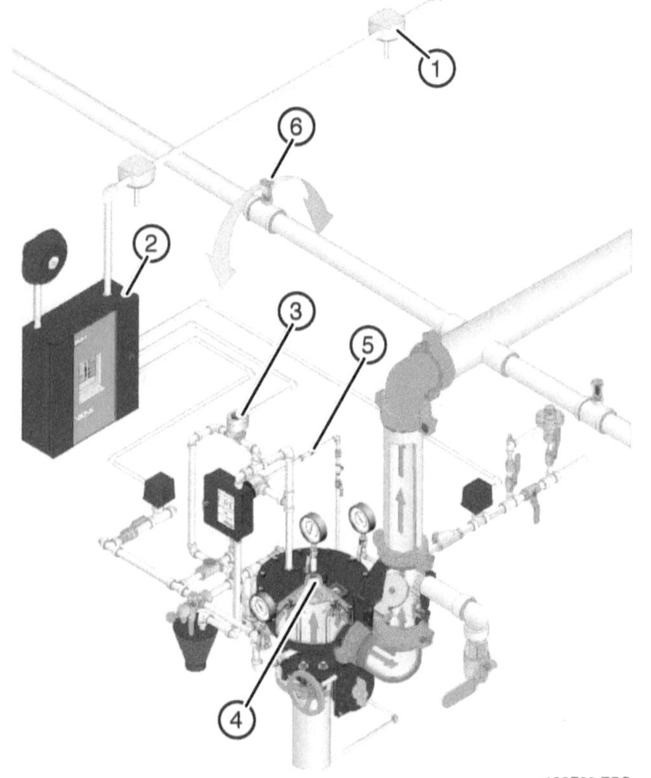

Figure 3 Preaction sprinkler system.

2.1.4 Deluge Sprinkler System

A deluge system (*Figure 4*) is a type of preaction sprinkler system that uses open sprinklers. When a fire triggers the detection system, the deluge valve is released, which produces immediate water flow through all sprinklers in a given area. These systems are typically found in places like aircraft hangers and chemical plants where high volume and fast reaction are required.

The purpose of a deluge system is to wet down an entire area in which a fire may originate. This is done by flowing water to open sprinklers and nozzles, rather than by using automatic sprinklers that would open independently as the fire spreads.

System operation is initiated by heat triggering the fixed temperature release (1). Pressure is released, resulting in the pneumatic actuator (2) opening. When the actuator opens, pressure is released from the priming chamber (3) of the deluge valve, causing the deluge valve to open.

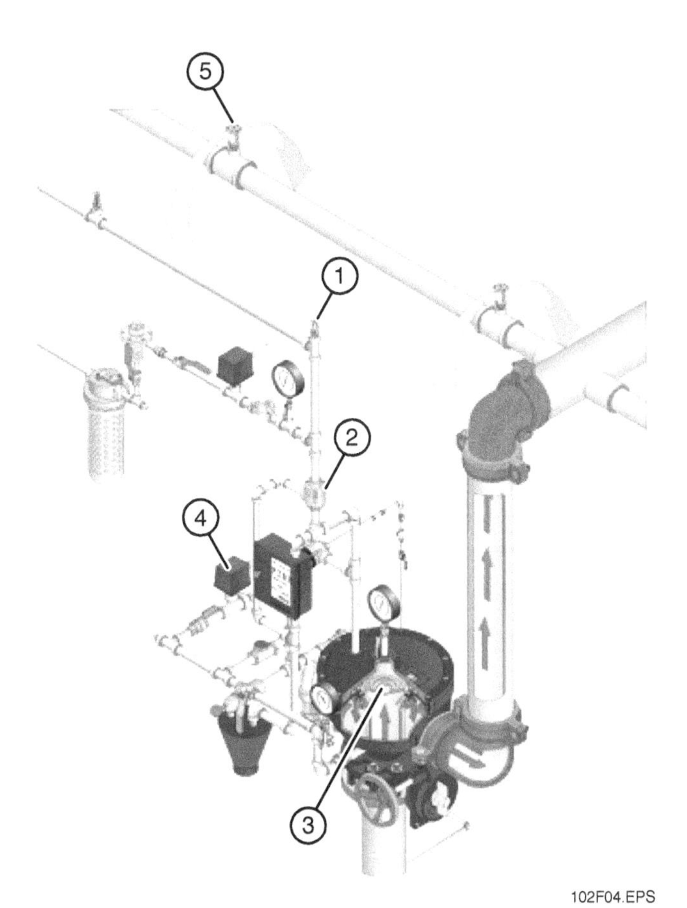

Figure 4 Deluge sprinkler system.

Water then flows into the system piping. This flow causes a pressure switch (4) to turn on an electric and/or water motor alarm. Water flows from all open sprinklers or nozzles (5). When the deluge valve opens, the pressure operated relief valve (PORV) opens (not shown). An open PORV routes priming water to drain, causing the deluge valve to latch in the open position.

2.2.0 Automatic Sprinklers

An automatic sprinkler is a fire suppression or control device that operates automatically when its heat-activated element is heated to its thermal rating or above, allowing water to discharge over a specified area.

Sprinklers are generally manufactured from brass with a plug held in place by either a metal link that melts or a glass bulb that breaks at a preset temperature. The water pressure in the system forces the plug out, allowing water to discharge.

2.3.0 Sprinkler Orientation and Spray Patterns

The orientation and spray pattern of a sprinkler is very important to the effective operation of the system and the application of the sprinkler. The spray pattern and orientation of the sprinkler are determined by the type of sprinkler used. Each of the following types of sprinklers is oriented in a particular manner to produce the desired spray pattern:

- Upright sprinklers
- Pendent sprinklers
- Sidewall sprinklers
- Conventional sprinklers

2.3.1 Upright Sprinklers

An upright sprinkler (*Figure 5*) shall be installed with the frame arms parallel to the branch line, unless specifically listed for other orientations. The weakest part of the pattern is in line with the frame arms, so lining them up with the pipe provides the strongest possible pattern. The upright spray sprinkler throws all of its water down and no water up toward the roof structure.

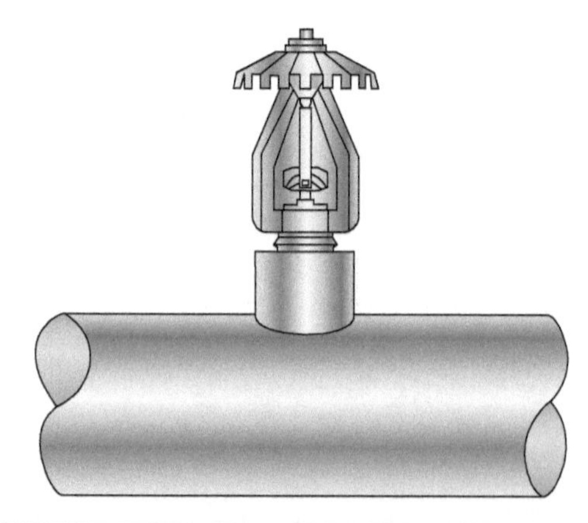

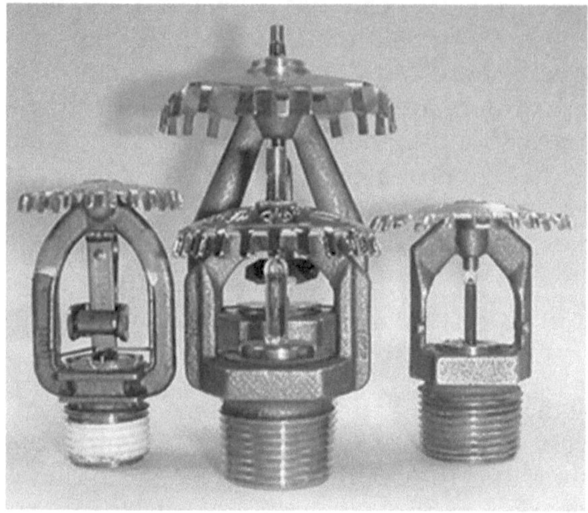

102F05.EPS

Figure 5 Upright sprinkler.

2.3.2 Pendent Sprinklers

A pendent sprinkler (*Figure 6*) is always located below the branch line or, in the case of a drop/suspended ceiling, at the ceiling below. The frame arms do not have to be parallel with the branch line because the pipe never interferes with the spray pattern. The weakest part of the pattern is in line with the frame arms, but lining them up in any specific manner is not required. The pendent spray sprinkler throws all of its water down and no water up toward the roof structure.

2.3.3 Sidewall Sprinklers

A sidewall sprinkler (*Figure 7*) is always located along a wall or soffit. There are horizontal sidewall and vertical sidewall sprinklers. Horizontal sidewall sprinklers are pointed straight out from a wall and are designed to throw water across a room and along the wall from which they extend. Vertical sidewall sprinklers are either upright or pendent. They do not extend from the wall, but are located in front of the wall. Unlike standard uprights and pendents, vertical sidewall sprinklers throw water across the room and along the wall in the same way as horizontal sidewall sprinklers.

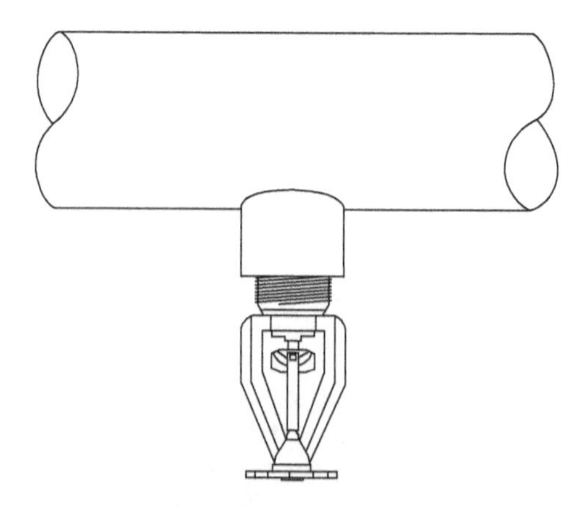

102F06.EPS

Figure 6 Pendent sprinkler.

2.3.4 Conventional Sprinklers

Conventional sprinklers (*Figure 8*) are generally called old-style sprinklers. They were used extensively until the advent of the spray sprinkler used today. Conventional sprinklers were very simple because they could be used as either uprights or pendants. They were designed to discharge approximately 40 percent of their water in the upward direction against the roof or ceiling. The other 60 percent of the spray was directed downward. They are still used in some European countries and in specific structures such as underneath piers, wharfs, and in fur storage vaults, per NFPA standards. Conventional sprinklers can also be used as replacement sprinklers in systems that were designed and installed with the old-style sprinklers, but they cannot be used in a new system or to replace standard spray sprinklers.

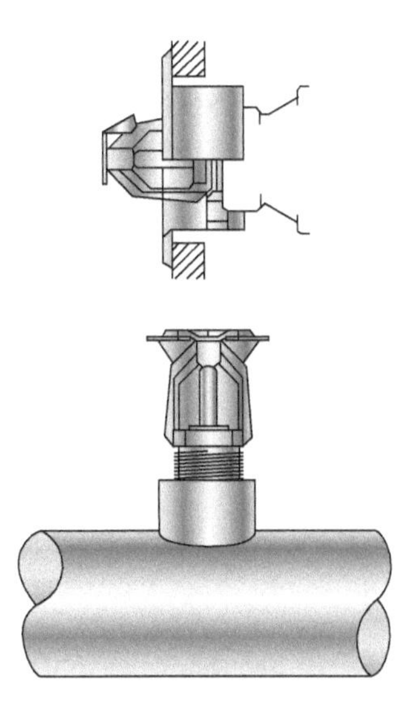

102F07.EPS

Figure 7 Sidewall sprinkler.

102F08.EPS

Figure 8 Conventional sprinkler.

2.4.0 How Sprinklers Are Defined

NFPA 13 defines the following characteristics of a sprinkler that describe its ability to control or extinguish a fire:

- Thermal sensitivity
- Temperature rating
- K-factor
- Installation orientation
- Water distribution characteristics
- Special service conditions

Remember that the term *fast response* (like the term *quick response* used to define a particular type of sprinkler) refers to thermal sensitivity within the operating element of a sprinkler, not the time of operation in a particular installation.

On Site

Adjustable Drop Nipple

The adjustable drop nipple allows for the final adjustment of the length of drop between a branch line and the pendent sprinkler without re-cutting the drop or disturbing the ceiling. This screw-type adjustable sprinkler fitting is designed for use in wet-pipe fire protection systems with suspended (lift-out) ceilings.

102SA01.EPS

As sprinkler manufacturers and researchers continue to investigate the effects of modifying sprinkler characteristics, sprinklers with faster operating times, broader spray patterns, and deeper water penetration capabilities are becoming more widely available.

2.5.0 Sprinkler Types

There are a large number of sprinklers in use today. They can generally be sorted into three categories: design and performance characteristics, installation orientation, and service conditions. *Table 1* lists sprinklers defined by design and performance characteristics. *Table 2* lists sprinklers defined by installation orientation.

Table 1 Sprinklers Defined by Design and Performance Characteristics

Sprinkler Type	Description
Control Mode Specific Application Sprinkler (CMSA)	Specific application control mode sprinkler capable of producing large water droplets; can provide fire control of specific high-challenge fire hazards.
Early Suppression Fast-Response (ESFR)	Has a thermal element with RTI of 50 (meters/seconds)$^{1/2}$ or less. Provides fast response fire suppression of specific high-challenge fire hazards.
Extended Coverage	Spray sprinkler with maximum coverage areas as specified in *NFPA 13* and the individual listing.
Nozzles	Used in applications requiring special water discharge patterns, spray direction, or other unusual discharge patterns.
Old-Style Conventional Sprinkler	Directs 40% to 60% of total water initially downward. Designed to be installed with deflector either upright or pendent.
Open Sprinkler	Does not have actuators or heat-responsive elements.
Quick Response Early Suppression (QRES)	Provides quick fire suppression of specific fire hazards. Has a thermal element with RTI of 50 (meters/seconds)$^{1/2}$ or less.
Quick Response Extended Coverage	Listed as having a thermal element with an RTI of 50 (meters-seconds)$^{1/2}$ or less and complies with extended protection areas as defined in *NFPA 13* Chapter 8.
Quick Response (QR) Sprinkler	Listed as a quick-response sprinkler. Has a thermal element with RTI of 50 (meters/seconds)$^{1/2}$ or less.
Residential Sprinkler	Fast-response type; has a thermal element with RTI of 50 (meters/seconds)$^{1/2}$ or less. Investigated for its ability to enhance survivability in the room of fire origin. Listed for protection of dwelling units.
Special Sprinkler	Tested and prescribed for use by its individual listing.
Spray Sprinklers	Listed for its capability to provide fire control for a wide range of fire hazards. Maximum coverage areas specified in individual listing.
Standard Spray Sprinkler	A spray sprinkler with maximum coverage area as specified in Sections 8.6 and 8.7 of *NFPA 13*.

102T01.EPS

Table 3 lists sprinklers defined by special service conditions. These devices are intended for use in specific environments. They are either covered with a special coating or are designed for a specific function. Only the sprinkler manufacturer can apply coatings.

A few of the many available sprinklers are illustrated in *Figure 9*.

For additional information on sprinkler types, refer to the *NFPA 13* sprinkler definitions.

Did You Know?

George Manby invented the portable fire extinguisher in the early 1800s. His idea was that a small amount of water applied early and in the right place would accomplish what even large amounts of water could not accomplish later on. The idea of putting out a fire in the earliest stages led to the invention of the automatic fire sprinkler in 1864 by Henry S. Parmalee, who needed to protect his piano factory.

Table 2 Sprinklers Defined by Installation Orientation

Sprinkler Type	Description
Concealed Sprinkler	A recessed sprinkler with a cover plate.
Flush Sprinkler	All or part of the body, including the shank thread, is mounted above the lower plane of the ceiling.
Pendent Sprinkler	Designed to be installed in such a way that the water stream is directed downward against the deflector.
Recessed Sprinkler	All or part of the body, other than the shank thread, is mounted within the recessed housing.
Sidewall Sprinkler	Has special deflectors that are designed to discharge most of the water away from the nearby wall in a pattern resembling ¼ of a sphere; a small portion of the discharge is directed at the wall behind the sprinkler.
Upright Sprinkler	Designed to be installed in such a way that the water spray is directed upwards against the deflector.

102T02.EPS

Table 3 Sprinklers Defined by Special Service Conditions

Sprinkler Type	Description
Corrosion-Resistant Sprinkler	Fabricated with corrosion-resistant material, or with special coatings or platings, to be used in an atmosphere that would normally corrode sprinklers.
Dry Sprinkler	Secured in an extension nipple with a seal at the inlet end that prevents water from entering nipple until the sprinkler operates.
Intermediate Level Sprinkler/ Rack Storage Sprinkler	Equipped with integral shields to protect its operating elements from the discharge of sprinklers installed at higher elevations.
Ornamental/Decorative Sprinkler	Has been painted or plated by the manufacturer.
Institutional Sprinkler	A sprinkler specifically designed for resistance to load-bearing purposes and with components not readily converted for use as weapons.
Pilot Line Detector	Standard spray sprinkler or thermostatic fixed-temperature release device used as a detector to pneumatically or hydraulically release the main valve, controlling the flow of water into a fire protection system.

102T03.EPS

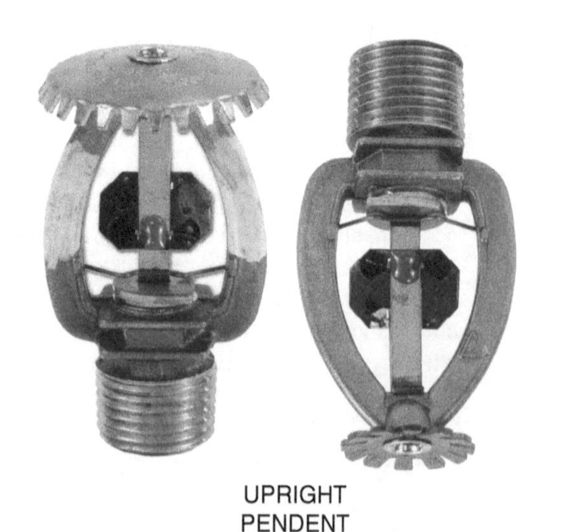

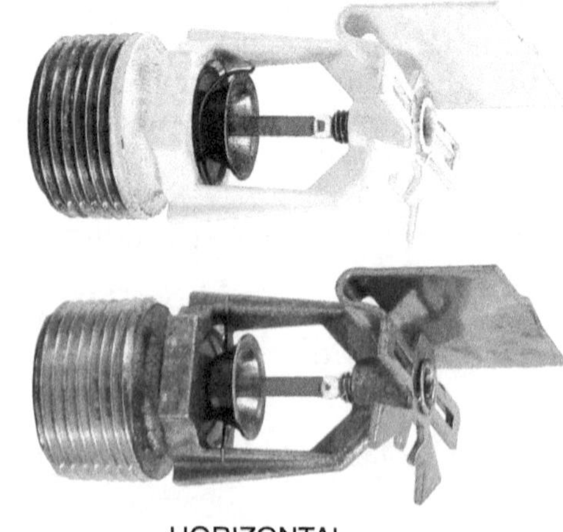

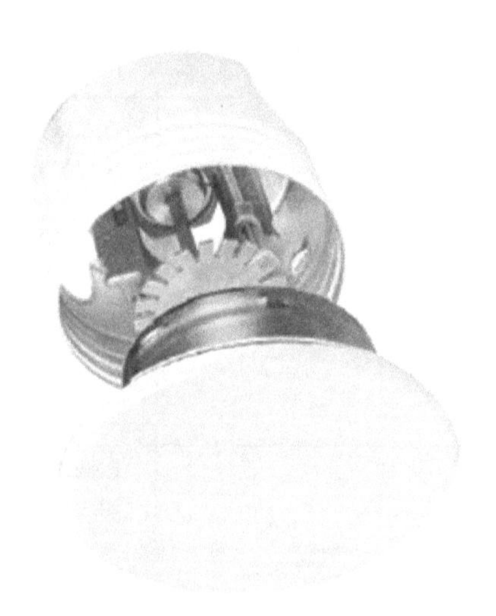

Figure 9 Typical sprinkler heads.

2.6.0 Sprinkler K-Factor

The opening (orifice) through which water is released from the sprinkler varies in size. The size varies from ¼" to 1", with the different sizes being assigned a K-factor. The larger the orifice, the larger the K-factor. You can see in *Table 4* that as the K-factor increases, so does the rated discharge from the sprinkler, with K-5.6 set as the 100 percent nominal discharge. Therefore, a sprinkler rated at a K-factor of 1.4 produces much less water (25 percent of nominal discharge) than a sprinkler rated at a K-factor of 11.2 (200 percent of nominal discharge).

2.7.0 Sprinkler Temperature Ratings

Sprinklers have a predetermined temperature at which they open or fuse. A color, either in the fusing element (frangible bulb) or paint on the sprinkler's frame, indicates the sprinkler's temperature range. *Table 5* lists sprinkler temperature ratings, classifications, and color codes.

3.0.0 Aboveground Pipe and Tube

To make sure that sprinkler systems perform adequately and are reliable, *NFPA 13* requires Listing of components crucial to system performance during a fire. Components that might malfunction during a fire, but would not adversely affect system performance, do not have to be Listed. The requirement for Listed products and materials is aimed at making sure the fire protection system will perform as it should during a fire.

Steel pipe and copper tubing do not have to be Listed if they meet the applicable American Society for Testing and Materials International (ASTM) standards. Otherwise, they must be Listed.

Table 4 Sprinkler Discharge Characteristics Identification (Table 6.2.3.1)

Nominal K-factor [gpm/(psi)$^{1/2}$]	Nominal K-Factor [L/min/(bar)$^{1/2}$]	K-Factor Range [gpm/(psi)$^{1/2}$]	K-Factor Range [L/min/(bar)$^{1/2}$]	Percent of Nominal K-5.6 Discharge	Thread Type
1.4	20	1.3–1.5	19–22	25	½ in. NPT
1.9	27	1.8–2.0	26–29	33.3	½ in. NPT
2.8	40	2.6–2.9	38–42	50	½ in. NPT
4.2	60	4.0–4.4	57–63	75	½ in. NPT
5.6	80	5.3–5.8	76–84	100	½ in. NPT
8.0	115	7.4–8.2	107–118	140	¾ in. NPT or ½ in. NPT
11.2	160	10.7–11.7	159–166	200	½ in. NPT or ¾ in. NPT
14.0	200	13.5–14.5	195–209	250	¾ in. NPT
16.8	240	16.0–17.6	231–254	300	¾ in. NPT
19.6	280	18.6–20.6	272–301	350	1 in. NPT
22.4	320	21.3–23.5	311–343	400	1 in. NPT
25.2	360	23.9–26.5	349–387	450	1 in. NPT
28.0	400	26.6–29.4	389–430	500	1 in. NPT

Note: The nominal K-factors for dry-type sprinklers are used for sprinkler selection. See *NFPA 13*, 23.4.4.9.3 for use of adjusted dry-type sprinkler K-factors for hydraulic calculation purposes.

102T04.EPS

Table 5 Sprinkler Temperature Ratings, Classifications, and Color Codings (Table 6.2.5.1)

Maximum Ceiling Temperature		Temperature Rating		Temperature Classification	Color Code	Glass Bulb Colors
°F	°C	°F	°C			
100	38	135–170	57–77	Ordinary	Uncolored or Black	Orange or Red
150	66	175–225	79–107	Intermediate	White	Yellow or Green
225	107	250–300	121–149	High	Blue	Blue
300	149	325–375	163–191	Extra High	Red	Purple
375	191	400–475	204–246	Very Extra High	Green	Black
475	246	500–575	260–302	Ultra High	Orange	Black
625	329	650	343	Ultra High	Orange	Black

102T05.EPS

3.1.0 Common Pipe Sizes

Steel pipe is manufactured in many sizes. The most common sizes in the industry range from ½" to 10". Steel pipe is manufactured with different wall thicknesses. This is referred to as a schedule. A given pipe size may be manufactured in different schedules. For example, a 2" pipe may be manufactured as Schedule 40 or even Schedule 10. With pipe, the outside diameter (OD) of a given pipe size is the same no matter what the schedule is. Thus, the thinner the wall, the larger the inside diameter (ID) of the pipe.

The most common form of steel pipe used in fire sprinkler systems is black steel. Black steel can also be galvanized (dipped in zinc during the manufacturing process) for corrosion resistance. Stainless steel pipe is rarely used because it is costly and difficult to thread. Steel pipe can be joined by various means, including the following:

- *Threading* – For use with threaded fittings.
- *Grooving* – For use with grooved fittings.
- *Drilling* – For use with mechanical tees or welded outlets.
- *Welding* – Either end-to-end; or welded, grooved, or threaded outlets.

3.2.0 Common Tube Sizes

Copper tube, like steel pipe, is manufactured in many sizes. The most common sizes in the industry range from ¾" to 2". Copper tube is not used as extensively as steel pipe.

Copper tube differs from steel pipe in that it is not manufactured in schedules. Rather, it is identified by types. The type refers to its hardness or flexibility. The three types used in fire sprinkler systems are K, L, and M.

3.2.1 Tube Sizes and Wall Thicknesses

Types K, L, and M copper tube are allowed by *NFPA 13* for use in sprinkler systems. Types K and L are flexible and come in straight lengths and in rolls. Type K is normally used in commercial plumbing systems and Type L is normally used in residential plumbing systems. Type M is hard-drawn copper and is available in straight sections only. Hard-drawn copper will kink before it will bend and care must be taken to maintain the quality of the tube during installation. The cost of copper tube is directly related to wall thickness and the thinnest wall permitted is normally used. The actual internal diameter of copper tube is very close to the nominal tube size. Copper specifications are often referred to as CTS, meaning copper tube size, as differentiated from IPS, originally standing for iron pipe size.

There are other types of copper tube available, such as drain, waste, and vent (DWV) and air conditioning and refrigeration (ACR). These are not acceptable under *NFPA 13* for use in fire sprinkler systems. Common methods of joining copper tube are soldering, compression, and brazing.

3.3.0 Chlorinated Polyvinyl Chloride (CPVC) Pipe

CPVC pipe must be Listed for aboveground fire sprinkler applications. It is typically identified by its bright orange color. The most common industry sizes of CPVC pipe are ¾" to 4". CPVC pipe can be joined by the solvent-cement method using CPVC fittings or with mechanical couplings.

3.4.0 C-Factor

When pipe sizes are determined for fire sprinkler systems, one of the things that must be determined is the C-factor of the pipe. The C-factor refers to the roughness of the inside of the pipe. The higher the C-factor, the smoother the pipe and the easier the water flows. This translates into smaller pipe sizes, which in turn save money. Typically, black steel has a C-factor of 120 and copper and CPVC have a C-factor of 150.

3.5.0 Fittings

Fittings used with steel pipe or copper tube do not have to be Listed, but must meet or exceed other nationally recognized standards. Fittings used with steel pipe in aboveground systems must be manufactured of cast, malleable, or ductile iron. Fittings used with CPVC must be Listed. *Figure 10* shows examples of various pipe fittings.

4.0.0 Underground Pipe

Underground piping must comply with the standards in *NFPA 24* or *NFPA 13*, as applicable. Pipe must be Listed or comply with American Waterworks Association (AWWA)or ASTM standards.

4.1.0 Piping Materials

Steel pipe is not typically used for underground service. When it is, such as between the check valve and the fire department connection, it must be externally coated and wrapped and internally galvanized.

Pipe must be designed to a working pressure of not less than 150 psi. Typical underground pipe materials are PVC, CPVC, ductile iron, and copper.

The type and class of pipe for a particular underground installation is determined by the following factors:

- Fire resistance of the pipe
- Maximum system working pressure
- Depth at which the pipe is to be installed
- Soil conditions
- Corrosion
- Susceptibility of the pipe to other external loads, including earth loads, installation beneath buildings, and traffic or vehicle loads

4.2.0 Fittings

Fittings must meet the standards of *NFPA 24* or *NFPA 13*, as applicable. Listed fittings are permitted for the system pressures as specified in their listing, but not less than 150 psi (10 bar).

4.3.0 Depth of Cover

How deep to lay the pipe (or depth of cover) depends on the maximum depth of frost penetration. The top of the pipe must be buried at least 1' (0.3 m) below the frost line for the locality. Where frost is not a factor, the depth of cover should not be less than 2½' (0.8 m) to prevent mechanical damage. Where pipe is installed below driveways and railroad tracks, the minimum depth to bury the pipe is 3' and 4' respectively.

STRAIGHT THREADED TEE

PLAIN END COUPLING WITH GRIPPERS

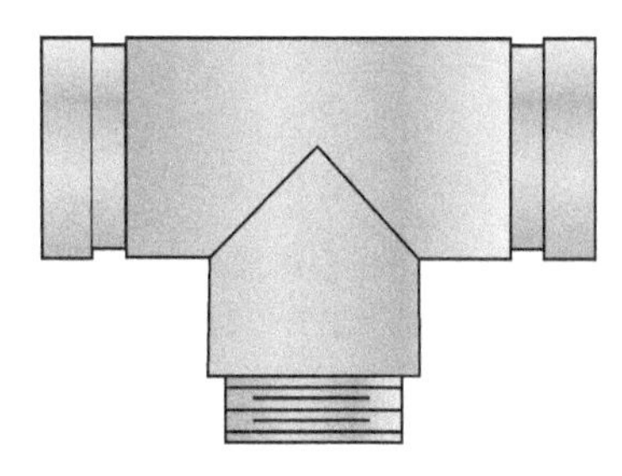

GROOVED TEE WITH THREADED OUTLET

THREADED MECHANICAL TEE

RIGID COUPLING

102F10.EPS

Figure 10 Pipe fittings.

4.4.0 Protection Against Freezing

Where pipe cannot be buried, it can be laid aboveground, provided that the pipe is protected against freezing and mechanical damage.

4.5.0 Protection Against Damage

Pipe should not be run under buildings. However, where pipe must be run under buildings, special precautions must be taken, including the following:

- Arching the foundation walls over the pipe
- Running pipe in covered trenches
- Providing valves to isolate sections of pipe under buildings

Fire service mains may enter the building adjacent to the foundation. Where adjacent structures or physical conditions make it impractical to locate risers immediately inside an exterior wall, such risers can be located as close as practical to exterior walls to minimize underground piping under the building.

Where a riser is located close to building foundations, underground fittings of proper design and type must be used to avoid locating pipe joints in or under the foundations. To prevent electrolysis when joining pipes of dissimilar metal, the joint must be insulated against the passage of an electric current using an approved method.

4.6.0 Requirement for Laying Pipe

Care must be taken to not drop or damage pipe, valves, and fittings. Small cracks and fractures can occur that may go unnoticed until the pipe is installed and pressure checked.

Bolted joints must be properly torqued. Pipes, valves, hydrants, and fittings must be clean inside and at their mating surfaces. Care must be taken to prevent stones and foreign materials from entering pipes, valves, and fittings.

Carefully lower all pipe, fittings, valves, and hydrants into the trench using appropriate

Did You Know?

Homeowners should never install their own fire sprinkler systems. A fire sprinkler system must be installed to meet applicable standards, local codes, and ordinances. This is not a project for the novice, and there may be local ordinances that prevent the homeowner from installing a fire sprinkler system.

Fire sprinkler design and layout is based on a number of issues related to occupancy. Fire sprinkler systems are installed by specialized contractors who understand the installation requirements set out by recognized standards. Many states require fire sprinkler contractors to be licensed and demonstrate competency in the trade.

equipment. Carefully examine each component for cracks or other defects while the pipe is suspended over the trench.

4.7.0 Joint Restraint

Large poured-in-place concrete blocks, called thrust blocks, are used to restrain movement of underground piping systems using push-on or mechanical-joint fittings.

Underground piping systems that use fused, threaded, grooved, or welded joints should not require additional restraining, provided that such joints can pass the hydrostatic test of *NFPA 24* or *NFPA 13*, as applicable, without shifting of piping or leakage in excess of permitted amounts.

4.7.1 Thrust Blocks

Thrust blocks are limited to use where the soil is suitable. Thrust blocks are installed between undisturbed earth and the fitting to be restrained, and must be able to keep the pipe and fitting from moving. Ideally, thrust blocks must be placed so that the joints are accessible for repair. Note that larger diameter piping creates greater thrust under pressure, requiring larger thrust blocks.

4.7.2 Restrained Joint Systems

Some underground piping systems use a restrained joint method of keeping the pipes, valves, and fittings from moving. With this type of system, thrust blocks are not necessary. The following are typical restrained joint methods:

- Locking mechanical or push-on joints
- Mechanical joints using setscrew retainer glands
- Bolted flange joints
- Heat-fused or welded joints
- Pipe clamps and tie rods
- Other approved methods or devices

5.0.0 Hangers, Bracing, and Restraint-of-System Piping

Because of water hammer (a shock wave in the piping caused by a sudden change in water flow) and vibration, sprinkler systems can move vertically and horizontally. Hangers, bracing, and restraints are installed at specified intervals to control this movement.

Hangers, bracing, and restraints are chosen by sprinkler contractors and installed by sprinkler fitters according to sprinkler system drawings. Correctly installed hangers, bracing, and restraints allow pipes to drain, expand, contract, and vibrate without damaging the sprinkler system.

5.1.0 Hanger Assemblies

The pieces of a hanger assembly that attach directly to the pipe or the building structure must be Listed. The other components do not have to be Listed.

Hangers (*Figure 11*) attach fire sprinkler piping to the building structure. The type of hanger is

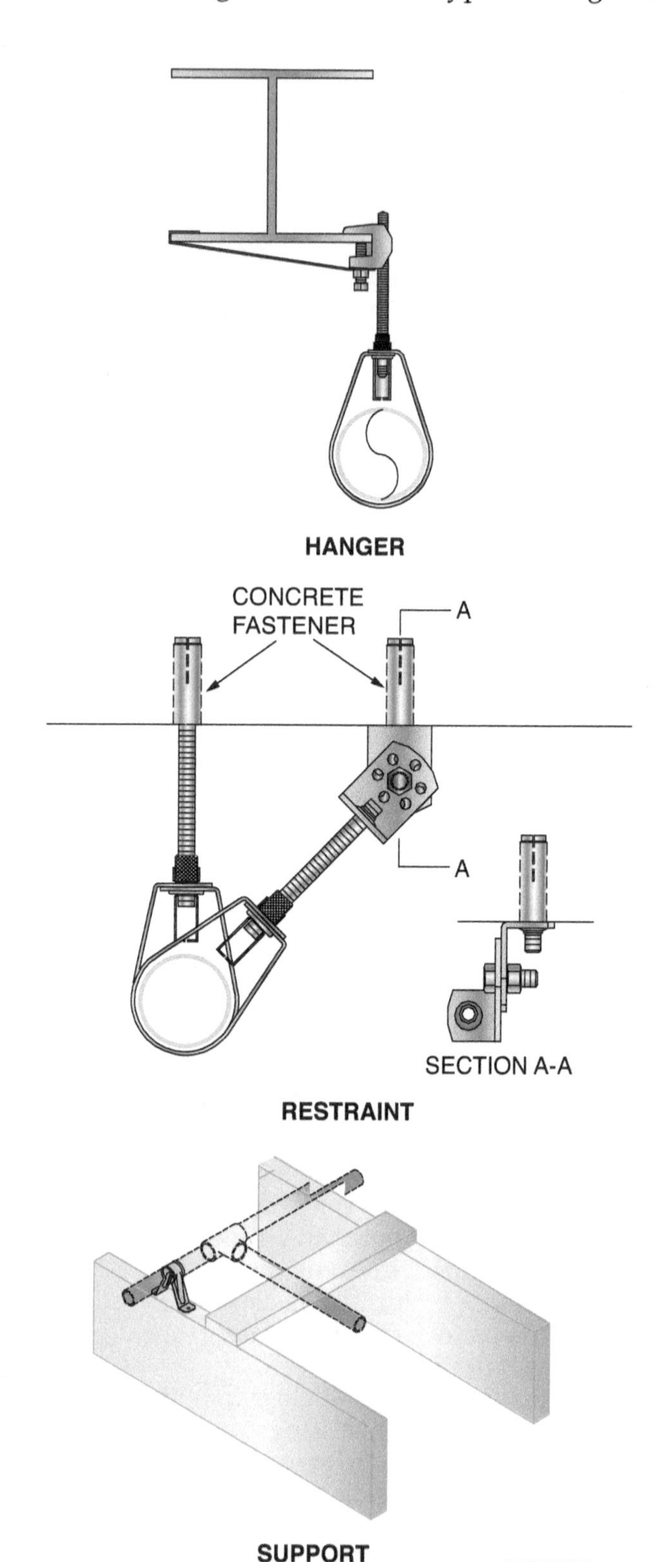

Figure 11 Hanger, support, restraint, and guide.

determined by the size of the pipe and the material of the structure.

5.1.1 Strength Requirements

Hangers must be capable of supporting five times the weight of the water-filled pipe plus 250 pounds at each point of piping support. When it is determined that a single point of connection cannot meet this requirement, a trapeze hanger (*Figure 12*) must be used.

The trapeze hanger is designed to distribute the load of water-filled pipe over at least two joists or purlins.

Make sure the pipe hangers used on the trapeze unit are capable of supporting the pipe load in accordance with *NFPA 13* requirements. Charts in *NFPA 13* show options and correct components.

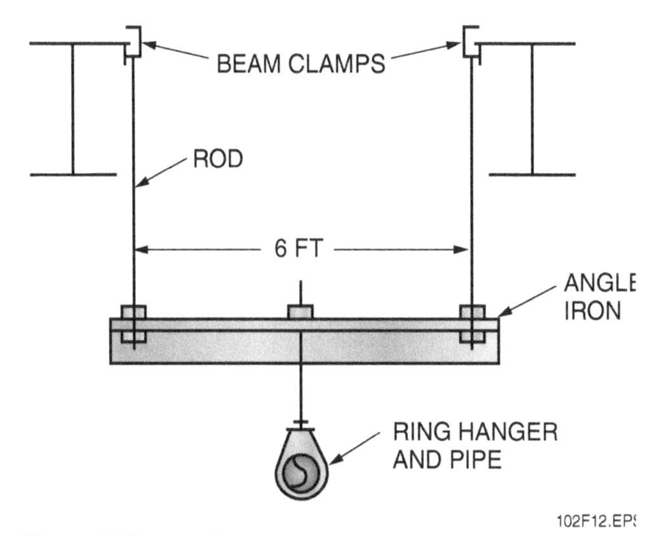

Figure 12 Trapeze hanger.

5.1.2 Hangers

Typical hanger assemblies (*Figure 13*) include the following:

- Side beam connectors
- Threaded screws with female connection
- U-hooks
- Drop-in anchors
- Ring hangers

5.2.0 Protection from Seismic Activity or Earthquakes

Certain areas of the country require that fire sprinkler systems be designed and installed with regard to protection from seismic activity or earthquakes. During an earthquake, piping systems tend to move in different directions from the structure to which they are attached; therefore flexibility of the pipe and clearance from the building structure must be taken into account. This is accomplished by the use of flexible grooved fittings and proper clearance around piping where it passes through walls and floors (*Figure 14A*).

The four-way brace (*Figure 14B*) should be attached above the upper flexible coupling required for the riser, and preferably to the roof structure, if the structure is suitable. The brace should not be attached directly to a plywood or metal deck. As with hangers, attachment points to the structure are very important. The sprinkler drawing must be followed very closely.

5.2.1 Sway Bracing

Sway bracing (*Figure 15*) of sprinkler pipe minimizes movement during an earthquake. Currently, there are two types of sway bracing:

- *Lateral sway bracing* – Resists the movement of sprinkler piping from side to side.

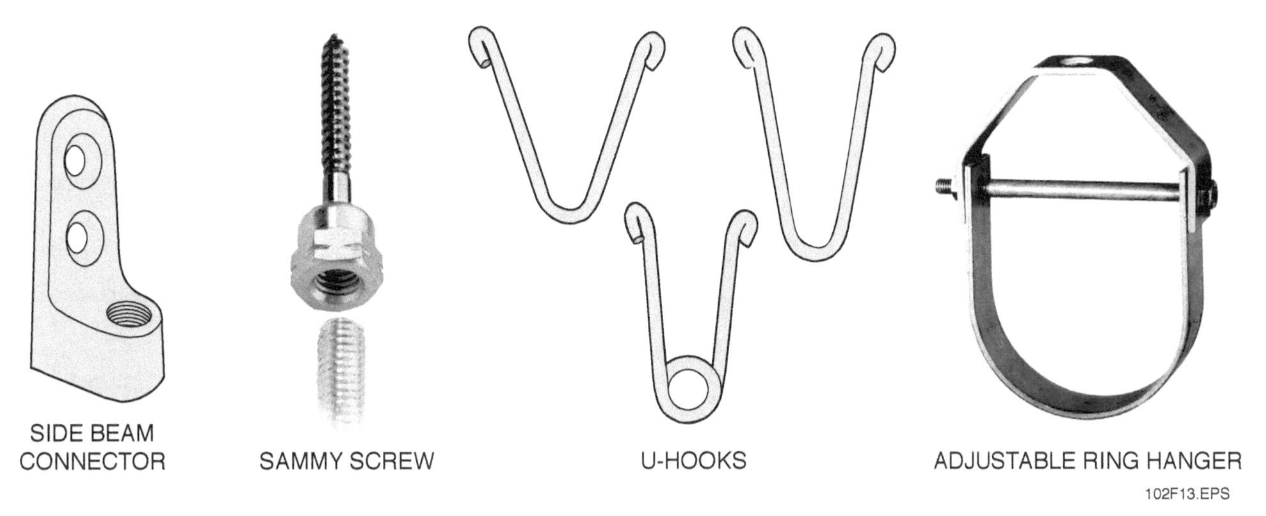

Figure 13 Examples of hangers.

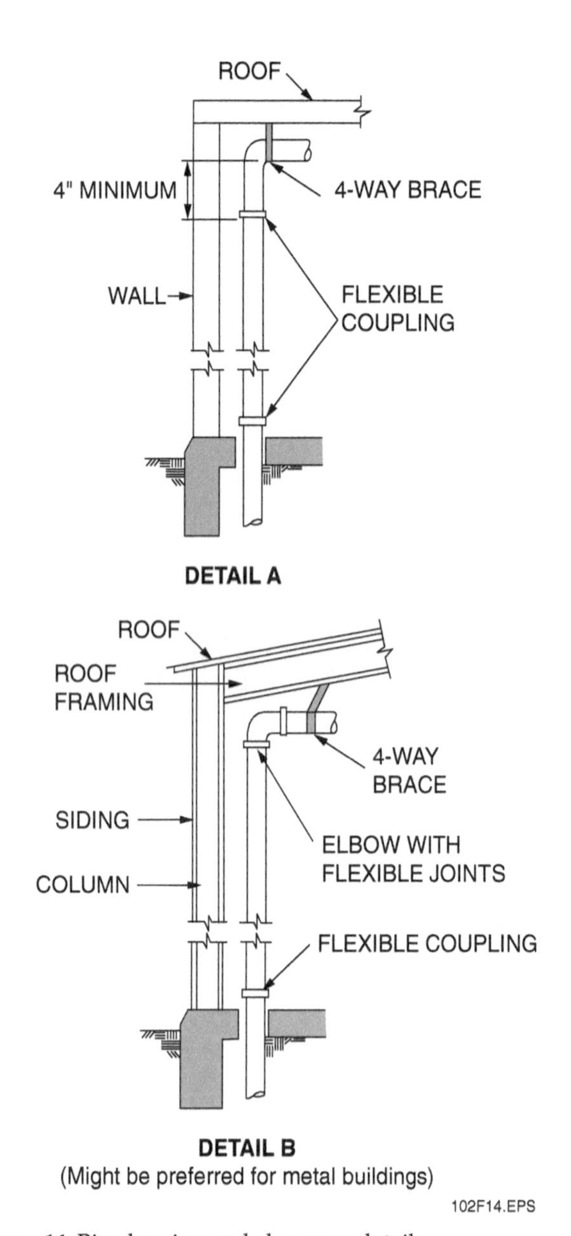

Figure 14 Pipe bracing and clearance details.

- *Longitudinal sway bracing* – Resists the movement of sprinkler piping along the length of the pipe run.

6.0.0 Valves

Fire sprinkler systems use various types of valves. They include control valves, check valves, drain valves, and test valves.

Control valves (*Figure 16*) control water supplies to sprinkler systems and must be Listed. They shall have a visual indicator that shows whether the valve is open or closed. Typical control valves are post indicator valves (PIV) and butterfly valves.

Check valves (*Figure 17*) allow water to flow into the sprinkler system, but not back out. Check valves must be Listed.

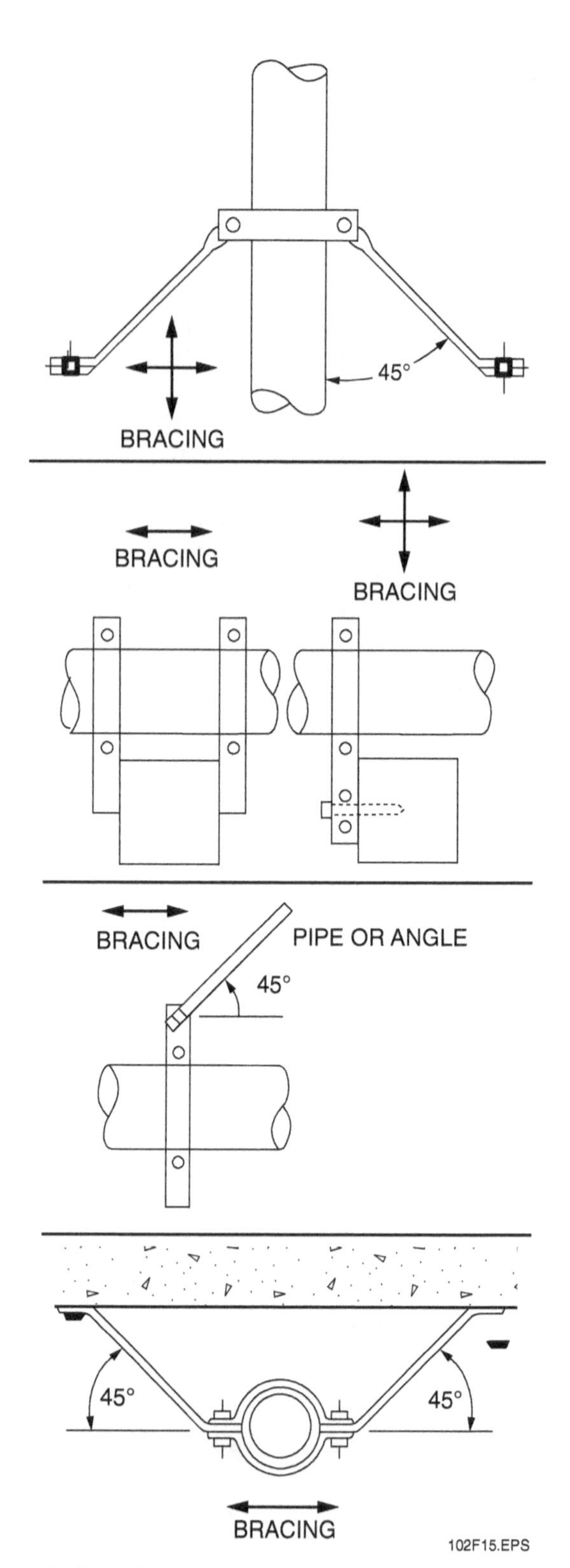

Figure 15 Sway bracing.

Drain valves (*Figure 18*) are used to drain the sprinkler system or portions of the system. They are usually 1", 1¼", or 2" in size.

Test valves are used to simulate a fire sprinkler discharge to test the water flow alarm. Drain valves and test valves must be approved, but need not be Listed. They are usually 1" in size.

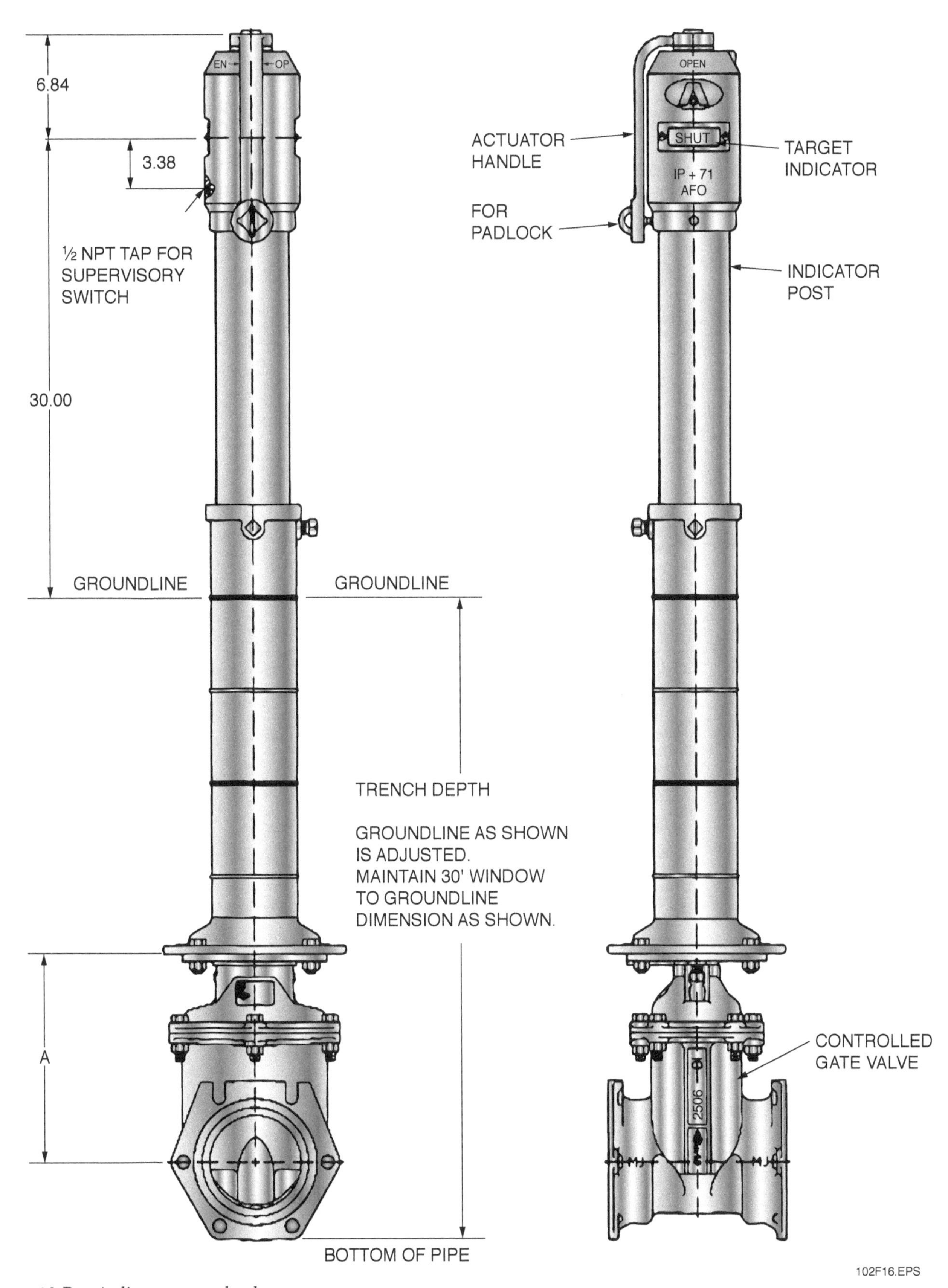

Figure 16 Post indicator control valve.

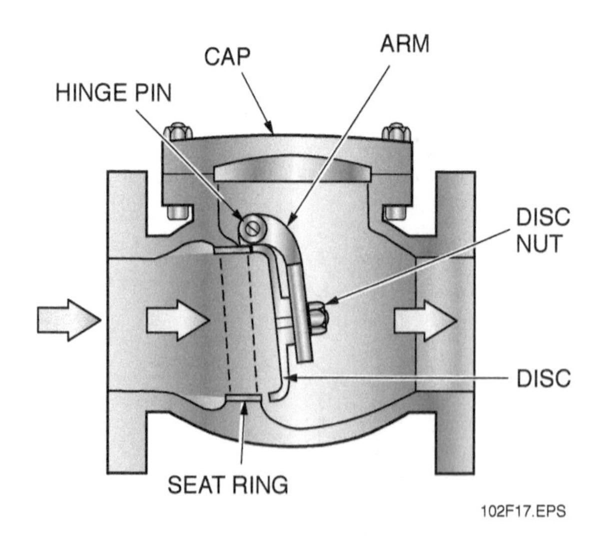

Figure 17 Flow through a swing check valve.

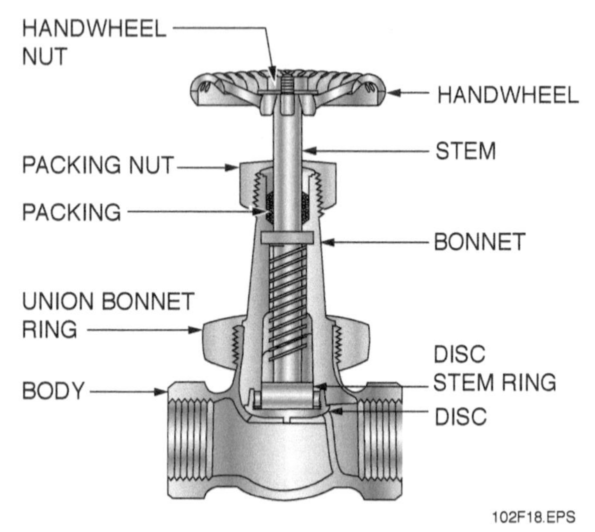

Figure 18 Drain (globe) valve.

7.0.0 Water Flow Alarms

Sprinkler systems are required to have an audible alarm (*Figure 19*) on the premises. Water flow alarms must be Listed. They may be electrically operated, such as a flow switch or pressure switch, or mechanically operated upon the flow of water into the system, such as a water motor and gong.

8.0.0 Fire Department Connections

Fire department connections (*Figure 20*), or FDCs, are inlets into which the fire department can use its pumper truck to supply water to, or boost the water pressure of, an operating sprinkler system. FDCs must be approved.

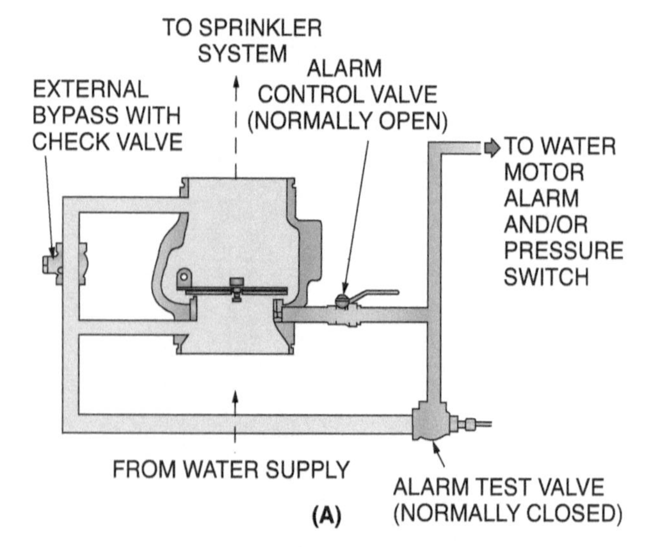

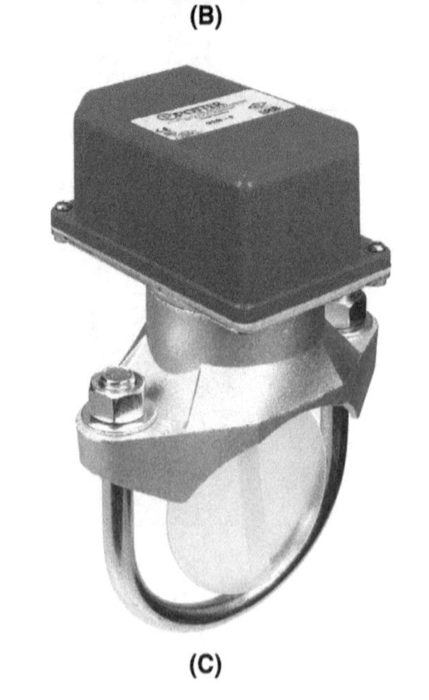

Figure 19 Water flow alarms.

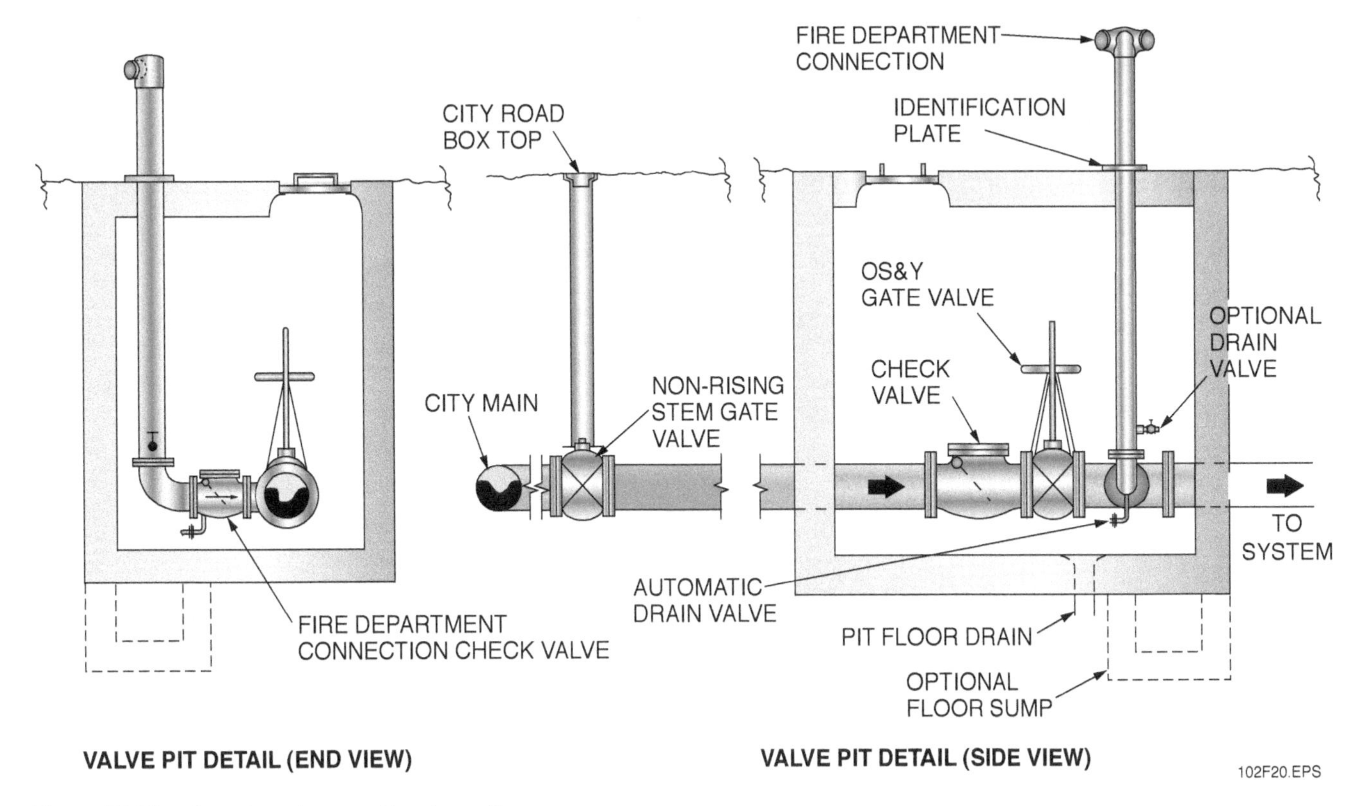

Figure 20 Fire department connection in a pit.

SUMMARY

The standards that govern sprinkler systems come primarily from The National Fire Protection Association. Codes and standards are developed using testing laboratories such as Underwriters Laboratories® (UL) and FM Global. When these agencies test a product and find that it meets code requirements and is suitable for use in fire sprinkler systems, that product is said to be Listed. Sprinkler components must be Listed for fire protection use.

Automatic sprinklers are designed according to certain component categories such as the sprinkler's capability for controlling a fire, its thermal response, spray pattern and coverage area, orientation, and occupancy. Also important are the sprinkler's K-factor rating, and the temperature rating at which the sprinkler discharges water.

There are a large number of sprinklers in use today. The sprinklers can be generally sorted into three categories by design and performance characteristics, installation orientation, and service conditions.

There are specific characteristics for pipe and tubing used above ground and for pipe used below ground. There are also specific code requirements for sprinkler system supports. Because of water hammer and vibration, sprinkler systems can move vertically and horizontally. Hangers, bracing, and restraints are installed at specified intervals to control this movement.

Sprinkler systems use various types of valves to control the flow of water through the system. You will learn more about these valves later in your training.

Review Questions

1. Underwriters Laboratories® is a type of _____.
   a. fire sprinkler association
   b. fire protection organization that writes regulations for the NFPA
   c. organization that evaluates products to determine compliance with safety standards
   d. Authority Having Jurisdiction over fire sprinkler systems

2. The UL product directory contains the names of companies that _____.
   a. donate funds to Underwriters Laboratories®
   b. buy only UL Listed equipment
   c. are qualified to use a UL Mark on products tested by UL
   d. have the most UL product listings

3. The deluge system is a variation of the _____.
   a. preaction system
   b. dry pipe system
   c. wet pipe system
   d. old-style conventional system

4. The frame arms must be parallel to the branch line to reduce the shadow effect of the frame in a(n) _____.
   a. upright sprinkler
   b. pendent sprinkler
   c. sidewall sprinkler
   d. conventional sprinkler

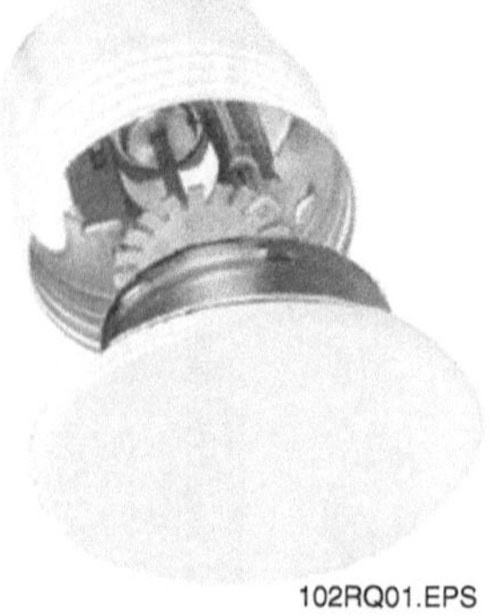

102RQ01.EPS

Figure 1

5. The sprinkler shown in *Figure 1* is an example of a(n) _____.
   a. combustible concealed space upright
   b. recessed horizontal sidewall
   c. recessed pendent
   d. extended coverage concealed pendent

6. A sprinkler rated at a K-factor of 11.2 discharges four times as much water as a sprinkler rated at a K-factor of _____.
   a. 1.4
   b. 2.8
   c. 5.6
   d. 22.4

7. A sprinkler color coded with white paint means that the sprinkler has a temperature classification of _____.
   a. ordinary
   b. extra high
   c. ultra high
   d. intermediate

8. A 2" Schedule 40 pipe has the same inside diameter as a 2" Schedule 10 pipe.
   a. True
   b. False

9. The roughness of the inside of a pipe is represented by the _____.
   a. K-factor
   b. color code
   c. inside diameter
   d. C-factor

10. Steel piping should *not* be used for general underground service unless specifically Listed for that service.
    a. True
    b. False

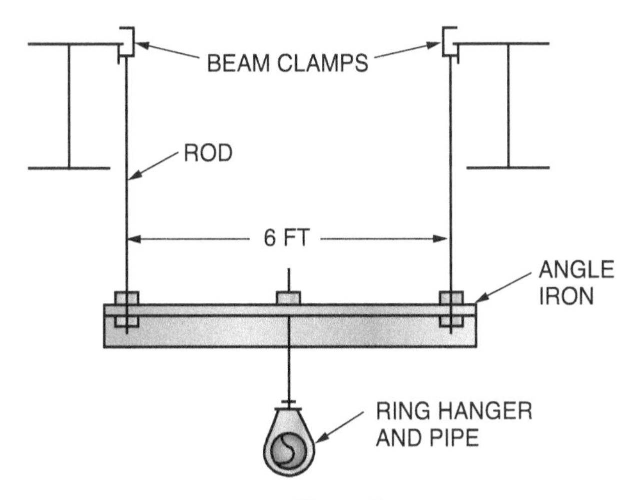

Figure 2

11. The hanger shown in *Figure 2* is an example of a _____.
    a. trapeze hanger
    b. single point hanger
    c. sway control brace
    d. side beam connector

12. Techniques used to protect piping in a building during an earthquake include adequate clearance, four-way bracing, and _____.
    a. trapeze hangers
    b. Schedule 10 pipe
    c. thrust blocks
    d. flexible couplings

13. Sway bracing is used primarily to minimize pipe _____.
    a. movement due to subsidence
    b. damage due to inferior installation
    c. movement during earthquakes
    d. damage due to freeze conditions

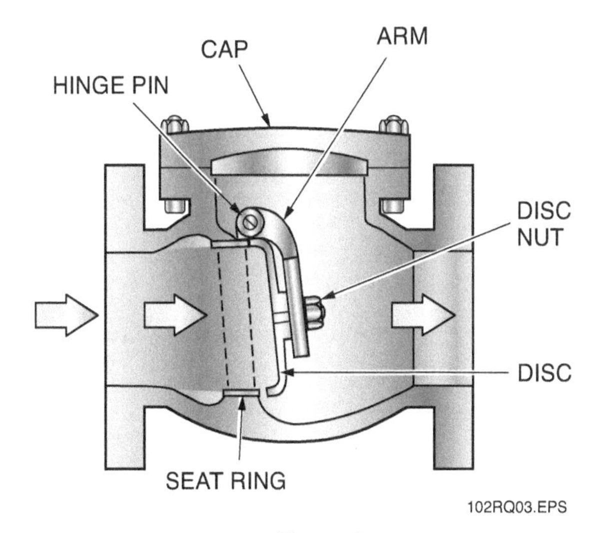

Figure 3

14. The valve shown in *Figure 3* is a type of _____.
    a. control valve
    b. check valve
    c. drain valve
    d. test valve

15. Sprinkler systems are *not* required to have an audible alarm on the premises.
    a. True
    b. False

Trade Terms Quiz

Fill in the blank with the correct trade term that you learned from your study of this module.

1. _________ is a testing laboratory and listing agency.
2. The water supply to a sprinkler system is controlled by a(n) _________.
3. A component that is tested by a recognized agency and found suitable for use in a fire sprinkler system for a specific purpose is said to be _________.
4. A ceiling that is not part of the structural framework of the building is called a(n) _________.
5. _________ is a measure of sprinkler orifice size.
6. The roughness inside the sprinkler pipe is referred to as the _________.
7. _________ causes a shock wave in the pipe and is due to a sudden change in water flow.
8. A product dipped in zinc during the manufacturing process is said to be _________.
9. Equipment or material which has attached a symbol or other identifying mark is said to be _________.
10. _________ allow water to flow in one direction only.
11. _________ are used to empty all parts of a sprinkler system.
12. Movement of sprinkler pipe during an earthquake is minimized by _________.
13. The decomposition of a substance caused by the passage of electricity through it is _________.

Trade Terms

C-factor
Check valve
Control valve
Drain valve
Drop/suspended ceiling
Electrolysis
Galvanized
K-factor
Labeled
Listed
Sway bracing
Underwriters Laboratories® (UL)
Water hammer

Cornerstone of Craftsmanship

Manning Strickland

Strickland Fire Protection, Inc.
Owner

Manning Strickland got his start in the industry unloading pipe for $2.65 an hour. He now has his own company, which now grosses over $12 million a year. Constant retrofitting of existing buildings and new construction in our nation's capital has provided steady work and economic opportunity. Manning employs 65 people, including his two sons who will run the business when he retires.

How did you choose a career in the sprinkler fitting field?
I was making 85 cents an hour working at a bakery. My cousin was working with a sprinkler company in Washington, DC. I had no desire to go to college. I was looking for a way to make more money, and I just went along with my cousin without thinking much about it.

What was your apprenticeship like?
I enrolled in the union apprenticeship program. It was tough. You'd work all day, hard physical labor, and then come home and study and do homework at night. It wasn't easy. Then you would take your exams at a proctored setting at the union. Your raises were based on completing the coursework and doing well on the tests.

What kinds of work have you done in your career?
My first job was unloading pipe. We unloaded tractor trailers and hauled the pipe up the stairs. We carried pipe all day long, eight hours a day. After that I started working with tools. I was cutting 4-inch pipe with hacksaws at giant airplane hangers at Andrews Air Force base. I worked for the same company for 22 years. I worked my way up from journeyman sprinkler fitter, to supervisor, job foreman, area superintendent, and then area manager. After 16 years in the field, I moved to an inside job and was responsible for the company's regional operations.

After my company was bought and sold three times in five years, I decided to start my own business. We have worked at the White House and the Capitol.

What factors have contributed most to your success?
My strong suit has always been working in the field. I was fortunate to work under men who taught me a trade. They taught me to make sure that when I did something, I did it right. I knew the importance of giving eight hours of work for eight hours of pay.

What advice would you give to those new to the sprinkler fitter field?
Availability is just as important as ability. Show up, be dependable. Do your best, always. Give your company and your customers as much as you can. Treat them as well as you want them to treat you.

Do the right thing; try to learn and to not make the same mistakes over and over. Everyone makes mistakes—the big thing is to learn from them. When you have problems, don't run from them. As I always say, run to the problem.

Trade Terms Introduced in This Module

C-factor: Refers to the roughness of the inside of the sprinkler pipe. The higher the C-factor, the smoother the pipe.

Check valves: Allow water to flow in one direction only.

Control valve: Controls the water supply to a sprinkler system.

Drain valve: Used to drain all or parts of a sprinkler system.

Drop/suspended ceiling: A ceiling that is not part of the structural framework of a building. This ceiling is installed suspended from the floor above, or from the roof, and is commonly used to provide space for services such as cables, recessed lighting, and piping.

Electrolysis: The decomposition of a substance by the passage of electricity through it.

FM Global: A testing laboratory and listing agency that certifies new products for service in particular locations and for specific applications.

Galvanized: Dipped in zinc during the manufacturing process.

Inside diameter (ID): The distance between the inner walls of a pipe. Used as the standard measure for tubing used in heating and plumbing applications.

K-factor: A measure of sprinkler K-factor. The larger the orifice, the larger the K-factor.

Labeled: Equipment or material which has attached a label, symbol, or other identifying mark that is acceptable to the Authority Having Jurisdiction.

Listed: Included in a list of products or services published by a recognized listing agency and acceptable to the AHJ for use in fire sprinkler systems for a specified purpose.

Outside diameter (OD): The distance between the outer walls of a pipe. Used as the standard measure for ACR tubing.

Sway bracing: Bracing of sprinkler pipe to minimize pipe movement during an earthquake.

Underwriters Laboratories® (UL): A testing laboratory and listing agency that certifies new products for service in particular locations and for specific applications.

Water hammer: Shock wave in the piping caused by a sudden change in water flow.

Additional Resources

This module presents thorough resources for task training. The following resource material is suggested for further study.

FM Global Approval Guide, Latest Edition. Norwood MA: FM Global.

NFPA 13, Standard for the Installation of Sprinkler Systems, Latest Edition. Quincy, MA: National Fire Protection Association.

The Pipefitters Blue Book, 2002. Graves. Webster, TX: W.V. Graves Publishing Company.

Underwriters Laboratories Fire Protection Equipment Directory, Latest Edition. Northbrook, IL: Underwriters Laboratories.

Figure Credits

Victaulic Company, Module opener, 102F10 (plain end fitting and rigid coupling)

The Viking Corporation, 102F01–102F04, 102SA01

Tyco Fire and Building Products, 102F05-102F09, 102F10 (threaded mechanical tee), 102F19 (A and B)

National Fire Protection Association, 102T04, 102T05

Reprinted with permission from *NFPA 13-2013, Standard for Installation of Sprinkler Systems*, Copyright © 2012, National Fire Protection Association, Quincy, MA. This reprinted material is not the complete and official position of the NFPA on the referenced subject, which is represented only by the standard in its entirety.

Mueller/B&K Industries, Inc., 102F10 (straight threaded tee)

ITW Buildex, 102F13 (threaded rod)

Anvil International, Inc., 102F13 (adjustable clevis hanger)

American Flow Control, 102F16

Potter Electric Signal Company, 102F19 (C)

MODULE 18102-13 — ANSWERS TO REVIEW QUESTIONS

Answer		Section
1.	c	1.2.1
2.	c	1.2.1
3.	a	2.1.0
4.	a	2.3.1
5.	d	2.5.0; Figure 9
6.	b	2.6.0
7.	d	2.7.0; Table 5
8.	b	3.1.0
9.	d	3.4.0
10.	a	4.1.0
11.	a	5.1.1; Figure 12
12.	d	5.2.0
13.	c	5.2.1
14.	b	6.0.0; Figure 17
15.	b	7.0.0

MODULE 18102-13 — ANSWERS TO TRADE TERMS QUIZ

1. UL
2. Control valve
3. Listed
4. Suspended/drop ceiling
5. K-factor
6. C-factor
7. Water hammer
8. Galvanized
9. Labeled
10. Check valves
11. Drain valves
12. Sway bracing
13. Electrolysis

NCCER CURRICULA — USER UPDATE

NCCER makes every effort to keep its textbooks up-to-date and free of technical errors. We appreciate your help in this process. If you find an error, a typographical mistake, or an inaccuracy in NCCER's curricula, please fill out this form (or a photocopy), or complete the online form at **www.nccer.org/olf**. Be sure to include the exact module ID number, page number, a detailed description, and your recommended correction. Your input will be brought to the attention of the Authoring Team. Thank you for your assistance.

Instructors – If you have an idea for improving this textbook, or have found that additional materials were necessary to teach this module effectively, please let us know so that we may present your suggestions to the Authoring Team.

NCCER Product Development and Revision

13614 Progress Blvd., Alachua, FL 32615

Email: curriculum@nccer.org
Online: www.nccer.org/olf

❑ Trainee Guide ❑ AIG ❑ Exam ❑ PowerPoints Other ____________________

Craft / Level: Copyright Date:

Module ID Number / Title:

Section Number(s):

Description:

Recommended Correction:

Your Name:

Address:

Email: Phone:

Steel Pipe

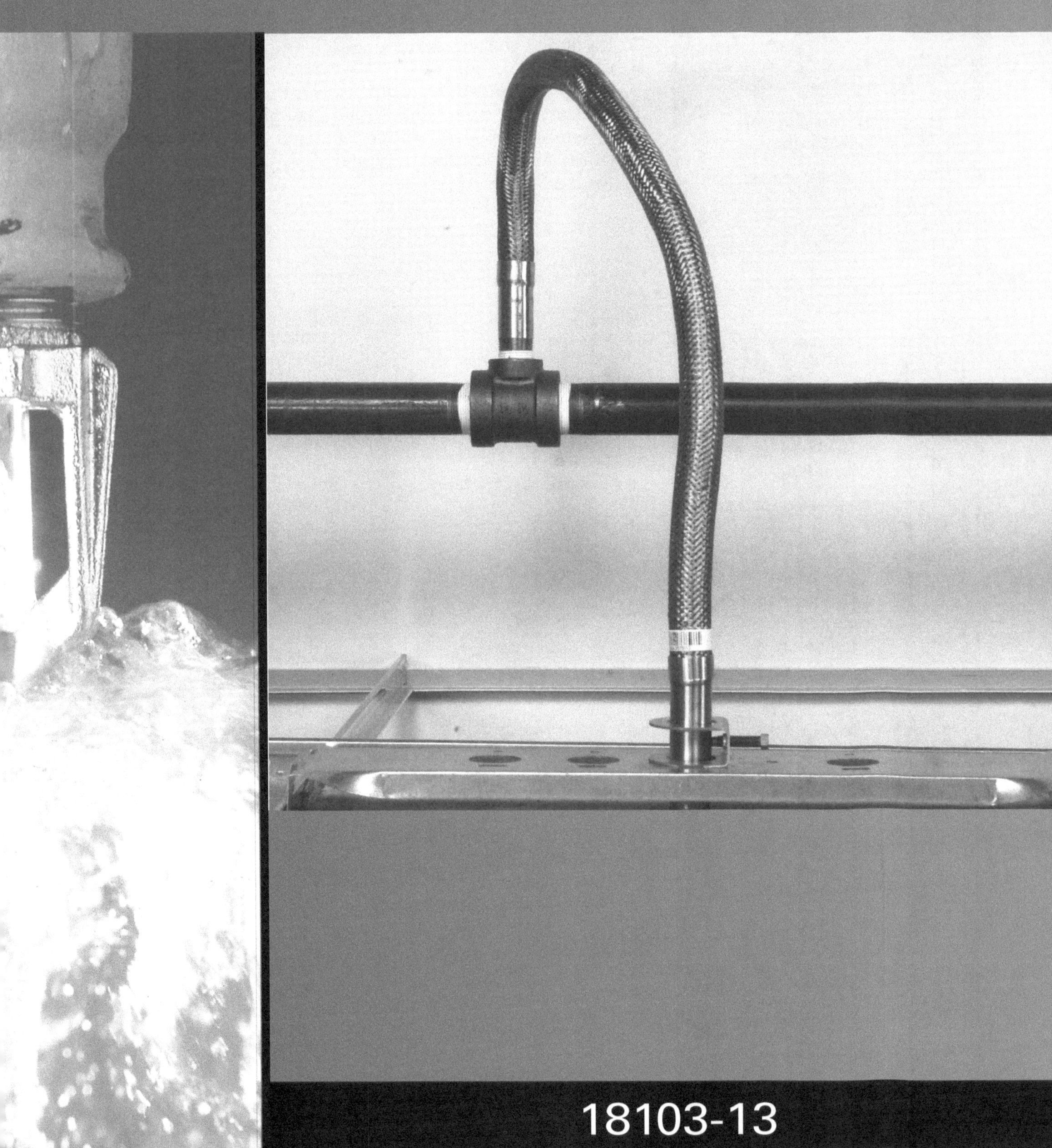

18103-13

SPRINKLER FITTING LEVEL ONE

18106-13 Underground Pipe

18105-13 Copper Tube Systems

18104-13 CPVC Pipe and Fittings

18103-13 Steel Pipe

18102-13 Introduction to Components and Systems

18101-13 Orientation to the Trade

Core Curriculum: Introductory Craft Skills

This course map shows all of the modules in *Sprinkler Fitting Level One*. The suggested training order begins at the bottom and proceeds up. Skill levels increase as you advance on the course map. The local Training Program Sponsor may adjust the training order.

Objectives

When you have completed this module, you will be able to do the following:

1. Follow basic safety precautions for preparing and installing steel pipe.
2. Identify types of steel pipe and fittings.
3. Recognize tools for cutting and threading steel pipe.
4. Calculate takeouts.
5. Set up equipment, including power threading machines.
6. Measure, cut, ream, and thread steel pipe.
7. Assemble threaded, grooved, flanged, and plain-end pipe.
8. Check for correctness of pipe-end preparation.
9. Read a fitting.

Trade Terms

Bushing
Cast-iron fitting
Close nipple
Companion flange
Concentric reducing coupling
Cross
Cut groove
Ductile
Eccentric reducing coupling
Elbow
Extra heavy
Flange-thread
Flexible drop
Gasket
Gib
Liquid Teflon®
Makeup
Malleable iron
Nipple
National Pipe Thread (NPT)
Plain-end
Prepared-end
Rolled grooves
Shoulder-to-shoulder
Standpipe
Street elbow
Tee
Teflon® tape

Prerequisites

Before you begin this module, it is recommended that you successfully complete *Core Curriculum* and *Sprinkler Fitting Level One*, Modules 18101-13 and 18102-13.

NOTE

Unless otherwise specified, references in parentheses following figure and table numbers refer to *NFPA 13*.

Contents

Topics to be presented in this module include:

Contents (continued)

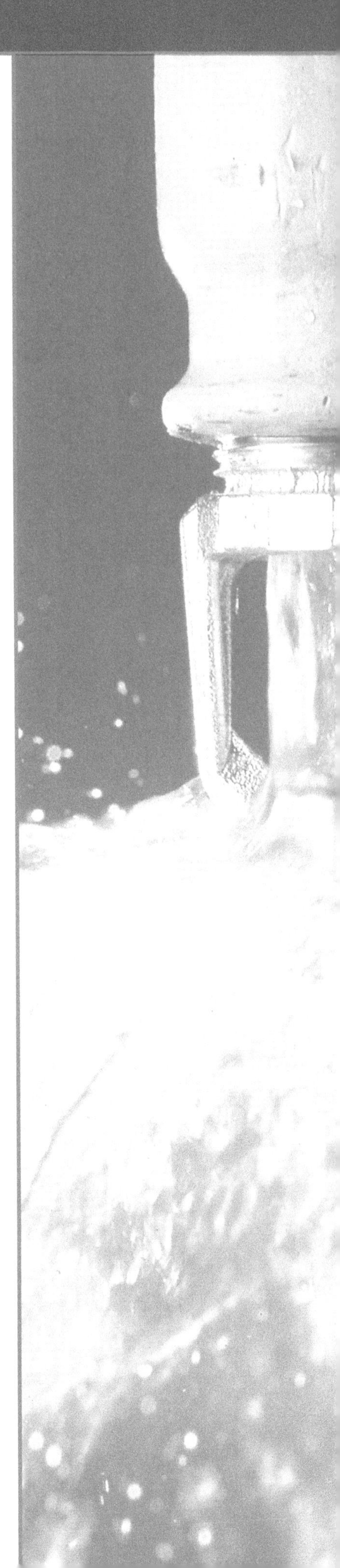

Figures and Tables

1.0.0 INTRODUCTION

Steel pipe is the most common type of pipe used in fire sprinkler systems. While most steel pipe is joined with threaded fittings, some systems use grooved joints, flanges, or plain-end joining methods. The sprinkler fitter must be familiar with all these joining methods, and must also know the conditions and limitations that apply to the different methods. For example, steel pipe below a specified wall thickness may not be threaded unless it is Listed for that purpose. This module will cover all three joining methods, starting with threaded pipe.

> **NOTE**
>
> The capitalized term *Listed* is used in this module to represent a listing such as UL.

There are several advantages to using threaded pipe. Threaded systems require no specialized in-shop fabrication, and can be easily fabricated on the job site using pipe and various fittings. Portable threading equipment can be used almost anywhere on the job site. Threaded pipe is much safer and faster than welded pipe for piping systems that must be installed near flammable liquids or gases, because it poses no fire hazards. Finally, threaded pipe can be cleaned on the inside before it is installed, reducing the possibility of metal particles becoming trapped and damaging valves or strainers.

Threaded pipe also has its disadvantages. Threaded pipe joints are more prone to leak than other types of joints, and the strength of the pipe is weakened by the threading process because the wall thickness of the pipe is reduced at the threads.

2.0.0 MATERIALS USED IN THREADED PIPING SYSTEMS

Various types of materials can be used in threaded piping systems. These materials are classified according to the size of the pipe, the schedule or wall thickness of the pipe, and the material of construction.

The material most commonly used in threaded piping systems is steel. Steel pipe is widely used in various applications because it is durable, machinable, and less expensive than most other types of pipe.

Galvanized pipe is steel pipe that has been dipped into a mixture of hot zinc. The zinc coats the pipe both externally and internally and protects the pipe from corrosion. Galvanized pipe is used to convey air, water, and other fluids.

Stainless steel can also be used in threaded systems; however, if at all possible, avoid using stainless steel pipe, because it is difficult to correctly cut the threads and to make up leakproof joints. Stainless steel pipe is also expensive, although it is good for special applications that require the maximum resistance to corrosion.

2.1.0 Pipe Sizes

Pipe is Listed in inches by its nominal size. For pipe sizes up to and including 12", nominal size is an approximation of the inside diameter. From 14" and larger, nominal size reflects the outside diameter of the pipe. *Figure 1* shows the inside and outside diameters of pipe.

There are times when the nominal size of a pipe and its actual inside or outside diameter differ greatly, but nominal size is used to describe the pipe. The important thing to remember is that the material the pipe or fitting is made from will affect the size. For example, a ¾" brass tube fitting will not fit a ¾" copper pipe. Brass is, however, interchangeable with steel pipe.

2.2.0 Schedules and Wall Thicknesses

Wall thickness can be described in two ways: by schedule and by manufacturer's weight. The larger the schedule number, the thicker the pipe wall. Therefore, pipe with higher schedule numbers is stronger and can withstand more pressure. Steel pipe with wall thicknesses less than Schedule 30 in sizes 8" and larger or Schedule 40 in sizes smaller than 8" typically must not be joined by threaded fittings, although there are

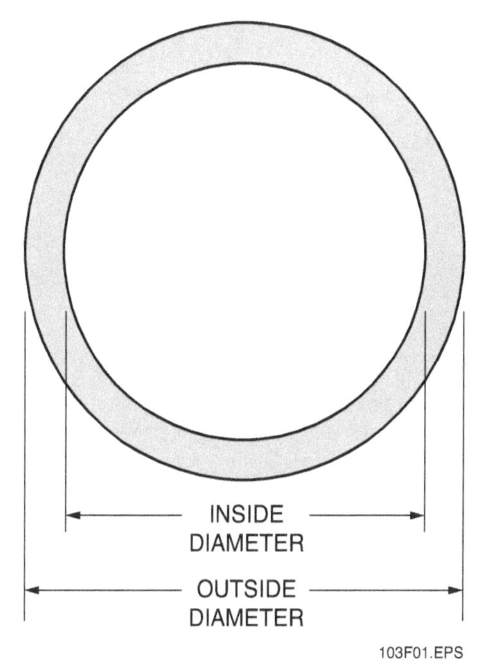

Figure 1 Inside and outside diameters of pipe.

exceptions to this rule made for Specially Listed thin-wall pipe that has been designated as threadable. Only if a threaded assembly is investigated for suitability in automatic sprinkler installations and Listed for this service can it be used. It is also important to remember that the schedule numbers refer only to the wall thickness of a pipe of a given nominal size. For example, a ¾" Schedule 40 pipe does not have the same thickness as a 1" Schedule 40 pipe.

Another way to describe pipe wall thickness is by manufacturer's weight. From smallest to largest, there are four classifications in common use today:

- Thin wall
- Standard wall (STD)
- Extra-strong wall (XS)
- Double extra-strong wall (XXS)

Schedule numbers and wall thicknesses are somewhat interchangeable. Schedule 40 galvanized pipe and standard-wall galvanized pipe have the same wall thickness for all sizes up to and including 10" nominal size. Schedule 80 and extra-strong wall galvanized pipe have the same wall thickness through 8" nominal size. Common wall thicknesses of steel pipe are Schedule 10, Schedule 30 (threadable thin wall), Schedule 40, Schedule 80, and Schedule 160. Common wall thicknesses of stainless steel pipe range from Schedule 5 to 160.

2.2.1 The Schedule System

Years ago, almost all sprinkler systems were installed using Schedule 40 and Schedule 30 black steel pipe. All sizes through 6" were Schedule 40. Sizes 8" and greater were Schedule 30. Schedule 30 and Schedule 40 pipe are called standard (STD) pipe.

According to *NFPA 13*, standard piping material is usable at up to 300 psi. Extra-strong, Schedule 80 pipe is also available, but is rarely used in fire sprinkler work.

Standard materials are specified by nominal outside diameter and wall thickness in American National Standards Institute (ANSI) *Welded and Seamless Wrought Steel Pipe*. ANSI is a volunteer standards group that is privately funded to provide uniformity standards for parts and materials in industry. *Table 1* shows the schedule system used in the fire sprinkler industry.

The outside diameter of standard wall, extra-strong wall, and double extra-strong wall is the same for all schedules, although the wall thicknesses and inside diameters change. Therefore, the same fittings can be used on the pipe regardless of the schedule. The size designation (1", for example) is an approximation; although it is called 1" pipe, the actual outside diameter is 1.315".

The actual diameters of the pipe used in fire sprinkler supply systems have evolved through practical experience. Sizes now available are those

Table 1 Steel Pipe Dimensions

Nominal Pipe Size	Outside Diameter	Schedule 5		Schedule 10[a]		Schedule 30		Schedule 40	
		Inside Diameter	Wall Thickness	Inside Diameter	Wall Thickness	Inside Diameter	Wall Thickness	Inside Diameter	Wall Thickness
½[b]	0.840	–	–	0.674	0.083	–	–	0.622	0.109
¾[b]	1.050	–	–	0.884	0.083	–	–	0.824	0.113
1	1.315	1.185	0.065	1.097	0.109	–	–	1.049	0.133
1¼	1.660	1.530	0.065	1.442	0.109	–	–	1.380	0.140
1½	1.900	1.770	0.065	1.682	0.109	–	–	1.610	0.145
2	2.375	2.245	0.065	2.157	0.109	–	–	2.067	0.154
2½	2.875	2.709	0.083	2.635	0.120	–	–	2.469	0.203
3	3.500	3.334	0.083	3.260	0.120	–	–	3.068	0.216
3½	4.000	3.834	0.083	3.760	0.120	–	–	3.548	0.226
4	4.500	4.334	0.083	4.260	0.120	–	–	4.026	0.237
5	5.563	–	–	5.295	0.134	–	–	5.047	0.258
6	6.625	6.407	0.109	6.357	0.134[c]	–	–	6.065	0.280
8	8.625	–	–	8.249	0.188[c]	8.071	0.277[d]	7.981	0.322
10	10.750	–	–	10.370	0.188[c]	10.140	0.307[d]	10.020	0.365
12	12.750	–	–	–	–	12.090	0.330[c]	11.938	0.406

NOTES:

All dimensions in inches.

(a) Schedule 10 defined to 5 inch (127 mm) nominal pipe size by *ASTM A 135, Standard Specification for Electric-Resistance-Welded Steel Pipe.*

(b) These values applicable when used in conjunction with 8.15.20.4 and 8.15.20.5.

(c) Wall thickness specified in 6.3.2.

(d) Wall thickness specified in 6.3.3.

103T01.EPS

that have been useful in the past. The nominal outside diameters, such as 1", 1¼", 1½", and 2" are used universally for reference in the mechanical trades, even though the actual diameters are different. *Figure 2* shows a comparison of wall thickness.

Standard pipe, which is threaded pipe used for sprinkler systems, uses Schedule 40 pipe for up to 6" pipe. Schedule 30 pipe is used in 8" and 10" sizes. The materials used to manufacture standard pipe and the specifications for standard pipe are further defined in the following specifications:

- *ASTM A53, Standard Specification for Pipe, Steel, Black and Hot-Dipped Zinc-Coated (Galvanized) Welded and Seamless*
- *ASTM A135, Standard Specification for Electric-Resistance-Welded Steel Pipe*
- *ASTM A795, Standard Specification for Black and Hot-Dipped Zinc Coated (Galvanized) Welded and Seamless Steel Pipe for Fire Protection Use*

2.3.0 Pipe Manufacturing Processes

Three different methods are used to manufacture pipe. The piping industry has typically supplied steel pipe in lengths of 21' and 25'. Some specifications require the use of domestic-made pipe as opposed to foreign-made pipe. This is a contractual requirement based on national economics. As long as the pipe is made to specifications, there should be no difference between the quality of foreign and domestic material.

2.3.1 Continuous-Weld Pipe

Continuous-weld pipe is manufactured from skelp, which is a coil of flat steel. In a furnace, the skelp is drawn through a series of forming rollers. The edges are heated to welding temperature. Rollers form the round section and also squeeze the edges together, producing the weld. Sizing rollers adjust the outside diameter, and saws cut the pipe into the desired lengths.

2.3.2 Seamless Pipe

Seamless pipe is made from round steel billets called tube rounds, which are heated and then pierced. A mandrel, which is a cylindrical bar of a precise diameter, is inserted into the shell or pierced tube round. The size of the mandrel determines the interior diameter of the pipe. Rollers form the outside to the proper diameter and wall thickness. Seamless tubing is more expensive than other kinds of pipe or tubing and is not frequently used in the sprinkler industry.

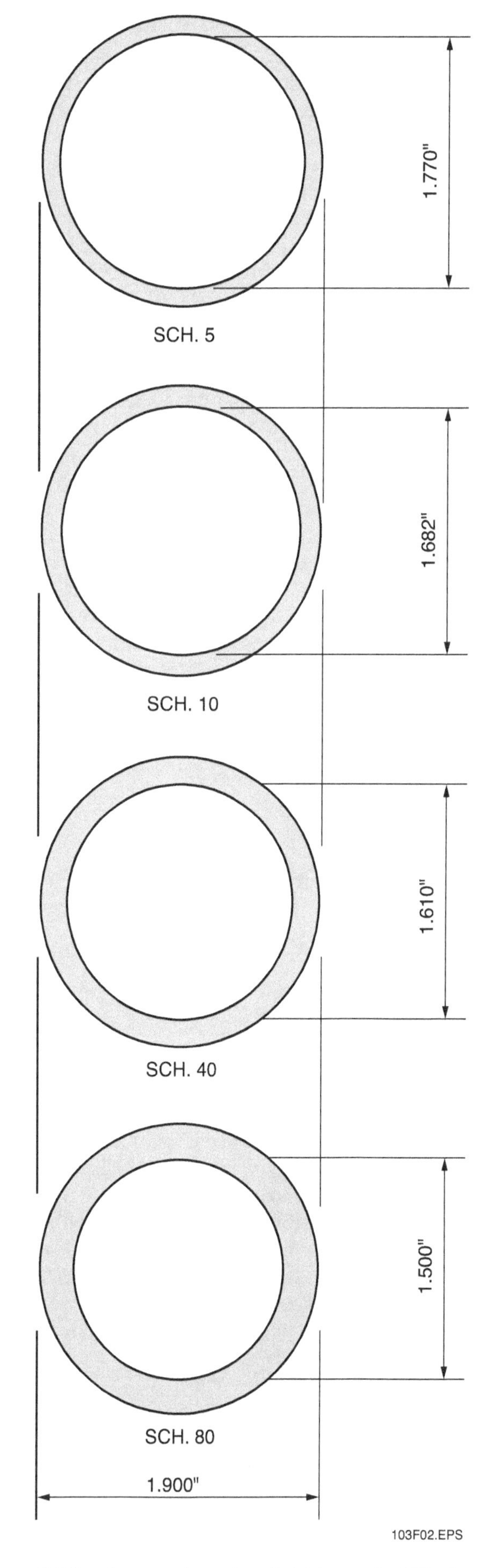

Figure 2 Visual comparison of wall thickness for 1½" pipe.

2.3.3 Electric Resistance Welded (ERW) Pipe

Electric resistance welded (ERW) pipe is formed by drawing the skelp through a series of rollers. The rollers gradually form the material into a round shape. It then passes through two copper welding wheels. One wheel has a positive charge and the other has a negative charge. The skelp edges are heated to welding condition by the resistance to electrical current in the joint. After the rollers pass, the material cools in a bonded condition.

2.3.4 Galvanized Pipe

The normal piping material used for sprinkler systems is black pipe. This pipe usually has a rust-delaying varnish or lacquer coat on the outside. However, this lacquer coat does not classify as corrosion-resistant. Hot-dipped galvanized pipe is black pipe that is submerged in molten zinc so that a layer of zinc (galvanize) attaches itself to both the inside and outside of the pipe.

Hot-dipped galvanized pipe is more expensive than black pipe and is never used unless there is a condition where corrosion is a problem or when specifications or standards require it. FM Global requires the use of galvanized pipe in dry systems. *NFPA 13* allows metallic corrosion-resistant material such as galvanized steel, brass, or copper for water motor-operated devices.

2.4.0 *ASTM A53* Piping

ASTM A53 is premium grade piping. The standard identifies the following types of *ASTM A53*:

- *Type F* – Furnace butt-welded, continuous-weld
- *Type E* – Electric resistance welded, Grades A and B
- *Type S* – Seamless, Grades A and B

ASTM A53 is sometimes specified for sprinkler systems because *ASTM A53* is chemically controlled to ensure ductility. Under *NFPA 13*, bending is permitted for Schedule 10 steel pipe or greater wall thickness, and Types K and L copper tube, provided the pipe has no kinks, ripples, distortions, reductions in diameter, or any noticeable deviations from round.

2.5.0 *ASTM A135* Piping

ASTM A135, Specification for Electric-Resistance-Welded Steel Pipe, describes another frequently used piping material. There are two different grades of *ASTM A135* piping, A and B. Like *ASTM A53* piping, Grade A piping may also be bent in accordance with NFPA requirements. Grade B piping is the more commonly used material.

2.6.0 Threadable Thin-Wall Pipe

The threading process removes metal from the pipe wall, thereby reducing the wall thickness. In fact, if the pipe is too thin, the threads could be visible from the inside of the pipe. If the pipe wall is thin to start with, the risk of failure is increased by threading. For that reason, *NFPA 13* has traditionally limited the size of steel pipe that can be threaded. Over the years, however, pipe manufacturers have developed thin-wall steel pipe that is suitable for threading. This has caused *NFPA 13* to be modified to incorporate the following exceptions:

Did You Know?

Producing Steel

Carbon steel is produced by one of four processes: Bessemer, open hearth, electric furnace, or basic oxygen. The Bessemer process, developed in the 1850s, produced steel by blowing air through molten iron at high pressure. The process used oxygen to draw off impurities, but it was efficient enough to remove all the unwanted elements. The process is no longer used in the United States.

The open-hearth process was the main method of producing steel for many years. An open-hearth furnace is a huge, shallow tank lined with firebrick. A fuel such as gas, oil, or powdered coal, is mixed with hot air in burners and blasted across the tank. Molten iron, recycled steel, and limestone are melted together in the tank. After the removal of impurities collected by the limestone, the steel is ready.

The electric-furnace process uses tremendous charges of electricity to produce steel. The furnace is loaded with scrap iron. An electrical arc is produced using electrodes. The resulting heat of 6,000°F (3,312°C) melts the scrap iron and produces the steel alloy.

The basic oxygen process is the method most often used today. Molten iron from a blast furnace (a giant oven lined with firebrick) is poured into the basic oxygen furnace. Pure oxygen is blown onto the metal through a tube called an oxygen lance. The oxygen combines with the impurities and is then removed from the furnace, leaving the steel.

After the liquid steel has been created in the furnaces, it must be further processed and cast into solid form before manufacturers can use it.

- Other types of pipe or tube may be used if investigated and Listed for fire protection service.
- Steel pipe with wall thicknesses less than Schedule 30 in sizes 8" and larger, or Schedule 40 in sizes smaller than 8", must not be joined by threaded fittings. A threaded assembly must be investigated for suitability in automatic sprinkler installations, as well as Listed for this service, or it cannot be used.

Several manufacturers now offer threadable thin-wall pipe approved by Underwriters Laboratories® (UL) for threading. One offering is approximately Schedule 20 pipe.

Quality control of threading is important. To enhance the lasting quality, some of this material may be zinc coated on the outside only. This is a plating process as opposed to the hot dip galvanized process that coats both the inside and outside of the pipe. The threading operation must follow the specifications of the manufacturer and the approving agencies if the installation is to be acceptable.

The annex of *NFPA 13* states that threaded thin-wall pipe should be checked by the installer using working ring gauges conforming to the *Basic Dimensions of Ring Gauges for USA (American) Standard Taper Pipe Threads, NPT* as per ANSI/ASME. Some Authorities Having Jurisdiction (AHJs) are enforcing this as a requirement. Not all AHJs will accept threaded light-wall pipe.

Threaded light-wall pipe is available in limited sizes. The UL *Fire Protection Equipment Directory* lists 1" through 3" sizes, and ¾" threaded light-wall pipe is produced on demand only.

On Site

Steel Pipe

The only difference between black iron pipe and galvanized pipe is the addition of a protective zinc coating on the galvanized pipe.

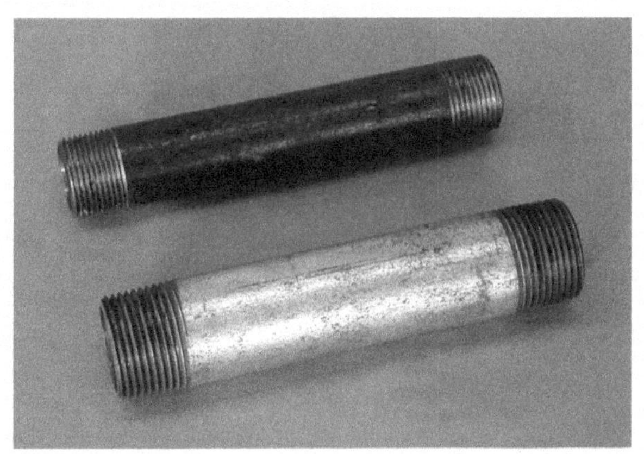

103SA01.EPS

2.7.0 *ASTM A795* Piping

ASTM A795, Specification for Black and Hot-Dipped Zinc-Coated (Galvanized) Welded and Seamless Steel Pipe for Fire Protection Use, is a more recent development in fire sprinkler piping. Other than those materials specified in *NFPA 13* by ASTM or ANSI standards, look for the listing by Underwriters Laboratories® or FM Global in the *Fire Protection Equipment Directory* and for the listing mark on the material.

Steel pipe remains the major installation material of fire sprinkler systems. There are other types of piping that are approved and cost-effective under specific sets of circumstances. They will be discussed in subsequent modules.

3.0.0 Tools For Cutting and Threading Steel Pipe

Both hand and power tools are used to cut and thread steel pipe. It is important that you learn to use these tools safely and to care for and maintain them properly.

The life of any tool depends on proper lubrication. If the frame is steel, a light coating of oil is essential to prevent rusting. If there are moving parts, the bearing surfaces require lubrication. Excess lubricants leave a trap for dust, dirt, grime, and abrasive action. If you follow the manufacturer's specified lubrication requirements for any tools you use, you will keep them functioning for a long time.

3.1.0 Hand Tools

Sprinkler fitters use a variety of hand tools for a variety of purposes, including cutting, reaming, and threading pipe. The most common hand tools used are steel pipe cutters, hinged cutters, reamers, and dies. In many cases, the tool selected to perform a job must be the exact size to perform the job or personal injury or equipment damage can occur.

WARNING

If you have not completed the *Basic Safety* module in the *Core Curriculum*, stop here. You must complete the *Basic Safety* module first. Also, you must wear appropriate protective equipment when you operate any power tool or when you are near someone else who is operating a power tool.

3.1.1 Steel Pipe Cutters

Steel pipe cutters are used for cutting steel pipe up to 8" in diameter. The standard steel pipe cutter is available with one, three, or four cutting wheels.

Steel pipe cutters operate like a vise clamped on a piece of pipe. The wheels are lined up on the cutting mark and tightened against the pipe. As the cutter is rotated around the pipe, the adjusting handle is turned to maintain uniform tightness on the pipe.

When using a steel pipe cutter, ensure that the pipe is clean. Clean pipe cuts easily, but dirt and grit act as abrasives on the cutting wheels and cause the cutters to wear rapidly.

Misalignment of the pipe cutter against the pipe causes different cutting wheels to cut on different lines. The cutting wheels crawl along the pipe and the cumulative effect of multiple cutters and multiple passes is lost. *Figure 3* shows examples of steel pipe cutters.

3.1.2 Hinged Cutters

Hinged cutters can be used to cut 1" to 12" diameter pipe. They work well in close quarters and are more safe than an abrasive saw or a milling cutter. *Figure 4* shows a hinged cutter designed for 2½" to 4" pipe.

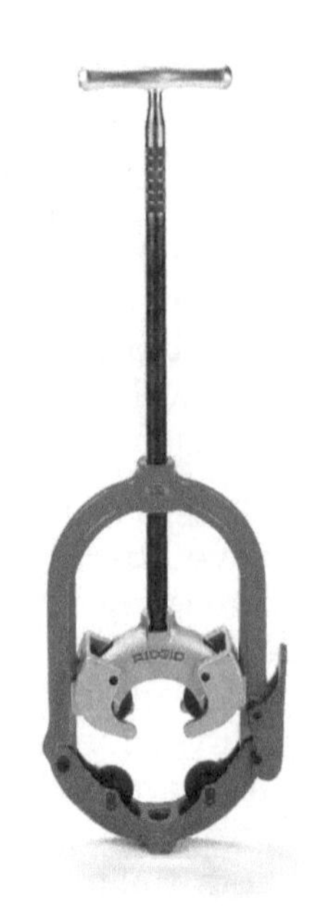

Figure 4 Hinged cutter.

3.1.3 Reamers

Whether pipe is cut to length by the manufacturer, in the shop, or on the job site, it must be reamed to clear the waterway. When the cutting process leaves burrs or fins at the cut end of a length of pipe, reaming removes these burrs and rough material from the end of the pipe. You can also use a plasma cutter, which does not produce burrs. *NFPA 13* requires that burrs and fins be removed whether the pipe is to be threaded, grooved or welded. Reaming helps ensure free flow, as well as a leakproof, strong joint.

Pipe sizes up to 2" can be reamed with a hand reamer (*Figure 5*). Hand-reaming requires the use of a ratchet ring and a ratchet handle to turn the reamer. On larger pipes, it is more economical to use a power reamer because power reamers are faster and more uniform than hand reamers.

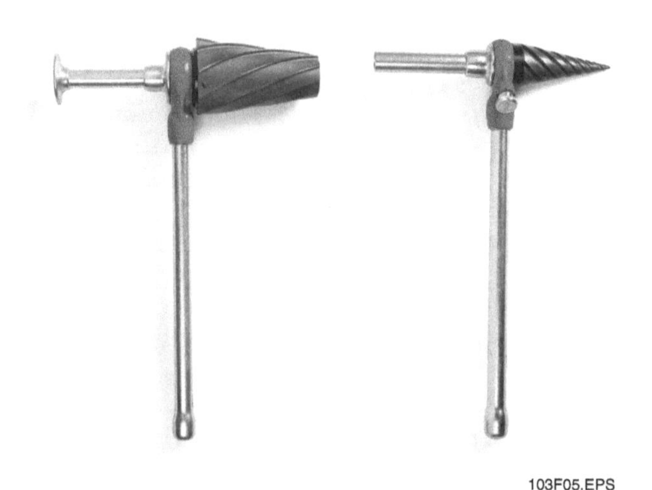

Figure 5 Reamers with ratchet housing attached.

When power reamers are not available, half-round files are often used to ream larger-size pipe.

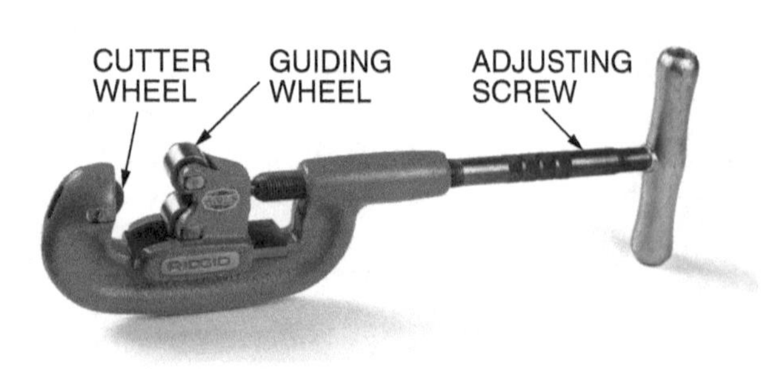

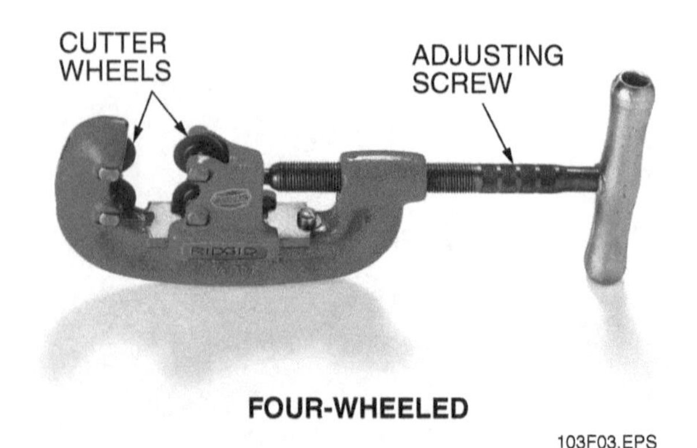

Figure 3 Steel pipe cutters.

Reaming must be done carefully to avoid oversized holes, bell-mouthed holes, or a poor finish. This is especially true with power reamers. If the reamer is out of alignment, the pipe end will be tapered, producing a mechanically weak, bell-mouth opening. To ensure a smooth finish, set the speed of the power drive according to the manufacturer's instructions.

3.1.4 Dies

Occasionally, you may have to hand-thread a piece of pipe or a hanger rod. A solid die with a holder is required for this procedure. Dies come in a wide range of sizes and thread patterns. Some holders are called three-way pipe threaders and hold up to three dies within a specified range. Each die set is locked in place and a guide keeps the threader straight for consistent, true threads. Most die sets include the commonly used die sizes as well as the holder. The die sizes most commonly used to thread hanger rods are shown in *Table 2*.

The die sizes most commonly used to thread pipe are shown in *Table 3*.

WARNING

Never operate a power tool until you have been trained to use it. You can greatly reduce the chance of accidents by learning safety rules for each task that you perform.

WARNING

Wear appropriate protective equipment, including safety glasses, gloves, and a hard hat when using a tool or when working near someone who is using a tool.

Table 2 Common Die Sizes for Hanger Rods

Hanger Rod Diameter	Die Size (threads/inch)
3⁄8"	16
1⁄2"	13
5⁄8"	11

103T02.EPS

Table 3 Common Die Sizes for Threaded Pipe

Nominal Pipe Size	Die Size (threads/inch)
1⁄16"–1⁄8"	27
1⁄4"–3⁄8"	18
1⁄2"–3⁄4"	14
1"–2"	11½
2½" and up	8

103T03.EPS

3.2.0 Portable Power Threading Machines

Portable power threading machines are used extensively in the sprinkler trade because they require less time than hand threading. The common portable models can thread pipe up to 2" in diameter. Always read the manufacturer's instructions for care, maintenance, and safety instructions before using a power threader. *Figure 6* shows examples of threading machines and accessories.

As a safety precaution, portable power threading machines are equipped with a foot pedal for controlling the power. Lifting your foot off the pedal automatically turns the power off. Accessories for cutting and reaming pipe can also be added to the same machine for performing these tasks. The portable power threading machine's standard components include the following:

- Chuck
- Ways and support bars
- Carriage
- Dies for power threading machines

3.2.1 Chuck

The chuck on a power threading machine centers the pipe and holds it in place for threading. The dual-jaw chuck rotates with the pipe as it is being threaded. On most models, the rear jaw does the actual centering and must be tightened only enough to ensure that the jaws are touching the pipe. The front jaw clamps the pipe in place and must be tightened so that the pipe does not slip.

3.2.2 Ways and Support Bars

A carriage moves along a bedway or support bar and holds the die head in place. Horizontal bars that are supported on both ends are called ways. Horizontal bars that are supported on one end are called support bars. The ways or support bars can be either flat, round, or V-shaped. It is important to keep them free of dirt, grit, and shavings to allow free movement of the carriage. The ways or support bars align the die head with the pipe being threaded. Do not use the ways as a tool rack

On Site

Die Sets

It is very important to keep the die sets together. Do not mix dies of one set with dies from another set. Usually, each die is marked with a serial number and the four dies that are made together have the same number.

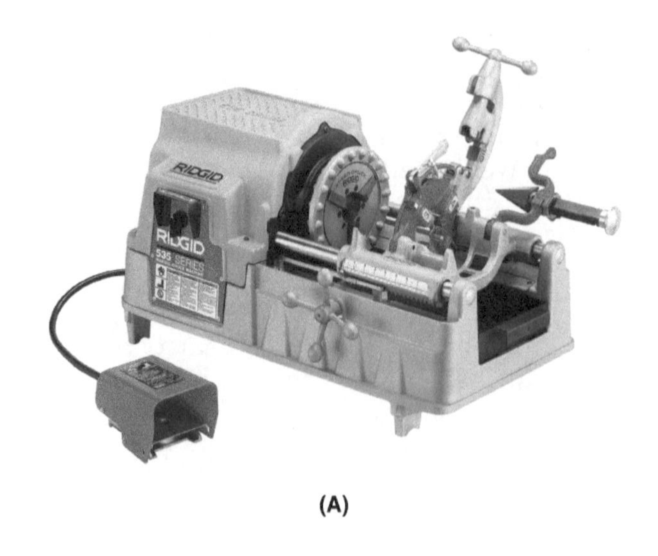

(A)

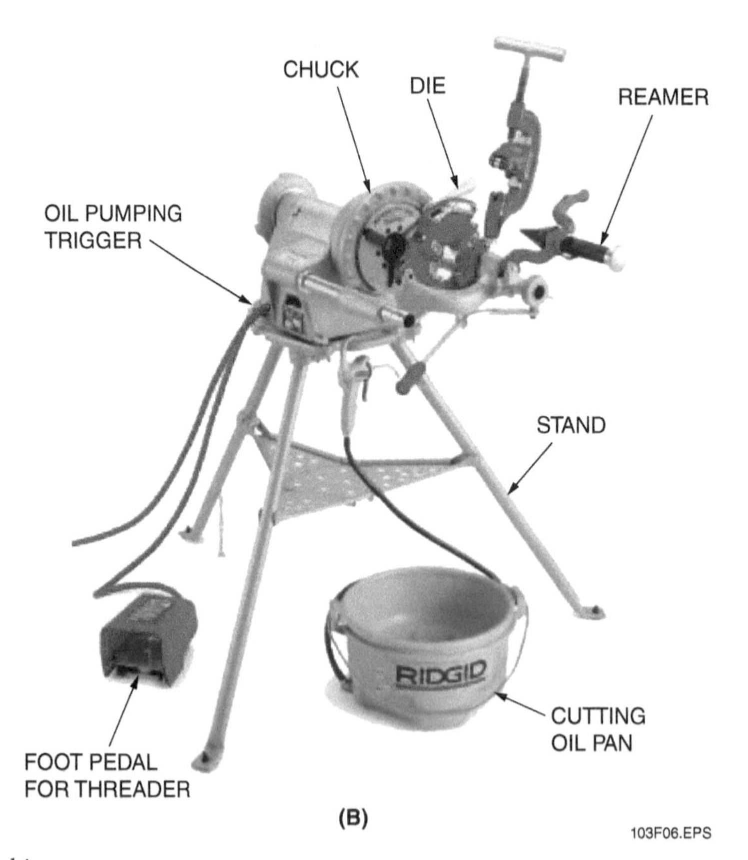

(B)

103F06.EPS

Figure 6 Pipe threading machines.

or handle and use care to ensure that the ways do not get dented or bent.

3.2.3 Carriage

Generally, there are two carriages on a power threader. One carriage carries the die head, and the other carriage carries the chuck. The carriages must also be kept free of dirt, grit, and shavings. The carriage normally has some side play, but if the carriage becomes excessively worn, there will be too much side play and the carriage will have to be replaced. A carriage on a flat bedway will sometimes have gibs that can be adjusted to eliminate excessive side play.

3.2.4 Dies for Power Threading Machines

The dies are the most important components of a power threading machine and if not maintained properly can cause a number of problems, from chipping to improperly sized threads. Dull dies can be sharpened or honed, and dies must be cleaned periodically using a manufacturer-approved cleaning solution and lightly oiled with lubricating oil. Before reinserting the die head, use a wire brush to clean away any foreign material that may have accumulated in the holding apparatus. Because 1" and 2" pipe have the same number of threads per square inch, some threading machines have adjustable heads that are set to size so that the dies do not have to be changed for different pipe sizes (*Figure 7*).

Torn threads are caused by dies that have chipped or dull teeth, or by using the wrong material, such as Schedule 10 thin wall. A bent die head hinge pin causes double threading, and die segments that are not properly ground will make it impossible to obtain a water-tight seal. Always consult the manufacturer's literature to troubleshoot threading problems.

3.3.0 Cutting Oil

The use of cutting oil (*Figure 8*) is essential when threading pipe and rod by hand or using power tools. Cutting oil serves the following purposes:

> **CAUTION**
>
> Make sure cutting oil is compatible with CPVC, if CPVC is being used.

- Keeps the pipe and die cool, preventing unnecessary wear on the die
- Speeds cutting
- Helps keep the threads clean, increasing the quality of the threads by preventing rough or torn threads

Figure 8 Cutting fluid.

When using tools that require cutting oil, make sure that water or other liquids do not get into the oil. The disposal of used oil may be a problem due to environmental requirements. Always dispose of oils and other chemicals in accordance with local, state, and federal requirements.

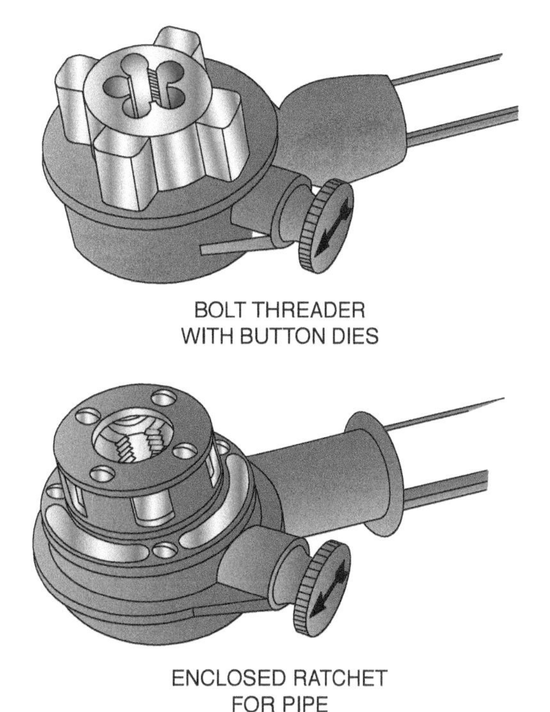

Figure 7 Universal die head.

CAUTION

Use only cutting fluid or cutting oil that is specifically designed for metal cutting. Otherwise, the equipment can be damaged. Automobile engine oil is not satisfactory. Review the MSDS for the cutting oil to identify any hazards to people or the environment before using or disposing of the oil.

3.4.0 Safety Precautions for Power Tools

Using power tools to cut, clean, ream, or deburr pipe requires specific safety precautions. Adhere to the following safety guidelines when using power tools:

- Perform the following to ensure that the power supply is adequate and safe:
  - Check the power rating of the tool relative to its power handling capacity.
  - Check all cords for signs of excessive wear or fraying.
  - Ensure that the plug has a proper grounding connection (three wires on 120-volt AC). If the grounding pin is missing or broken, discard the cord or have it repaired.
  - Ensure that there is a ground fault circuit interrupter (GFCI) on the line.
- Do not use power tools near water, while standing in water, or in wet, damp, or foggy weather.
- Ensure that extension cords are safe for the application. Discard cords that are cracked, frayed, or cut. Ensure that extension cords are properly grounded.
- Wear the appropriate personal protective equipment, including safety goggles and gloves.

WARNING

When you are working with a power threading machine, do not wear loose clothing, jewelry, or gloves that could easily be caught in the rotating parts of the machine. Be sure to wear appropriate personal protective equipment. Be sure to tie back long hair to prevent if from being entangled in the machine during operation. Use a piece of cloth or cardboard to absorb any oil that spills during the machine's operation, so the oil does not get on the bottom of your work shoes and you do not track the slippery oil around the job site.

3.5.0 Power Threading Machine Setup

It is important to read the manufacturer's instruction manual thoroughly before operating the equipment. Make sure you understand how the machine works, what its capabilities are, and what hazards are associated with using the equipment.

Step 1 Make sure the machine's power switch is in the OFF position before connecting the power.

Step 2 Make sure the machine is properly grounded. Insist on installation of a GFCI on the line, or use a GFCI extension cord (*Figure 9*).

Step 3 Remove any tools that might be on the machine.

Step 4 Check the die heads, support bars, or bedways. They must be free of dirt, grime, shavings, or any other foreign material.

Step 5 Wear the appropriate safety equipment.

Step 6 Use a floor pipe stand or support for long lengths of pipe (*Figure 10*).

Step 7 Make sure the threading machine and pipe stand are level and aligned.

3.5.1 Alignment

The components of a power threading machine must be properly aligned and adjusted to achieve quality threads. If the machine is out of alignment or adjustment, the threads will be defective. Misalignment may prevent the die from starting on the pipe, break the die, cause off-center threads, or cause spiraled threads. Follow these steps to check the alignment of a power threading machine:

Step 1 Ensure the power is off.

Step 2 Place a pipe in the chuck.

Step 3 Close the die head so that it touches the pipe.

Step 4 Ensure the pipe touches all the die segments simultaneously.

If the pipe does not touch all points of the die segments simultaneously, the machine is out of

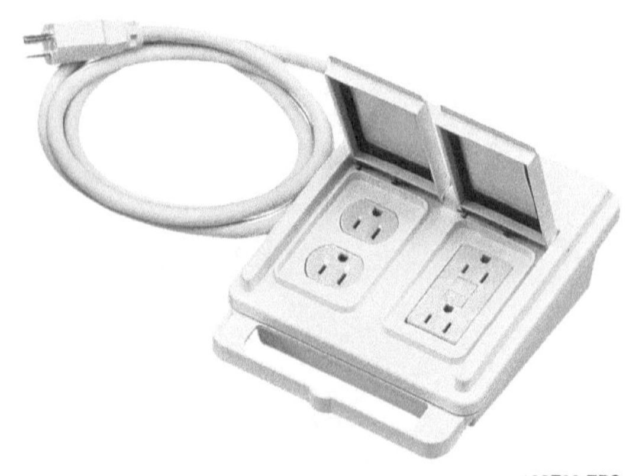
103F09.EPS

Figure 9 Extension cord with GFCI.

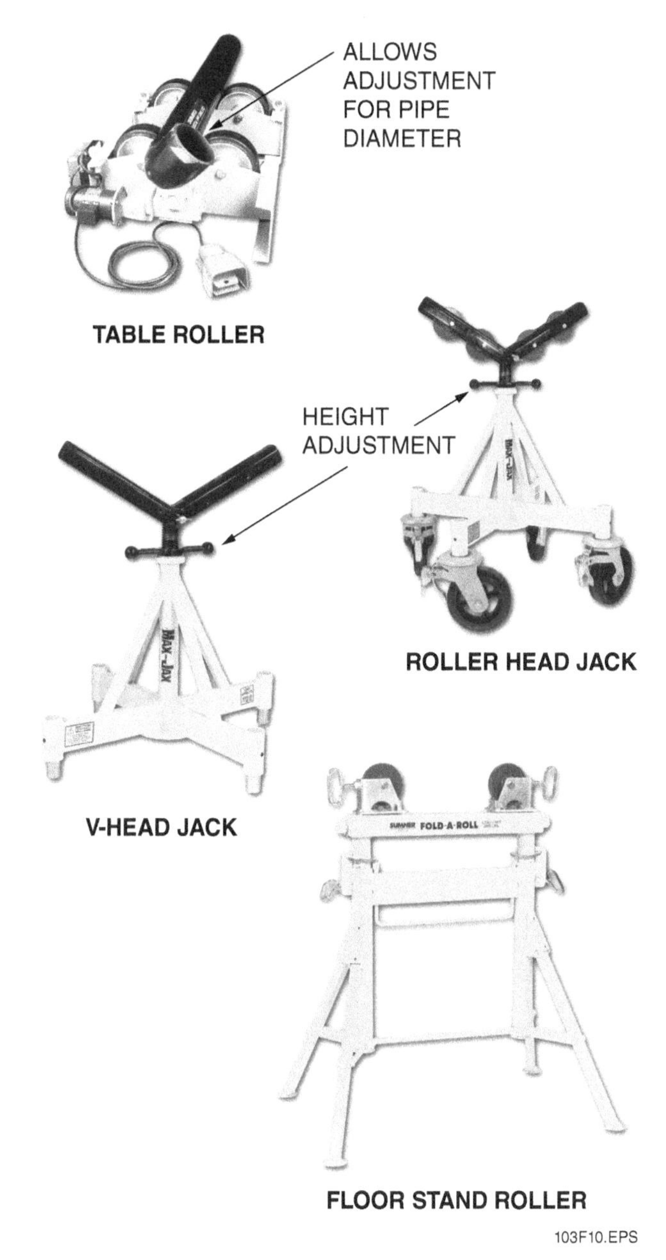

Figure 10 Pipe supports.

alignment. Follow the manufacturer's alignment procedure to align the machine before going ahead with the threading operation.

3.5.2 Leveling

It is also critical that the power threading machine be as level as possible and be set on a stable surface. In the field, this is the fitter's responsibility. Make sure that the machine is level and stable before the power is turned on.

After ensuring that the threading machine is level, insert the pipe into the chuck and secure the other end of the pipe with a pipe stand or support. Some models allow the pipe to move backward and forward only and other models allow the pipe to rotate freely for operations such as cut- or roll-grooving. A pipe support or stand keeps the pipe aligned with the machine.

> **NOTE** After leveling the machine, add a slight tilt to the machine so that cutting oil will not run up the pipe.

4.0.0 THREADS

A pipe thread is an inclined plane wrapped around a cylinder. The thread angle is the angle formed by two inclined faces of the thread. The thread angle varies depending on the specific thread standard used. The thread pitch is the distance between the adjacent screw threads, measured from center line to center line. Pitch can also mean inch when referring to pitch as threads per inch. The crest of the thread is the outermost part of the thread, or the top of the thread when viewing it from the side. The root of the thread begins at the innermost part of the thread, or from the bottom if viewed from the side.

4.1.0 Types of Threads

The National Pipe Thread (NPT) is the standard thread for pipe connections. The National Pipe Thread has a threaded angle of 60 degrees and a slight flat at the crest and the root. While National Pipe Threads can be either straight or tapered, the type normally used has a tapered internal and external thread. The taper used is 1⁄16" per inch of threads. *Figure 11* shows the taper used in cutting pipe threads.

In addition, there are several other standard pipe threads for specific applications. All National Pipe Threads are designated by the nominal size of the pipe, the number of threads per inch, and the symbols for the thread series, respectively. For example, 3⁄8–18–NPT is the designation for a 3⁄8" nominal size pipe that has 18 threads per inch and uses National Pipe Threads.

4.1.1 Threads Per Inch

The proper number of threads on the end of a piece of pipe is specified in *Standard USAS (ASME) B1.20.1.* From 1" pipe through 2" pipe, the number of threads per inch is 11.5. This is determined by the dies or die segments that are used in the pipe threader. For larger sizes of commonly used pipe, there are only 8 threads per inch.

The wall thickness of the pipe and the setting of the dies affect how many threads are present. If the dies are set improperly, too many threads can be cut on Schedule 40 and Schedule 80 pipe. In such a case, the die continues down the pipe making threads until you decide to stop, and the pipe will make up too short. On the other hand, improper settings can also result in too few threads and the pipe will make up long and have reduced joint strength.

4.1.2 Pipe Thread Identification

The number of threads per inch varies with different sizes of pipe. The number of threads is determined in trade practice by *USAS (ASME) B1.20.1, Pipe Threads (Except Dryseal).* An abbreviated version is shown below. These threads are identified as NPT, with the following designation and meanings:

¾ - 14-NPT

Where:

¾ = Nominal pipe size in inches
14 = Number of threads per inch
N = USA Standard
P = Pipe
T = Taper

4.2.0 Taper Thread Connection

The joining, or makeup, of a taper-threaded pipe connection is performed in two distinct operations known as hand engagement and wrench makeup (*Figure 12*). Hand engagement is how far the fitting can be tightened by hand. Wrench makeup is the additional turning of the fitting, using a wrench to completely make up the joint. For an average size pipe, hand engagement on properly cut threads is normally between three and four complete revolutions. Hand engagement is a good way to check the quality of threads after cutting. Wrench makeup is usually an additional three turns, for a total of approximately seven revolutions for complete makeup.

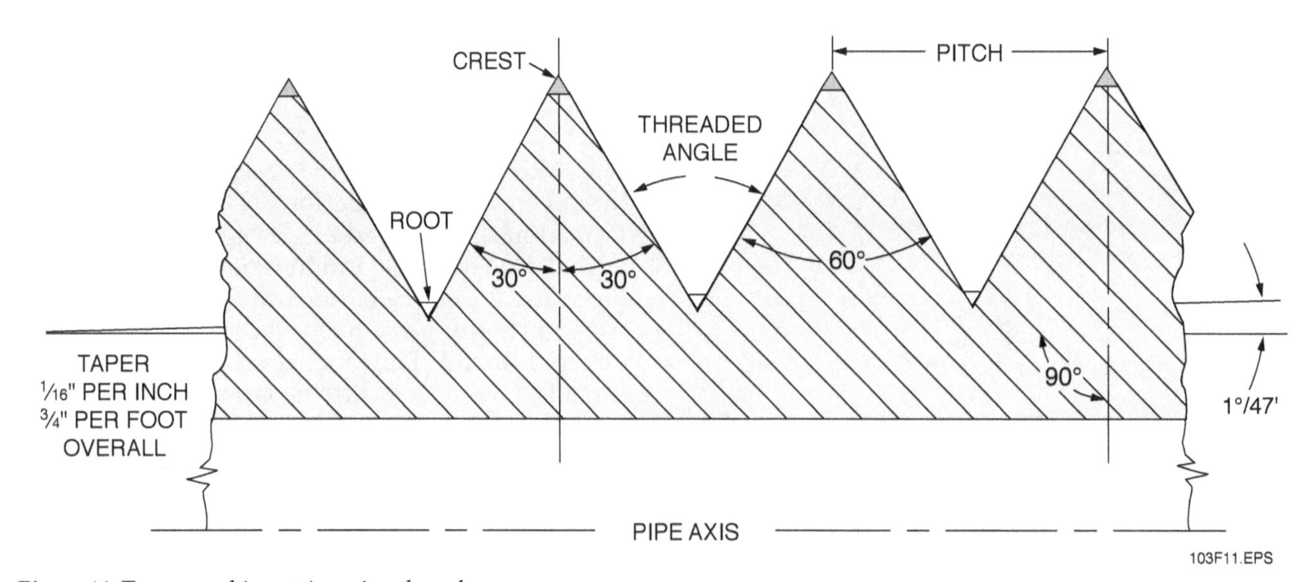

Figure 11 Taper used in cutting pipe threads.

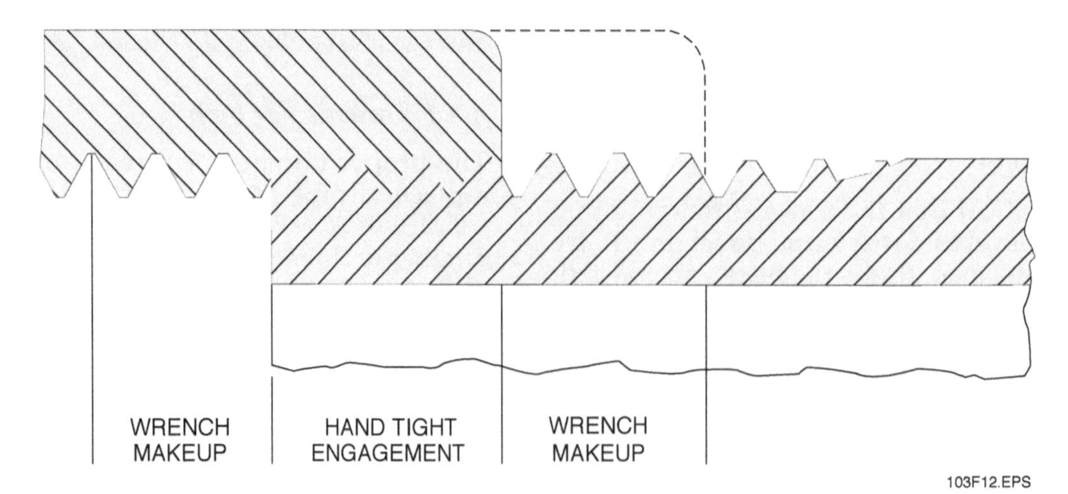

Figure 12 Hand engagement and wrench makeup.

On Site

Pipe Thread Standard

At its inception, this standard was known as the Briggs Standard. Robert Briggs originally formulated this basic thread schedule in England and published the Briggs Standard between 1820 and 1840. At that time, all manufacturers had their own standards and one manufacturer's threads would not fit those of another.

In 1886, manufacturers in the United States adopted the Briggs Standard jointly with the American Society of Mechanical Engineers (ASME). That standard resulted in the ability to interchange pipe and fittings made by various manufacturers and is the same basic standard that is in use today.

The standard for taper thread pipe makeup gives basic thread dimensions and the number of rotations for hand engagement and wrench makeup. The taper thread makeup may vary, plus or minus one turn, from the standard thread makeup allowances.

The overall external thread length is divided into three sections. The first section, from left to right, identifies the number of threads that engage with hand tightening. The second group consists of those additional threads that engage when a wrench is properly applied. The third group consists of those threads, called vanishing threads, that are essential in the threading process but which do not engage. If there are too many threads showing, the joint needs to be made up further. If there are too few threads showing, the joint is made up too tightly. In either case, leaks are likely and the system will not fit the way it was designed.

5.0.0 Threading Pipe

In order to create a strong, leakproof, threaded joint, the threads must be clean and smoothly cut. They must have the correct pitch, lead, taper, form, and size. If these conditions are not met, the connection will leak.

5.1.0 Threading Light-Wall Pipe

Threading of Schedule 30, 40, and 80 pipe in the sizes approved in *NFPA 13* normally presents no problem; however, the threading of light-wall pipe must be done carefully.

NFPA 13 calls for the use of "working ring gauges conforming to the Basic Dimensions of Ring Gauges for USA (American) Standard Taper Pipe Threads, NPT as per ANSI/ASME."

5.2.0 Threading Pipe Using Manual and Power Threading Machines

Manual and power threading machines are used extensively by sprinkler fitters. Manual threading machines (*Figure 13*) are used when you do not have many pieces of pipe to thread or if you must put threads on pipe that is already installed or being repaired. It's easier and quicker to use a manual threader than it would be to get the power threader and set it up to thread a few pieces of pipe.

Power threading machines are used when there is a lot of threading to be done. It is also easier to thread larger size pipe with a power threading machine than it would be to perform manual threading. These machines are cost effective because they save a lot of labor time. The following sections describe how to use manual and power threading machines.

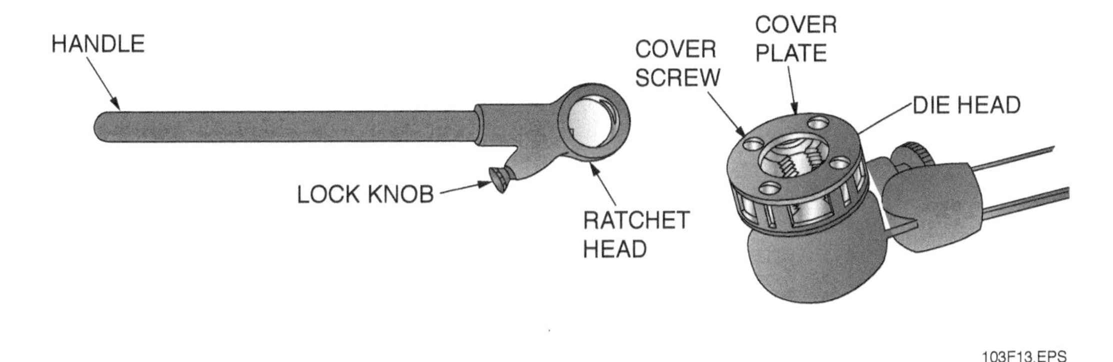

Figure 13 Hand threader.

5.2.1 *Threading Pipe Using a Manual Pipe Threader*

Follow these steps to thread pipe using a manual pipe threader:

Step 1 Identify the kind of material being threaded.

Step 2 Identify the diameter of the material being threaded.

Step 3 Select the die head best suited for the threading process.

Step 4 Insert the die head into the pipe threader handle.

Step 5 Inspect the die and die handle to ensure that all parts are clean and in good condition.

Step 6 Secure the pipe in a vise.

Step 7 Make sure that the end of the pipe is square, clean, and deburred.

Step 8 Position your body so that you are balanced, facing the work area, and within easy reach of the work area.

Step 9 Position the die flush on the end of the pipe.

Step 10 Press the die against the end of the pipe and turn the die handle clockwise to start the threading process.

Step 11 Turn the die handle two full rotations around the pipe.

> **NOTE**
> Add cutting oil as necessary to keep a thin coat on the threads while cutting.

Step 12 Stop the work, and back the dies off the pipe.

Step 13 Check the threads to ensure that they are properly started and clean the cutting threads from the die if necessary.

Step 14 Rethread the die onto the pipe and continue the process until the threading is complete. The threading is complete when one full thread extends past the back edge of the dies.

> **CAUTION**
> Do not thread the pipe too far. This can cause an improper fit between the fitting and the pipe, which results in leaks.

Step 15 Back the die off the pipe.

Step 16 Inspect the new threads and use a rag to remove any debris from the threads.

> **WARNING**
> Freshly cut threads are very sharp. Use care in handling them to avoid cuts.

5.2.2 *Threading Pipe Using a Power Threading Machine*

> **WARNING**
> Wear appropriate protective equipment, including safety glasses, gloves, and a hard hat when using a tool or when working near someone who is using a tool.

Be sure to check the cutting oil level and to prime the cutting oil pump before threading pipe. Follow these steps to thread pipe using a pipe threading machine:

Step 1 Load the pipe into the threading machine (*Figure 14*).

Step 2 Place pipe stands under the pipe as needed.

Step 3 Cut and ream the pipe to the required length.

Step 4 Make sure that the proper dies are in the die head.

Step 5 Loosen the clamp lever.

Step 6 Move the size bar to select the proper die setting for the size of pipe being threaded.

Step 7 Lock the clamp lever.

Step 8 Swing the die head down to the working position.

Step 9 Close the throw-out lever.

Step 10 Lower the lubrication arm and direct the oil supply onto the die, or prime the hand pump.

On Site

Correctly Cut Threads

If the threads are cut correctly, you should be able to screw a fitting three and a half revolutions by hand onto the new threads. If the threads are not deep enough, adjust the die using the clamp lever and repeat the threading operations.

Figure 14 Pipe mounted in a threading machine.

Step 11 Turn the machine switch to the forward position.

Step 12 Step on the foot switch to start the machine.

Step 13 Turn the carriage to bring the die against the end of the pipe (*Figure 15*).

Step 14 Apply light pressure on the handwheel to start the die.

Step 15 Release the handwheel once the dies have started to thread the pipe. The dies feed onto the pipe automatically as they follow the newly cut threads.

CAUTION

Make sure that the die is flooded with oil at all times while the die is cutting to prevent overheating the die and the pipe. Overheating can damage the die and the pipe threads.

Step 16 Open the throw-out lever as soon as one full thread extends from the back of the dies.

Step 17 Release the foot switch to stop the machine.

Step 18 Turn the carriage handwheel or arm to back the die off the pipe.

Step 19 Swing the die and the oil spout up and out of the way.

On Site

Handheld Power Pipe Threaders

A handheld power pipe threader can be used in conjunction with a portable stand when a number of pipes must be threaded in the field. The rotating power head of this threader turns at about 30 rpm and uses the same dies as an equivalent manual threader. When pipe over 2½" is being threaded, a support arm (not shown) is clamped to the pipe and the threader is rested against it to counteract the torque of the threader.

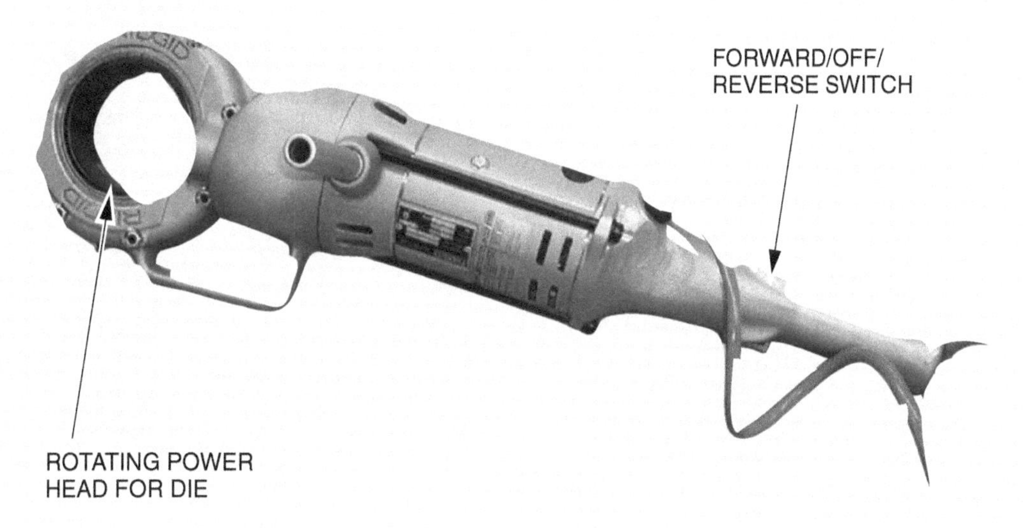

MOVABLE TOOL CARRIAGE
103F15.EPS

Figure 15 Starting the die on the end of the pipe.

Step 20 Screw a fitting onto the end of the pipe to check the threads. If the threads are correct, you should be able to screw a fitting three and a half revolutions by hand onto the new threads. If the threads are not deep enough, adjust the die with the clamp lever and repeat the threading operation.

Step 21 Turn the machine switch to the OFF position.

Step 22 Open the chuck and remove the pipe.

5.3.0 Thread Compounds

A good grade of pipe thread compound must be applied to the male threads to lubricate the joint. This will prevent galling of the threads and ease the makeup of the joint. Be careful about which compound you use to avoid product contamination or dangerous combinations of materials. Depending on the type of liquid being transferred through the pipe, the use of pipe compound may not be permitted. Always check with the sprinkler system design specifications and requirements before applying any type of pipe thread compound. Some compounds that can be used include Teflon® tape and Liquid Teflon®. Pipe thread compound is used to fill the gap produced by the truncated thread profile. The gap is a spiral flue space through which water or air will leak unless thread compound is used. Use caution when applying thread compound on dry systems. If the thread compound is too fluid, the air pressure of a dry system can literally push the thread compound out of the threads.

Apply thread compound properly and uniformly to the outside threads that screw into the fitting, as required by *NFPA 13*. If using a paste-type thread compound, placing it inside the fitting results in a lot of wasted sealant that becomes foreign material in the sprinkler system. Liquid Teflon® is applied to the outside threads, using a small brush. It must be applied to fill the threads. Teflon® tape must be wrapped on the threads clockwise around the outside thread if the thread is a right-hand thread and counterclockwise if the thread is a left-hand thread. This prevents the tape from coming off when the pipe and fitting are joined.

5.3.1 Teflon® Tape

Teflon® tape is a special tape that is wrapped around threads to prevent metal-to-metal contact and to create a leakproof joint. Follow these steps to use Teflon® tape:

Step 1 Remove all excess cutting oil from the threads to improve the grip of the tape on the threads.

On Site

Pipe Thread Compound

Pipe thread compound is available in a wide variety of materials. The most common pipe thread compound is used for domestic water lines. Joint compound is designed to resist interaction with substances moving through the system. Depending on the type of gas or liquid moving through the pipe system, manufacturers recommend different kinds of joint compound. Always read the manufacturer's recommendations before using the product.

Four or five layers of Teflon® tape being applied, clockwise, one thread back from the end of a pipe.
Pipe thread compound (dope) applied one thread back from the end of a pipe and just covering the threads.
103SA03.EPS

WARNING

Never use Teflon® tape on pipe that is to be welded or pipe that carries steam or other high-temperature service. When heated, Teflon® tape emits a highly toxic gas that could be fatal.

NOTE

Use ½" wide Teflon® tape for pipe that is ¾" nominal size and smaller. Use ¾" wide tape for pipe that is 1" nominal size or larger.

Step 2 Start the tape at the end of the pipe, leaving the first full thread bare to prevent the tape from bunching up at the beginning of the thread.

Step 3 Wrap the tape around the pipe in the direction in which the joint is to be assembled. Wrap the tape around right-hand threads in a clockwise direction and around left-hand threads in a counterclockwise direction. *Figure 16* shows wrapping a normal right-hand thread with Teflon® tape.

Step 4 Continue to wrap the tape around the joint, overlapping the edges of each wrap until all remaining threads have been covered.

Step 5 Press the tape against the threads to seal it to the threads and to prevent the tape from slipping off the threads once you start to make up the joint.

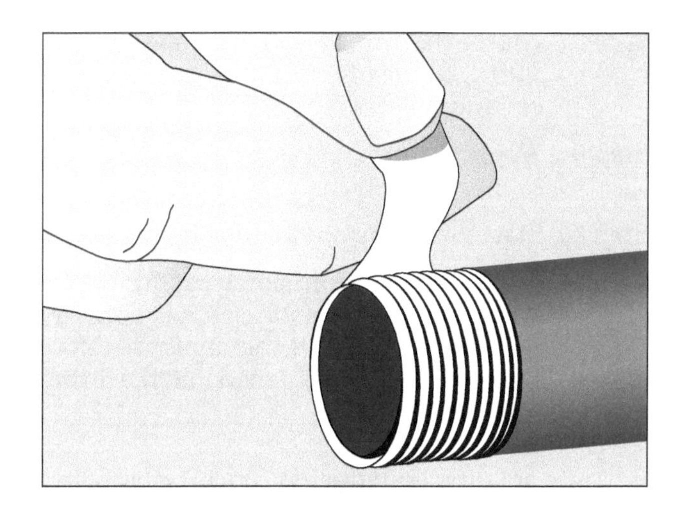

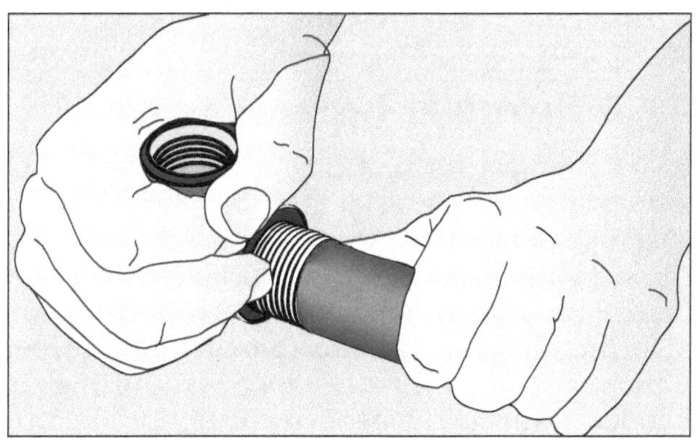

103F16.EPS

Figure 16 Wrapping right-hand thread with Teflon® tape.

5.4.0 Fitting Threaded Pipe and Fittings

The proper fitting of pipe and fittings is a skill that must be learned through experience. It is crucial to fabricating a quality piping system. Poor fitting practices can result in leaks and other defects in the system. When tightening fittings onto a piece of pipe, the sprinkler fitter must be aware of the end position that the fitting must face. If the fitting is turned past the desired end position and then turned back, the fitting will leak. Follow these steps to fit threaded pipe and fittings:

Step 1 Determine the type, size, and schedule of pipe being used in the system.

Step 2 Determine the length of pipe needed.

Step 3 Cut the pipe to the desired length.

Step 4 Ream the pipe to remove all internal burrs.

Step 5 Thread the pipe, taking caution not to cut the threads too deep.

Step 6 Select the proper size, shape, and type of fittings needed.

Step 7 Clean the pipe and fittings thoroughly inside and out. All sand, dirt, and oil must be removed from the inside of the pipe and fittings to avoid contaminating the system or clogging the line. The fittings and pipe ends may be cleaned with a clean rag soaked in nonflammable solvent.

On Site

Keep Piping Fittings to a Minimum

Piping fittings add resistance and reduce flow in piping systems. All piping systems should be planned prior to installation to eliminate as many fittings as possible. In some large industrial or commercial installations, bending pipes in curves instead of using elbows may be preferable in order to reduce flow resistance.

Step 8 Check the threads on the pipe and the fitting to ensure that they are properly cut and not damaged.

Step 9 Apply thread compound or Teflon® tape to the pipe threads.

Step 10 Start the fitting on the pipe by hand.

Step 11 Tighten the fitting slowly using a pipe wrench. When prefabrication is done, the fitting is already on the pipe. Therefore, the pipe must be tightened into the fitting.

> CAUTION
>
> Do not allow the threads to bottom out into the fitting. This can damage the threads and the fitting.

6.0.0 Selecting Threaded Fittings

In its simplest form, a fitting connects two or more pieces of pipe. A large variety of fittings are available to the fire sprinkler installer. Generally, the kind of fitting to use is specified on the drawings. Pipe fittings come in various sizes, materials, strengths, and designs to match the various piping systems. They are either plain fittings or banded fittings. The banded fittings are ductile iron, malleable iron, and cast-iron fittings. The plain fittings are forged steel.

Pipe fittings are attached to pipe to change the direction of fluid flow, to connect a branch line to a main line, to close off the end of a line, or to join two pipes of the same or different sizes.

The specific types of fittings used in installations may be chosen by the specifying engineer and written out in the contract specifications. Sometimes there are restrictions against using certain kinds of fittings. Fitters must inspect their own work to be certain the installation conforms to the project requirements.

6.1.0 Approvals

The following are the general approval standards for steel pipe fittings:

- Underwriters Laboratories Fire Protection Equipment Directory Listings
- FM Global Approval Guide
- ANSI and ASTM identified standards in *NFPA 13*

Fittings that do not conform to the standards in *NFPA 13* must be Listed by a testing laboratory to meet the requirements of *NFPA 13*. Be certain of the acceptability of the materials you are installing.

6.2.0 Types of Threaded Fittings

Threaded fittings are cast and then machined. All the threaded iron and steel fittings used in sprinkler systems are made using the American Standard Taper Thread.

Sometimes the castings are imperfect and will leak. The most common problem is caused by sand holes. Whether a leak is created from a sand hole, a crack, or a distorted fitting, the fitting must be removed and replaced. Report all problems to the job supervisor.

Other problems may occur if the shop is pulling the fittings onto the pipe improperly, causing too few or too many threads. Notify the shop supervisor of these problems so that they can be corrected. If fittings are being cracked in the shipping process, those responsible for shipping must be notified.

6.2.1 Cast-Iron Fittings

Cast-iron fittings are the most commonly used fittings in the supply system. They are generally used in water applications and have a classification of 125 or 250. The patterns are very old and the dimensions are well established. There are many uses for cast-iron fittings and they are manufactured in large quantities. For these reasons, cast-iron fittings are the least expensive type of fitting.

One disadvantage of cast-iron fittings is that they break more easily than other fittings. Dropping one on the floor can crack the fitting and cause a leak. One advantage is that the walls are thicker than those of malleable fittings, giving a greater weight per item.

Under *NFPA 13*, the standard-weight pattern cast iron fittings are allowed to be used with working pressures up to 300 psi in sizes 2" or smaller. Above 2", the standard weight cast iron fitting is permitted up to only 175 psi. Beyond these ratings, the extra-heavy pattern must be used; however, there are fewer extra-heavy pattern fittings than there are standard weight fittings.

6.2.2 Malleable Iron Fittings

Malleable iron fittings are produced by slowly reheating and cooling cast-iron fittings. The result is a change in the crystalline structure of the carbon, giving the fitting increased strength, ductility, and shock resistance. Because of this, malleable fittings are usually more expensive than cast fittings. Although malleable fittings are more expensive, they weigh less than comparable cast-iron fittings. In some instances, the higher cost of the fitting is more than offset by the reduction in shipping costs, which are based on weight.

Instead of cracking, malleable fittings stretch. Fittings that must be removed are often not reusable. On the other hand, there are times when the flexibility of a malleable fitting is important, such as in cases which the piping is subject to severe water hammer. If a malleable fitting accidentally freezes, it stretches rather than cracks apart. The potential water damage is reduced. *NFPA 13* allows the use of standard weight pattern malleable fittings of 6" or smaller, up to 300 psi.

Some manufacturers supply galvanized fittings only in the malleable condition. They are hot-dip galvanized, and the threads are chased after the fitting is dipped. Therefore, the threads are untreated.

6.2.3 Ductile Iron Fittings

Ductile iron fittings are similar in appearance, design, and function to those made from other types of iron. Ductile iron fittings are available in the same class thicknesses, pressure ratings, linings, and joint designs as standard ductile iron pipe. The standard class numbers of the fittings must match those of the pipe being used.

6.3.0 Standard Fittings in Cast, Malleable, and Ductile Iron

Many types of fittings, in a wide variety of types and sizes, are available in cast, malleable, and ductile iron. The fittings include the following:

- Elbows
- Tees
- Crosses
- Plugs and caps
- Couplings
- Bushings
- Mechanical tees
- Nipples

6.3.1 Elbows

The three types of elbows are standard, reducing, and street elbows (*Figure 17*). A standard elbow has internal threads with the same nominal pipe size on both outlets. Standard elbows are available in 90-degree and 45-degree patterns in both

Figure 17 Threaded elbows.

black and galvanized finish. A reducing elbow has internal threads with different nominal pipe sizes on the two ends. When specifying these, the largest opening is listed first. Street elbows have an internal thread on one end and an external thread on the other end. The main disadvantage of street elbows is the reduction in size of the flow path on the end with outside threads. These elbows are rarely used in fire sprinkler work. *Figure 18* shows the various elbow configurations.

6.3.2 Tees

Tees (*Figure 19*) allow three pieces of pipe to be joined. Tees can be straight tees with three outlets of the same nominal pipe size, or reducing tees, with outlets that are not the same size. The largest outlet on the straight-through waterway is read first, regardless of how the tee fits into the system. In the sprinkler industry, when a tee is installed with the largest outlet on the side, it is called a bullhead tee. *Figure 20* shows a tee listing.

Local supply houses stock tee sizes that are frequently demanded. Others can be obtained from the foundry by special order. A lead time of as much as 6 months may be required to obtain some types of tees.

6.3.3 Crosses

Crosses allow four pipes to be joined. *Figure 21* shows sizing of reducing crosses. A cross may also be a straight cross, as illustrated in *Figure 22*.

Figure 19 Threaded tees.

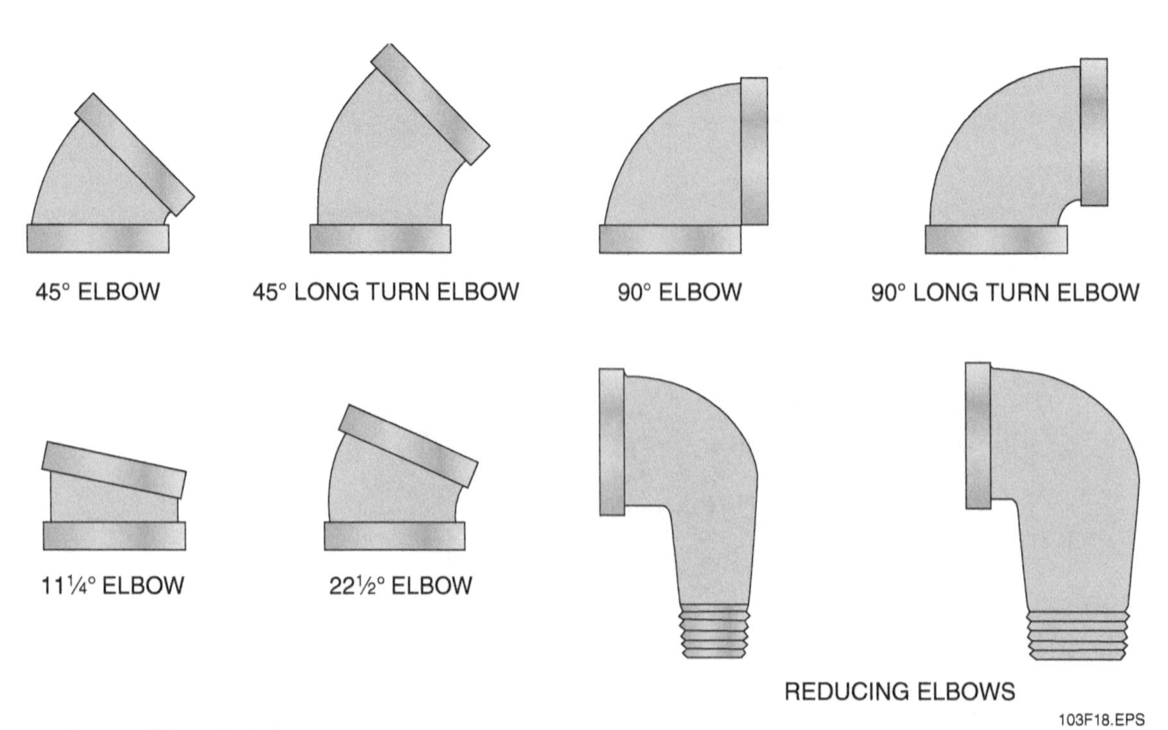

Figure 18 Types of threaded elbows.

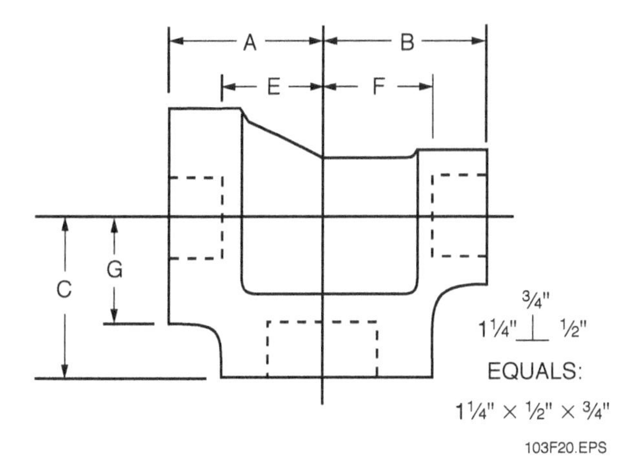

Figure 20 Tee listing.

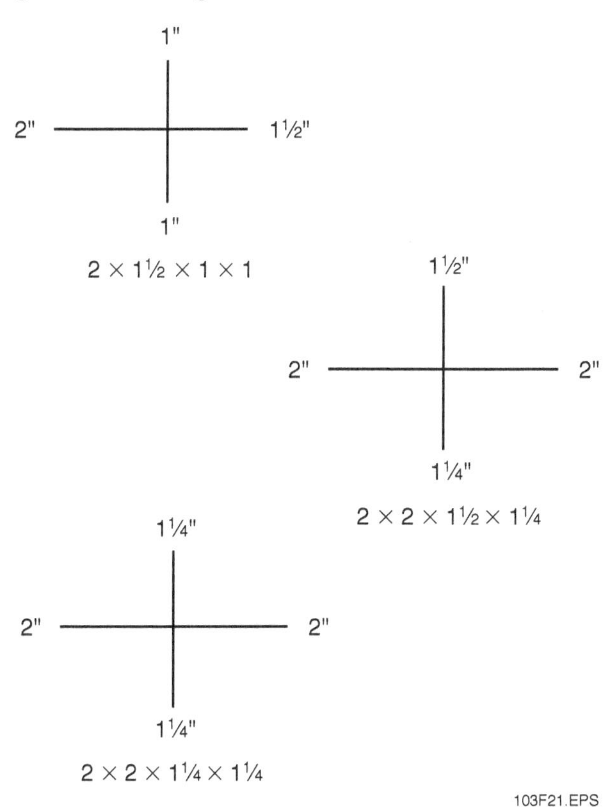

Figure 21 Sizing of reducing crosses.

Figure 22 Threaded cross.

Crosses, especially reducing crosses, are frequently unavailable and must be ordered as long-lead items. Some sprinkler installations require a large number of crosses. Careful scheduling must be performed to ensure that they are available.

6.3.4 Plugs and Caps

Caps and plugs are fittings that are used to close off lines. Caps and plugs must have the same pressure ratings as other fittings. A plug fitting (*Figure 23*) screws into the fitting end. A cap (*Figure 24*) screws onto the pipe end. Caps can either be plain caps with round heads or caps that have a square projection on the body to allow for tightening with a wrench. Plugs can either be solid square-head plugs or countersink plugs. The latter have a square depression where a square key is inserted for tightening.

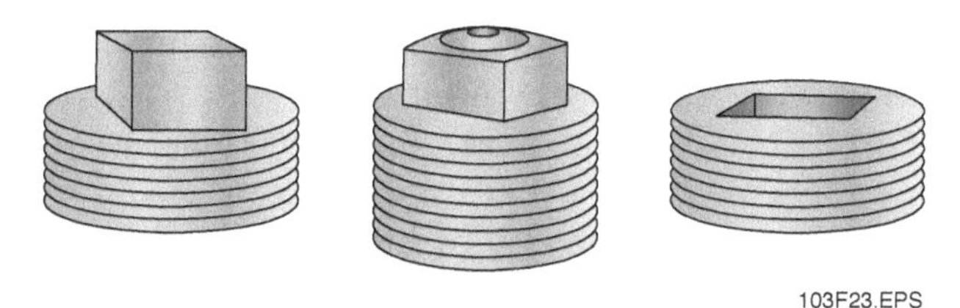

Figure 23 Threaded plugs.

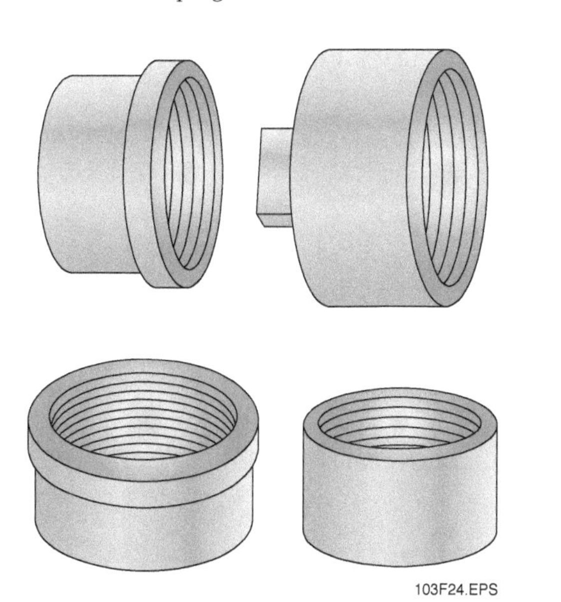

Figure 24 Threaded caps.

Did You Know?

The monkey wrench is a general-purpose wrench with an adjustable jaw for turning nuts of various sizes. Charles Moncky invented this tool around 1858. Monkey wrench also has an informal meaning: to disrupt, as in "He threw a monkey wrench into our plans." Monkey comes from the last name of the wrench's inventor.

6.3.5 Threaded Couplings

Couplings align two pipes on a straight run. Straight couplings have the same nominal pipe size on both ends. Concentric reducing couplings maintain one center line through the fitting and are more commonly used in sprinkler systems. Eccentric reducing couplings have different center lines for the inlet and outlet openings. A ground joint union fitting assembly has matched pieces that are ground together to form a metal-to-metal seal. In the sprinkler industry, ground joint union fittings are used mostly in valve trimmings where a point of joining is needed. The parts are matched and must be kept together. *Figure 25* shows types of couplings.

6.3.6 Bushings

Bushings (*Figure 26*) are occasionally used in fittings when a reducing fitting is not available. However, *NFPA 13* calls for one-piece reducing fittings, prohibiting the use of bushings unless the required size fitting is not available. Some AHJs interpret "not available" to mean "not manufactured." This would prohibit the use of bushings under any circumstances. One exception specifically allowed in *NFPA 13* is when the sprinkler head is initially upright, but whose sprinklers will be turned pendent with a drop nipple at a future date.

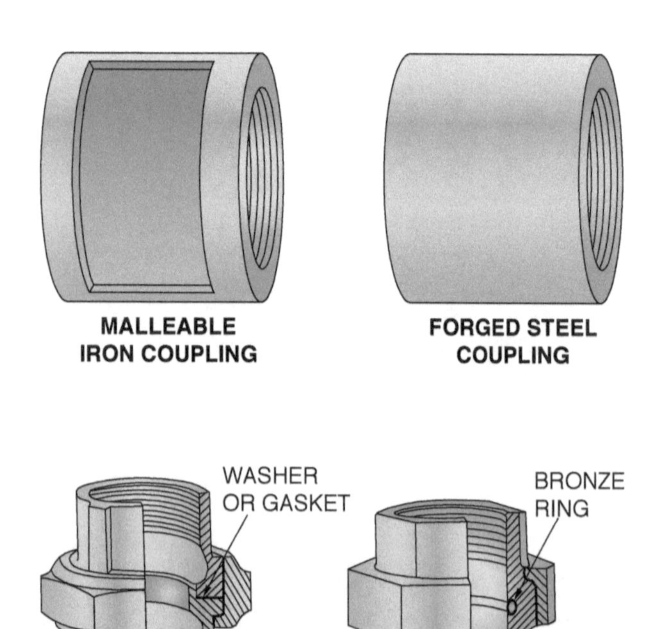

Figure 25 Types of threaded couplings.

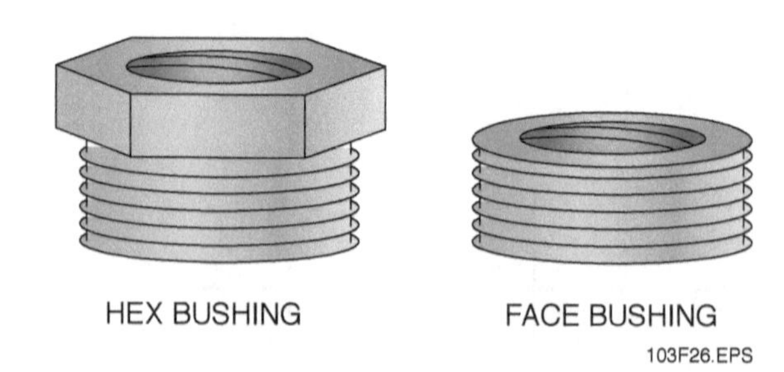

Figure 26 Types of threaded bushings.

Bushings add an extra set of threads and an extra potential for leaks to a fitting assembly.

6.3.7 Mechanical Tees

A mechanical tee provides a threaded or grooved side outlet without having to make a joint in the pipe. A hole of the proper size is cut into the pipe, using a hole saw. The mechanical tee is then clamped to the pipe so that the rubber gasket in the fitting aligns properly with the hole. The U-bolt is then tightened to the pipe to make a watertight connection.

Mechanical tees are very useful in modifying old systems or in making additional cut-ins on new systems without having to remove the pipe to be cut and threaded. The coupon, or cut-out, must be retrieved to avoid plugging the pipe. Some AHJs require that the coupon be pierced and wired to the pipe so the inspector can see that it has been removed. A mechanical cross consists of two such outlets that clamp together around the pipe. Outlets may be threaded or grooved. *Figure 27* shows mechanical tees.

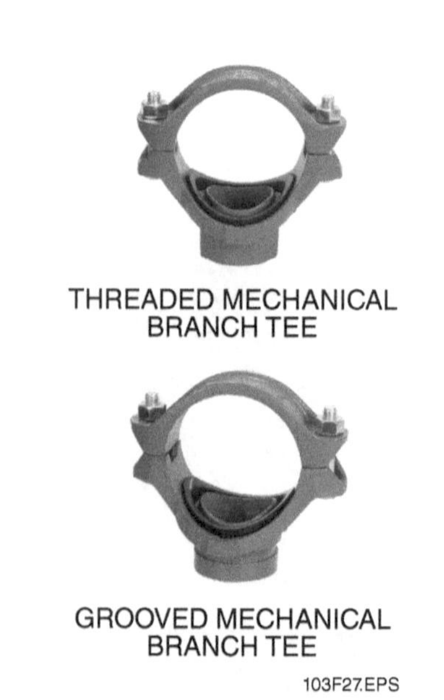

Figure 27 Threaded and grooved mechanical tees.

6.3.8 Nipples

Pipe nipples are short pieces of pipe (less than 12") that are threaded on both ends. *Figure 28* shows types of nipples.

The shortest available nipple is called a close nipple. Close nipples consist almost entirely of threads on the outside, making them difficult to install. Their wall is weak, contributing to flattening, they are prone to leak, and there is no safe place on which to grip them because of all the threads. The next-to-close, or shoulder nipple, is more desirable to use. Longer nipples provide more insurance against leaks. *Table 4* shows examples of nipples.

Pipe Nipples

Pipe nipples are available from suppliers in a variety of short lengths. It generally makes more sense to buy commonly used nipples in bulk, especially the close and shoulder nipples, rather than to cut and thread them in the field. Most threading equipment cannot be used to make close or shoulder nipples. However, short and long nipples can be fabricated using pieces of scrap piping gathered from the field installations. This type of task can be accomplished in the shop during slow periods.

6.4.0 Flexible Drops

Flexible fittings called flexible drops (*Figure 29*) were developed to span the distance from the branch line to the sprinkler. This fitting is usually made up of a threaded pipe nipple, a flexible connection that may be a corrugated tube or braided steel, a reducing fitting, and a mounting bracket that connects to the sprinkler and attaches to the T-bar or framing.

Table 4 Close and Next-to-Close Nipple Lengths

Nominal Pipe Size	Length of Close Nipple	Length of Shoulder Nipple
1	1½	2
1¼	1⅝	2
1½	1¾	2
2	2	2½
2½	2½	3
3	2⅝	3
3½	2¾	4
4	2⅞	4
5	3	4½
6	3⅛	4½
8	3½	5

NOTE: All dimensions in inches.

103T04.EPS

Flexible drops have become very popular over the last few years. They typically vary in length from 24" to 72". The greater the length, the more friction loss; this loss must be accounted for in the hydraulic calculations. In addition, there is a maximum number of bends and a minimum bend radius, neither of which can be exceeded.

Typical applications include intake or exhaust duct protection, hard ceiling and T-bars, and clean rooms. Currently not all drops are both UL and FM Global Listed. Be sure the flexible drop you are using is approved for the application.

NOTE

Some AHJ's will not allow flexible drops even though they are UL and/or FM Listed.

Some benefits to the use of flexible drops are:

- The system can hydrostatically tested with the drop and sprinkler head in place.
- The sprinkler can be relocated to accommodate small floor plan or ceiling changes without shutting and draining the system.

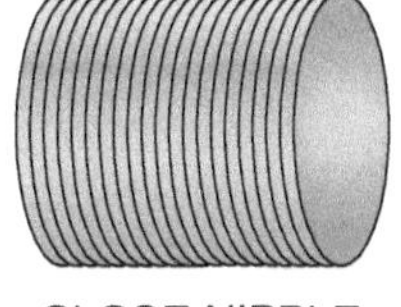
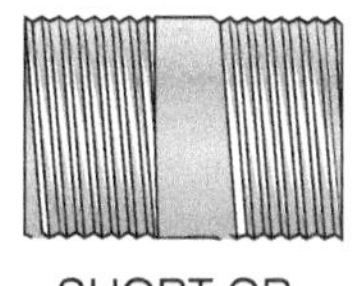
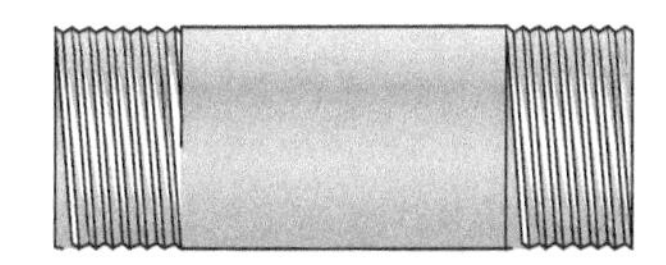

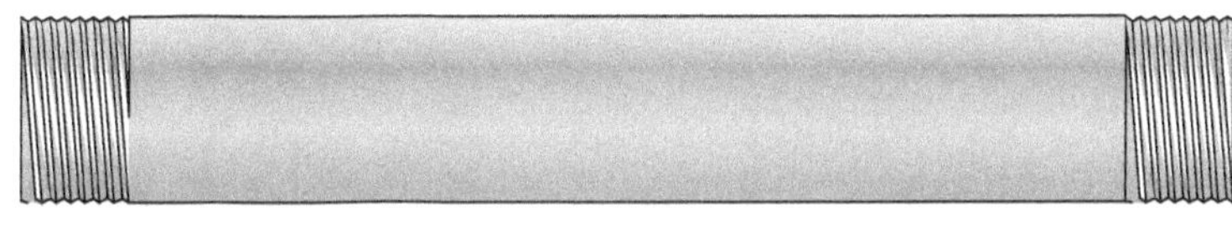

Figure 28 Types of nipples.

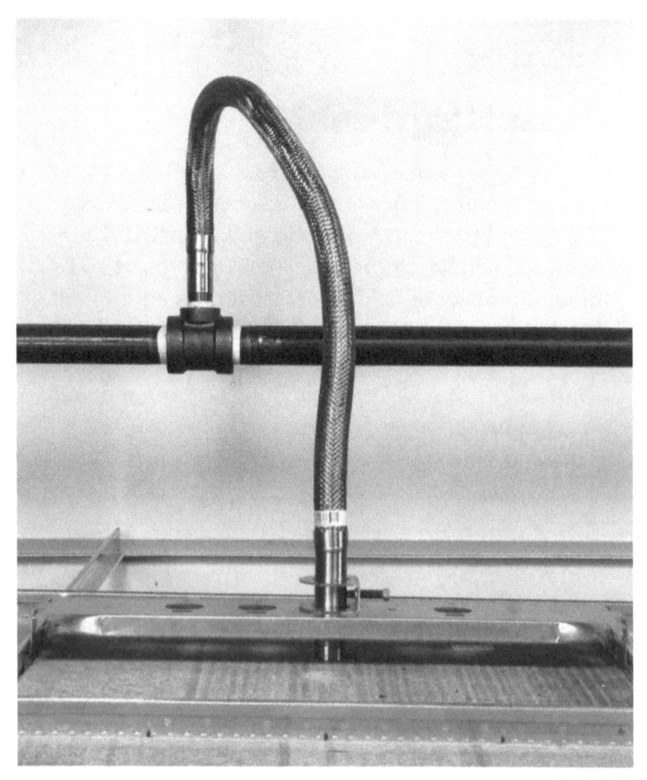

103F29.EPS

Figure 29 Flexible drop.

7.0.0 Plain Ends

There are a number of fittings designed for plain-end pipe. These usually require end preparation by cleaning and deburring. All coatings, notches, bumps, weld beads, and score marks 1" to 1½" from the pipe end must be removed so the gasket will seat properly. Internal chamfers, normal abrasive wheel cuts that leave burrs, and dull wheel cutters will produce unusable surfaces. Special grinders are available that reliably prepare the pipe end.

Listings for plain-end fittings must be checked to ensure that the application is correct. Some of the older products Listed require significant peripheral penetration of the pipe wall with grippers.

Some plain-end fittings use a quarter-turn locking lug that has a small-diameter tooth pattern on the end of the lug that grips the pipe. Others have a hex-head locking bolt. These fittings come in various sizes and reducing combinations. Refer to the manufacturer's catalog to determine application limits.

7.1.0 Couplings with Grippers

Couplings with grippers have curved grippers that bite into the pipe to create a mechanical joint (*Figure 30*). This type of plain-end fitting is suitable for use on square-end or beveled-end pipe. Other than making a square cut and deburring the pipe, no special pipe preparation is necessary.

Couplings with grippers are designed for Schedule 40 pipe. They are most useful when cutting into old systems to extend or modify them. In areas where there is significant expansion and contraction of piping due to temperature changes, couplings with grippers will not hold. These fittings may only be used in accordance with their listing.

7.2.0 Miscellaneous Plain-End Fittings

A number of plain-end fittings are designed to be used with gripper-type plain-end couplings. These fittings include elbows, tees, crosses, adapter nipples, adapter flanges, and swaged nipples. Be sure to reference the manufacturer's instructions before installing plain-end fittings.

7.3.0 Fittings for Prepared-End Piping

Prepared-end fittings were popular for a time, but they are no longer in common use because of leakage problems. They were used on Schedule 10 thin-wall and Schedule 40 pipe. A special pipe cleaning machine was used to prepare the pipe ends for the fitting, then a coupling similar to the plain-end fitting was installed.

7.4.0 Poz-Loc®

Poz-Loc® was a piping system that used a specialized ductile iron fitting and a U-shaped key to secure the piping. Special galvanized light-wall piping would be cut, leaving a rolled down or chamfered end. The fitting had an O-ring to provide a seal against the outside of the pipe. After preparation of the pipe end and lubrication of the O-ring, the pipe would have been inserted into the fitting 1⅜" until it met the pipe stop. The special U-shaped key was driven into a slot in the end of the fitting using a ball peen hammer. The key could be removed and replaced for disassembly

103F30.EPS

Figure 30 Plain-end fittings with grippers.

and reassembly of the connection. This material is no longer manufactured, though you will encounter it in the field.

7.5.0 Press-Fit

Another piping system that has been discontinued, but which you will also encounter in the field, is a press-fit system that uses Schedule 5 light-wall pipe and a steel fitting. The fitting has an O-ring to seal against the outside of the pipe. The pipe was cut and deburred and then inserted into the fitting. The pipe would be inserted until it contacted the pipe stop. A portable hand-held electric tool was placed over the fitting and an indentation made around the circumference of the fitting and pipe to secure the connection. Lubrication of the O-ring was not necessary. This system is no longer recommended for fire sprinkler work.

8.0.0 Grooved Pipe

Grooved piping systems provide an easy alternative to welded, flanged, or threaded joints. Each joint in a grooved piping system serves as a union, allowing easy access to any part of the piping system for cleaning or servicing. Grooved piping systems have a wide range of applications and can be used with many types of piping materials.

The grooved piping system offers varied mechanical benefits, including the option of rigid or flexible couplings. Rigid and flexible couplings can be incorporated as needed into any system to take full advantage of the characteristics of each. Rigid couplings create a rigid joint that is useful for risers, mechanical rooms, and other areas where positive clamping with no flexibility within the joints is desired. Flexible couplings allow for controlled pipe movement that occurs with expansion, contraction, and deflection. Rigid and flexible couplings have specific installation requirements for earthquake bracing as referenced in *NFPA 13*. *Figure 31* shows examples of rigid and flexible grooved pipe couplings.

Another advantage of grooved piping is the ability to join thin walled pipe. Schedule 5 pipe has a much thinner wall than Schedule 40 pipe and can be connected by grooved couplings only. Schedule 5 piping has been Listed by UL for use in sprinkler system installations. This is the thinnest wall steel pipe allowed to be used for sprinkler systems. Listed sizes in the UL *Fire Protection Equipment Directory* are 1¼" through 6".

8.1.0 Grooved Connections

The standard grooved pipe consists of a rubber gasket and two housing halves that are bolted together. The housing halves are tightened together

> **On Site**
>
> **Grooved Pipe**
>
> Like many products we now take for granted, the grooved pipe joining systems was developed in 1925 by Victaulic Corporation, then known as the Victory Pipe Grooving Company, to solve a military problem. The objective was to create a system that would allow troops in the field to rapidly assemble systems for water and fuel distribution.
>
> *Source: Victaulic Corp.*

HOUSING
GASKET
GROOVE
BOLTS AND NUTS
PIPE
RIGID

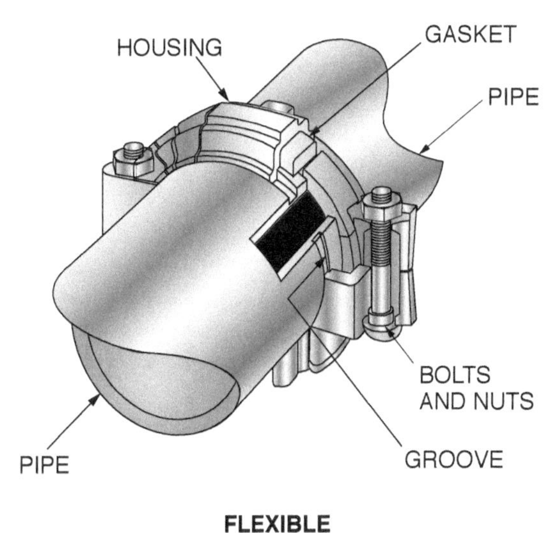

Figure 31 Rigid and flexible grooved pipe couplings.

in accordance with the manufacturer's instructions. This area must be clean and smooth for proper sealing to take place. The design of the gasket allows the internal pressure of the pipe to press on the back side of the gasket, forcing the gasket down on the pipe and forming the seal. The pipe can flex in the grooved joint without leaking. *Figure 32* shows standard cut and roll grooves in pipe ready for couplings.

The dimensions pointed out by the letters in *Figure 32* must comply with engineering specifications. The A dimension is the distance from the pipe end to the groove and provides the gasket seating area. This area must be free from indentations, projections, or roll marks to provide a leak-proof sealing seat for the gasket. The B dimension is the groove width. It controls expansion and angular deflection based on its distance from the tube end and its width in relation to the width of the coupling housing key. The C dimension is the proper diameter of the base of the groove. This dimension must be within the given diameter tolerance and concentric with the outside diameter (OD) of the pipe. The D dimension is the nominal depth of the groove and is used as a trial reference only. This dimension must be changed if necessary to keep the C dimension within the stated tolerances. The F dimension is used with the standard roll only and gives the maximum allowable pipe end flare. The T dimension is the lightest grade or minimum thickness of pipe suitable for roll or cut grooving.

8.1.1 Cut Grooves Versus Rolled Grooves

Grooves may be formed by cutting or rolling. The original method of grooving a pipe was by cutting. The cutting method was designed for Schedule 40 pipe. Schedule 10 pipe does not have enough wall thickness to allow for a cut groove. Only rolled grooves may be used on pipe whose walls are thinner than Schedule 40 (sizes less than 8") or Schedule 30 (8" and greater). On most machines, Schedule 40 is the maximum safe limit. All lesser schedules of pipe materials that are acceptable in *NFPA 13* can be roll-grooved.

CAUTION

Never cut-groove any pipe with a wall thinner than Schedule 40.

Cut grooving removes metal and thins the pipe wall. It is seldom used in the fire sprinkler industry. A cut groove removes less metal than a standard thread; however, the strength of a threaded joint is greater for two reasons. First, in the threaded joint, the fitting reinforces the pipe in the area where a groove would exist otherwise. Second, the edges of the cut groove introduce greater stress concentrations than threads. Rolled grooves, on the other hand, can actually increase the strength of the pipe at the joint location. Cold-working of the metal by displacing it can produce grain structure and increased tensile strength in a manner similar to that obtained with forging. However, there is a reduction of approximately 50 percent in available movement with a rolled groove versus a cut groove. If flexibility in the joints is a specific issue, cut grooves may be necessary.

CAUTION

Exercise care when roll grooving pipe to avoid damaging the galvanized coating.

8.1.2 Groove Specifications

Groove depth (diameter) is critical. If the groove is not deep enough, the coupling will not fit over the pipe. If the groove is too deep, the coupling will not grip the pipe. These depths may be gauged automatically by a stop on the grooving machine, or the sprinkler fitter may have to gauge each groove as it is prepared.

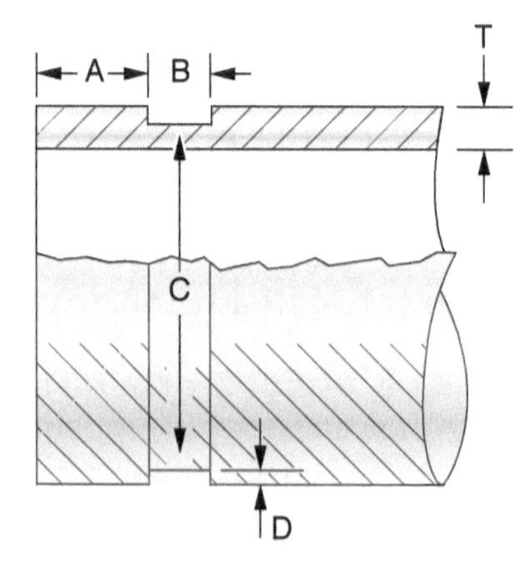

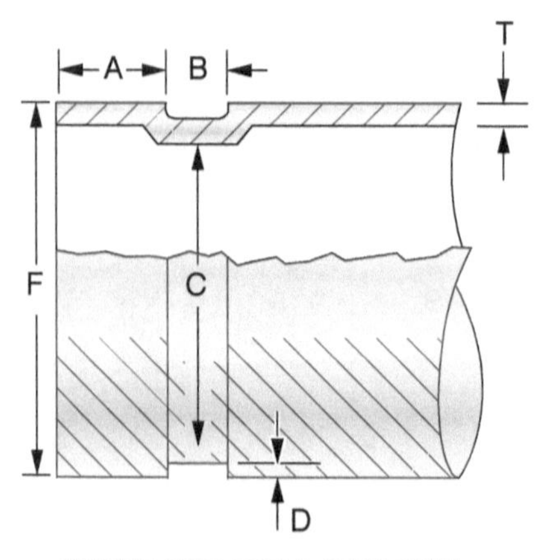

Figure 32 Grooved pipe.

The sprinkler fitter is responsible for quality control in the field. If the fitter rolls a groove and it leaks or does not make up, it is the fitter's responsibility. It only takes a moment to check the accuracy of each groove. It is a lot easier to correct the groove on the ground in the assembly process than it is to disassemble an already installed portion of a system when it leaks or will not make up.

Misalignment of the groove rollers on a machine can cut the pipe in the process of making the groove. Whether using pipe that you have grooved, or pipe grooved in the shop, check the grooves before installing the pipe to be certain it is correct. Another possible malfunction is splitting at the weld joint on electric resistance welded (ERW) pipe.

8.2.0 Grooving Tools

Different types of tools are used for cut-grooving and roll-grooving. Schedule 40 pipe in 1" through 6" and Schedule 30 pipe in 8" and 10" can be cut-grooved. Any pipe of lesser wall thickness must be roll-grooved. However, Schedule 40 and 30 pipe also can be roll grooved.

8.2.1 Roll Groover

Power roll-grooving machines and hand-roll groovers are used to roll grooves at the ends of pipe to prepare the piping for grooved fittings and couplings.

Power grooving machines are available to groove ¾" to 30" standard and lightweight steel pipe, aluminum pipe, stainless steel pipe, and PVC plastic pipe. Note that different rolls are required for roll-grooving PVC plastic pipe and aluminum pipe.

Many power roll-grooving machines are standalone units with their own power drives. Some models require a power drive or equivalent as a power source. The roll-grooving machine presses a groove into the pipe, using upper and lower die heads. The lower die head, or female die, rotates the pipe. The upper die head, or male die, is powered either by hydraulic pressure or a handwheel and pushes a groove into the pipe as the pipe is rotated. An adjustable stop controls the groove depth. Roll grooving removes no metal from the pipe but forms a groove by displacing the metal. Since the groove is cold-formed, it has rounded edges that reduce the pipe movement after the joint is made up.

Follow the manufacturer's suggested procedures on the proper setup and operation of the roll grooving tool, including roll selection for the appropriate material to be grooved. *Figure 33(A)* shows a power roll-grooving machine.

8.2.2 Portable In-Air Groover

Another tool that is used to roll grooves is the portable in-air (in place) groover. The portable in-air groover is a manually powered machine used to roll-groove piping that is already installed. Portable in-air grooving machines are available to groove 2½" to 16" copper tubing types K, L, M and DWV, as well as 1" Schedule 10 and Schedule 40 steel pipe. In-air groovers

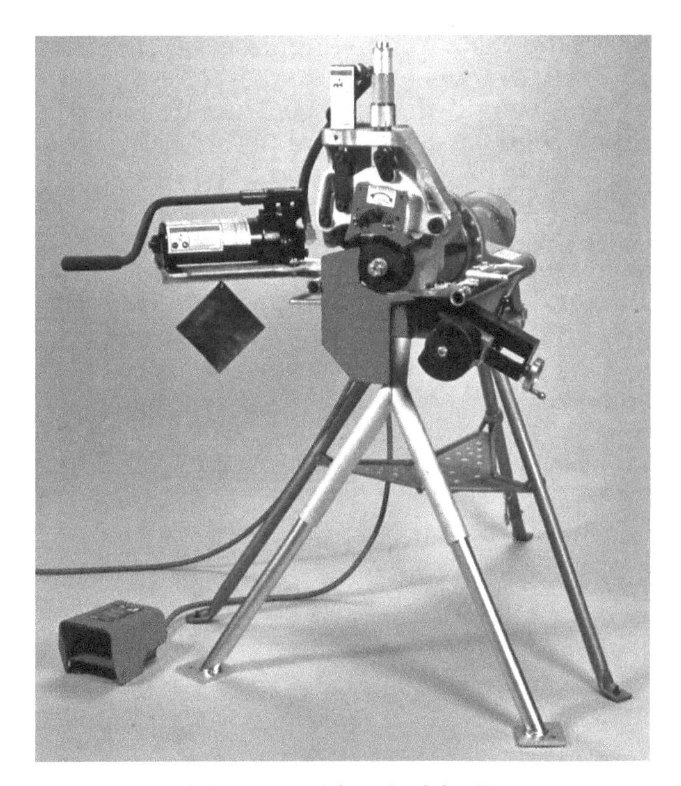

(A) POWER ROLL GROOVER

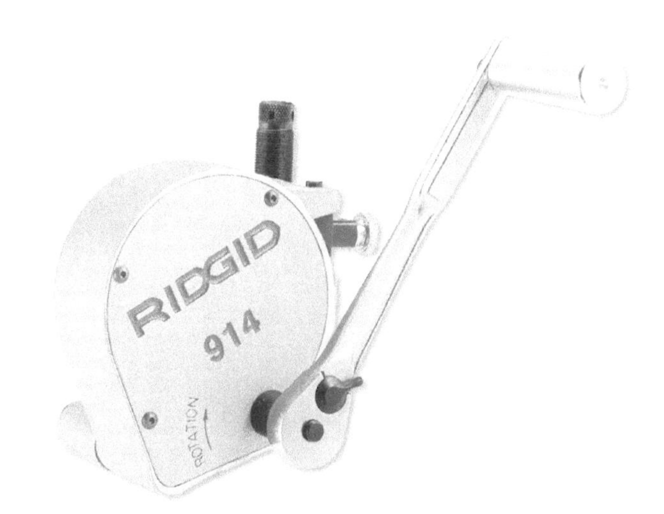

(B) PORTABLE IN-AIR GROOVER

103F33.EPS

Figure 33 Roll-grooving machines.

are capable of producing standard rolls for 1¼" to 6" Schedule 10 and 1¼" to 3" Schedule 40 pipe. In-air groovers require a different roll groove and drive roll to maintain the correct gasket seat and groove width dimension when grooving different types and sizes of pipe. *Figure 33(B)* shows a portable in-air groover.

8.2.3 Cut Groover

Cut grooving is rarely done in sprinkler fitting work because of the amount of wall material removed by the cut grooving process. There are a variety of tools on the market for cut grooving pipe from 1" through 24". The cut groover is a tool driven by an external power source. It is designed to be driven around a stationary pipe. This ensures that the groove is concentric with the outside diameter of the pipe. Most models have an adjustable pipe stop to ensure proper groove depth. Cut grooving differs from roll grooving in that a groove is cut into the pipe instead of being cold-formed. The cut groover removes less than one-half the pipe wall, which is less than the depth of thread cuts. Cut grooving is intended for standard weight or heavier pipe. *Figure 34* shows two types of cut grooving machines.

8.3.0 Hole-Cutting Tools

Hole-cutting tools are used for cutting fast, clean, and precise holes in steel pipe for steel mechanical tees and special sprinkler fittings. These types of tools expedite the installation process. Other types of hole-cutting tools are available for larger pipe sizes and for pipe being tapped under pressure. When cutting holes in pipe be sure to remove the coupons or discs from the piping so that the system will function properly. To verify removal of the discs, wire or attach the discs to the outside of the pipe. Always consult the tool manufacturer's instructions before using a hole-cutting tool. *Figure 35* shows a hole-cutting tool.

103F35.EPS

Figure 35 Hole-cutting tool.

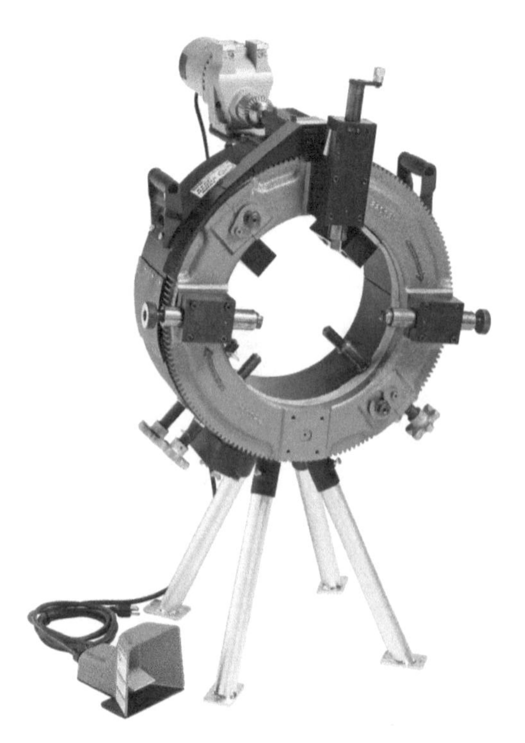

(A) POWER CUT GROOVER

(B) MANUAL CUT GROOVER

103F34.EPS

Figure 34 Cut grooving machines.

8.4.0 Grooved Fittings

Joining pipe using grooved fittings saves time as well as money. Grooved fittings are more expensive than threaded fittings, but fewer labor hours are needed to join pipe with grooved fittings. In large-diameter pipe, the labor savings in making joints with grooved fittings is greater than the material costs savings of using threaded fittings. Grooving pipe in the field is simple, whereas cutting threads on pipe larger than 2" usually requires a trip back to the shop. In addition, if the light-wall or thin-wall pipe being used is not approved for threading, one approved means of connection in the field is with a grooved coupling.

Lubrication of the rubber gaskets in accordance with manufacturer's instructions for these fittings is important. Use only a nonpetroleum-based lubricant or a pre-lubricated gasket recommended by the manufacturer. Petroleum products will attack the gasket and cause early failure. Removal of burrs is also very important. Failure to do either of these can result in cut gaskets and leaks. The manufacturer's instructions must be consulted to maintain the UL approvals and the warranty status. The types of fittings and couplings used with grooved pipe include the following:

- Flexible couplings
- Rigid couplings
- Reducer couplings
- Grooved flanges
- Elbows
- Tees
- Crosses and laterals
- Reducers
- Caps
- Valves

8.4.1 Flexible Couplings

The flexibility of standard grooved couplings can be a great advantage in areas where the pipe run needs a slight deviation from a straight line. They also have specific applications in earthquake areas.

Flexible grooved couplings are manufactured for standard, lightweight, and heavy-duty applications. They are rated according to working pressure. A rubber gasket provides the watertight seal.

There are a number of flexible coupling manufacturers Listed in the UL Equipment Directory. Each has its minor differences, and the gaskets from one manufacturer should not be used with couplings made by another. Some flexible couplings are approved for dry systems and some are not. It is important to determine which couplings work with rolled grooves, cut grooves, standard gaskets, and tri-seal gaskets.

Consult the manufacturer's literature to ensure that the application is suited for the use of flexible couplings. In addition, check the couplings and fittings to ensure they are compatible.

8.4.2 Rigid Couplings

Rigid couplings are particularly useful in dry pipe, freezer, standpipe, and fire pump installations. They also have applications in earthquake areas. The rigid coupling was developed for use where the growth of a pressurized line connected by grooved couplings is not desired. There is no flexibility in rigid couplings. A tri-seal (flexseal), Grade E EPDM (green color code) pre-lubricated gasket is used in freezer and dry pipe applications. *Table 5* shows application data for gaskets used in fire protection systems. *Figure 36* is an example of a rigid grooved coupling. Installation-ready couplings are

Table 5 Gasket Applications

Grade	Temp. Range	Compound	Color Code	General Service Application
A Pre-lubricated	Ambient to +150°F (+66°C)	EPDMA	Violet	Fire protection systems (not recommended for hot water systems)
E	−30°F (−34°C) to +230°F (+110°C)	EPDM	Green	Fire protection systems
E Tri-Seal	−30°F (−34°C) to +230°F (+110°C)	EPDM	Green	Fire protection systems (dry pipe or freezer systems)
L	−30°F (−34°C) to +350°F (+177°C)	Silicon	Red	Fire protection systems (dry pipe or freezer systems)

103T05.EPS

103F36.EPS

Figure 36 Rigid grooved coupling.

103F37.EPS

Figure 37 Reducing coupling.

103F38.EPS

Figure 38 Grooved flange.

available in rigid style with flush-seal or tri-seal gaskets. They are easy to assemble: simply push, join, and tighten.

8.4.3 Reducing Couplings

Grooved reducing couplings save costs in both material and labor because they allow the pipe size to be reduced on the run with no takeout. A typical reducing coupling is rated for 350 psi working pressure. *Figure 37* shows a reducing coupling.

8.4.4 Grooved Flanges

Grooved flanges use the hinge point and the rubber gasket principle to attach a flange to grooved pipe. *Figure 38* shows a grooved flange.

8.4.5 Elbows

The variety of available grooved elbows is much larger than the variety of threaded fittings. Grooved elbows are available in 90-degree, 45-degree, 22½-degree, and 11¼-degree models. In addition, there are adapter elbows, which are grooved on one end and threaded on the other. These are available in 90-degree and 45-degree configurations. An additional selection is available in long-radius elbows. Models in the range of 90 degrees down to 11¼ degrees are available in radii of 3D, 5D and 6D, where D is the diameter of the pipe. Another type of 90-degree elbow is the Add-a-cap®. An Add-a-cap® is a grooved cap with an outlet. An Add-a-cap® allows a sprinkler to be installed at the end of a grooved piping run and also functions as a cap. *Figure 39* shows types of elbows.

8.4.6 Tees

There are several tees available for use in sprinkler systems (*Figure 40*). There are mechanical tees, standpipe tees, bullhead tees, grooved tees, and outlet couplings. The mechanical tee previously discussed is available with a grooved outlet.

The outlet coupling is a special kind of tee. It serves as both a coupling and a tee in one. The outlet is available either threaded or grooved. Grooved tees are standard or reducing and are available in straight grooved form or with a threaded side outlet. Standpipe tees have a 2½" side outlet threaded internally. They are used specifically for outlets on standpipes that use grooved connections instead of flanged connections. The bullhead tee has a large outlet on the side.

8.4.7 Crosses

Crosses are available in the standard pattern only; there are no reducing models. *Figure 41* shows a grooved cross.

8.4.8 Reducers

Grooved reducers are used to join two different sized grooved pipes available in the grooved configuration. There are concentric reducers and eccentric reducers (*Figure 42*).

8.4.9 Caps

A type of end closure is a standard cap (*Figure 43*). A grooved cap is commonly used on grooved piping.

Figure 39 Types of grooved elbows.

8.5.0 Installing Grooved Pipe Couplings

Follow these steps to install a typical grooved pipe coupling. Refer to *Figure 44*.

Step 1 Ensure that the gasket to be installed is the proper size and type. Check pipe ends for imperfections or sharp edges.

Step 2 Apply a thin coat of lubricant (or silicone for freezer applications) to the gasket lips and to either the outside of the gasket or the inside of the coupling housing, or use a pre-lubricated gasket.

Step 3 Place the gasket over the pipe end, arranging it so that it does not hang over the pipe end.

Step 4 Align and bring the two pipe ends together.

Step 5 Center the gasket between the grooves on each pipe.

Step 6 Ensure that no portion of the gasket extends into the groove on either pipe.

Step 7 Place housings over the gasket.

Step 8 Engage the housing keys into the grooves.

Step 9 Insert the bolts, then tighten the nuts finger-tight.

Step 10 Tighten the nuts with a wrench, alternately and equally until the manufacturer's torque specifications are met.

CAUTION

Excessive nut tightening is not necessary. Uneven tightening may cause the gasket to pinch.

9.0.0 Flanged Pipe

Flanged pipe has many applications. Flanged pipe and fittings are used on all valves larger than 2", as well as other applications. Flanged fittings are particularly useful in riser assemblies, cold storage facilities, fire pump installations, and division boundaries of systems.

9.1.0 Flanged Fittings

Flanged fittings are rarely used today. Takeouts for flanged fittings can be calculated more accurately than takeouts for threaded fittings. Flanged fittings consistently use a ⅓"-thick gasket for all flange sizes. With threaded fittings the makeup varies and the fit is not as uniform as flanged makeup. Flanged fittings also appear in extensions of existing systems. *Figure 45* shows a typical flanged joint with the gaskets and bolts used to make it mechanically stable and leakproof. Whether a flanged fitting, a grooved fitting, or a plain-end fitting is used depends primarily on the preferences of the installing contractor and on the device or devices used in the piping assembly. Common types of flanged fittings include the following:

- Elbows
- Tees
- Crosses
- Flanged reducers
- Companion flanges
- Blind flanges
- Floor flanges
- Extra-heavy flanges

Figure 41 Grooved cross.

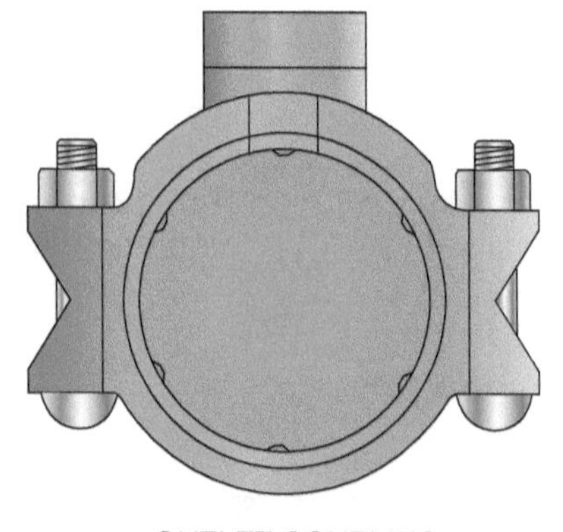

Figure 40 Types of grooved tees.

Figure 42 Grooved reducers.

Figure 43 Standard grooved cap.

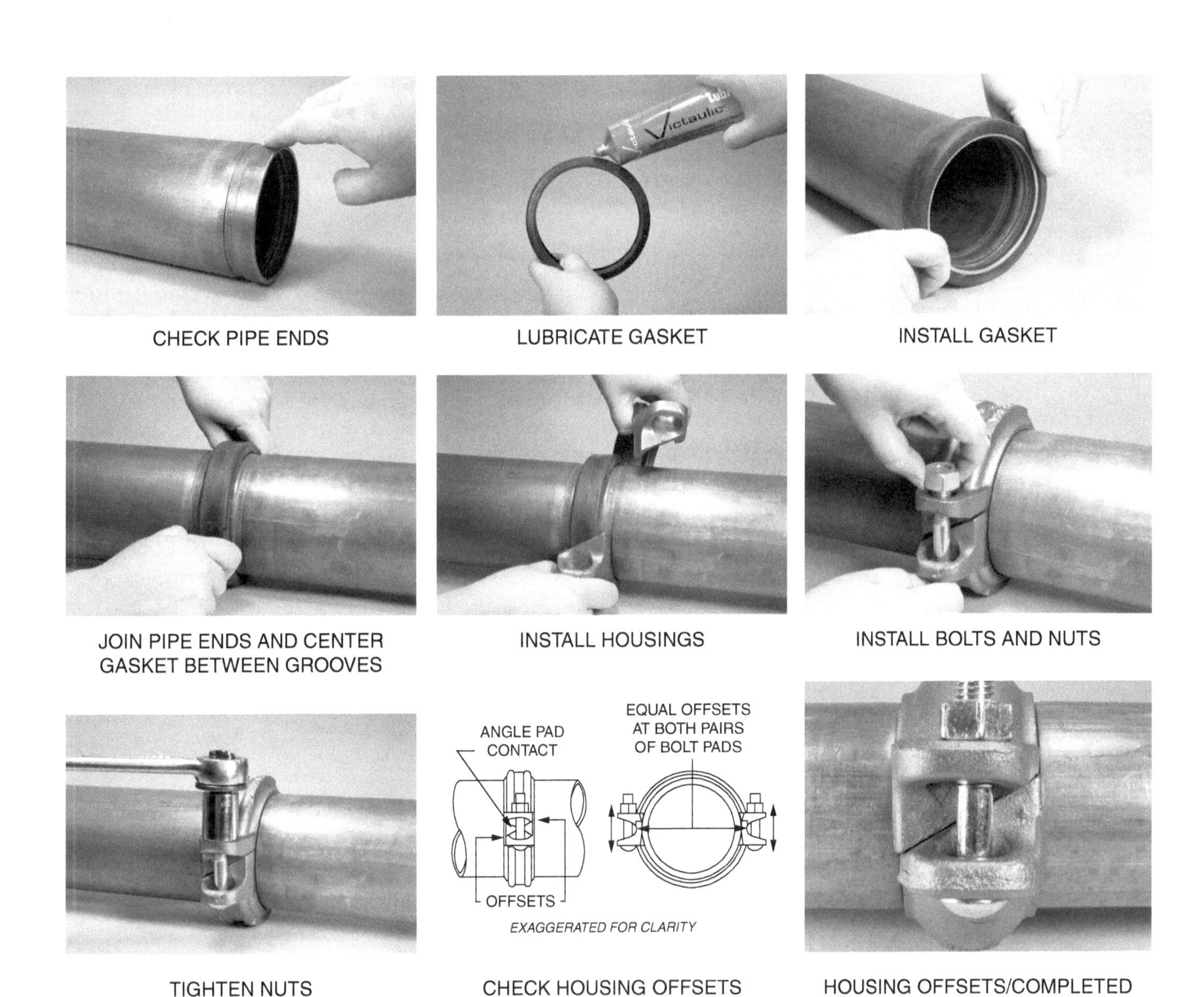

Always refer to instructions supplied with the product for complete information regarding pipe preparation, installation, product inspection, and safety requirements.

103F44.EPS

Figure 44 Joining grooved pipe.

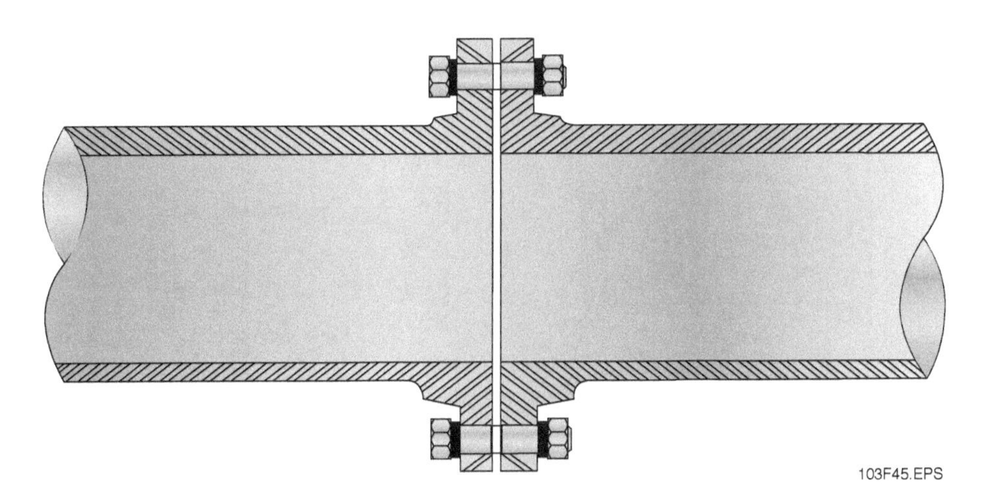

Figure 45 Typical flanged joint.

9.1.1 Elbows

Elbows commonly used in fire department connections are the 45-degree lateral and 90-degree flange-thread elbows. Some 90-degree straight elbows have flanges on both ends and others have flanges on only one end. Flanged-base elbows, although not often used, are especially useful in the horizontal run of riser assemblies where the end riser is fed by an elbow and support below the riser is needed. The cast-on base saves a lot of difficulty otherwise involved in supporting an elbow from below. The base may be round or square. Long-radius elbows also are available for fire sprinkler use. These are used to reduce friction losses or to make a connection in an installation that has special geometry.

Elbows are long lead-time items and must be ordered early enough so that they will be available when they are needed. *Figure 46* shows types of flanged elbows.

9.1.2 Tees

In addition to the flanged-base and long radius elbows discussed, there are two special drain fittings that are used in fire department connections. One is the base-flanged tee which is used when support is required at a connection, and the other is the reducing standard flanged tee that is used when larger drain piping needs to connect with smaller drain piping. Flanged fittings, like elbows, are long lead-time items, and unless your company keeps them in inventory, they may be difficult to acquire on short notice. *Figure 47* shows types of flanged tees.

9.1.3 Crosses

Flanged crosses are fittings used to connect four pieces of pipe. As with many of the flanged fittings, allow enough lead-time for the acquisition of crosses. Flanged crosses are rare and may have to be specially cast. Installers are not normally involved in the ordering process unless they happen to break or lose materials. *Figure 48* shows a flanged cross.

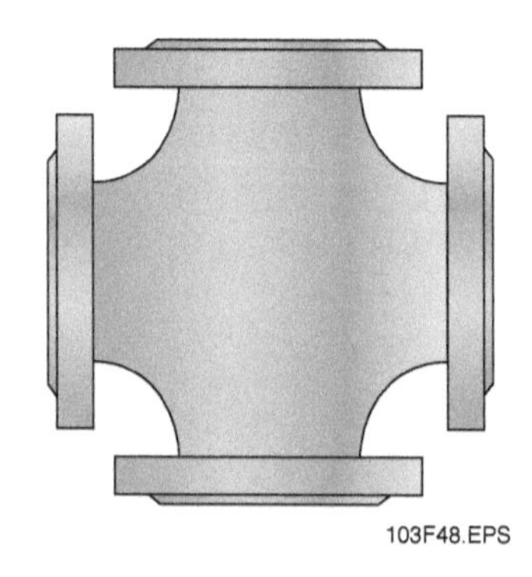

Figure 48 Flanged cross.

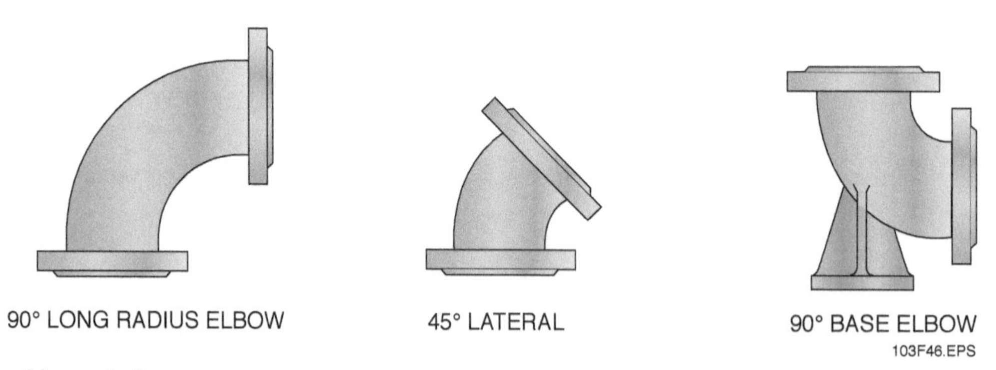

Figure 46 Types of flanged elbows.

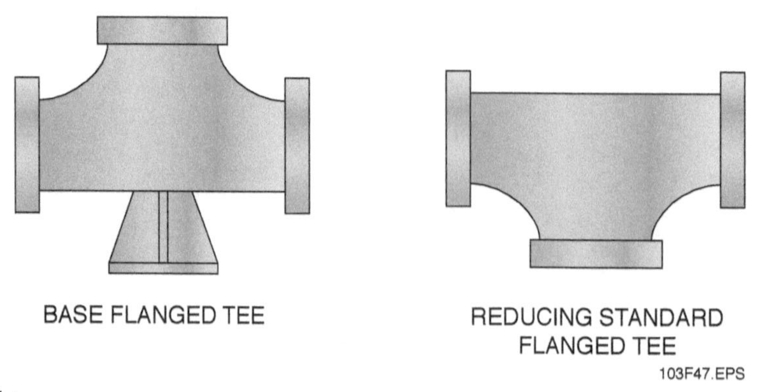

Figure 47 Types of flanged tees.

9.1.4 Flanged Reducers

Flanged reducers are available in concentric, threaded, and flanged configurations and in all sizes and ratings. Concentric flanged reducers are used in 8" dry pipe risers where a 6" dry pipe valve is used. Concentric flanged reducers have the same center line. Eccentric flanged reducers are specialized flanged reducers that may also be used. These reducers do not have the same center line. Concentric and eccentric flanged reducers are used primarily in fire pump installations. *Figure 49* shows a concentric flanged reducer and an eccentric flanged reducer.

9.1.5 Companion Flanges

Standard companion flanges are flanges that are mated with a fitting, device, or flange of the same nominal pipe and screwed to the end of a piece of pipe. *Figure 50* shows a companion flange.

Companion flanges are specified differently than threaded fittings. When determining the sizes of pipe to be mated using companion flanges, the first number is the nominal pipe size. The second number is related to the actual outside diameter of the flange.

To relate the sizes of the incoming pipe to the nominal pipe size of the companion flange and the reducing flange, cross-reference the actual outside diameter of the reducing flanges in *Table 6* with the actual outside diameter of the standard straight-through companion flange shown in *Table 7*. For example, if a 4" pipe must be mated to a 6" pipe using flanges, a reducing flange for the 4" pipe must be matched with a standard companion flange for the 6" pipe. To do this, find the 6" pipe flange size, which is 11", in *Table 6*. Then, examine the 11" flange selections in *Table 7* to determine if the 4" pipe can be accommodated. *Table 7* indicates that a 4" pipe can be mated to a flange size of 11".

9.1.6 Blind Flanges

Blind flanges (*Figure 51*) are used to close off the ends of pipes, valves, and pressure vessel openings. From the standpoint of internal pressure and bolt loading, blind flanges are the most highly stressed of all flange types since the maximum stresses in a blind flange are bending stresses at the center. Blind flanges are available in the seven primary ratings, and their dimensions can be found in manufacturer's catalogs. A blind flange is used in place of a cap or plug to blank off a flanged fitting. Sometimes blind flanges are used with a ¼" NPT outlet and a test gauge as a temporary means of testing a flanged spigot.

9.1.7 Floor Flanges

Although not used in waterways, floor flanges (*Figure 52*) are useful for pipe stands. Floor flanges are less expensive and lighter in weight than companion flanges and are designed for mounting to a supporting surface with flat-head wood screws or bolts.

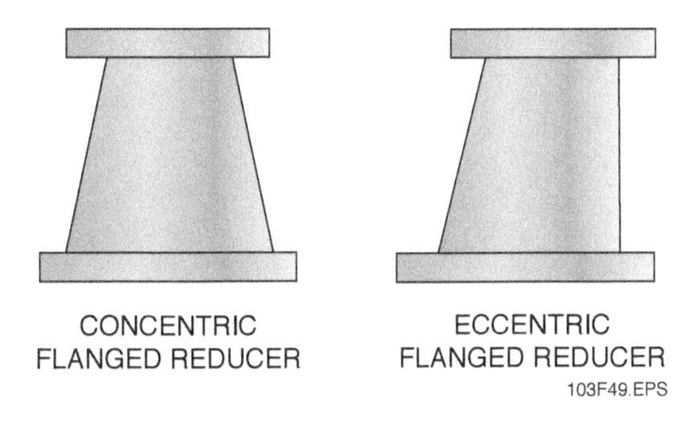

Figure 49 Concentric and eccentric flanged reducers.

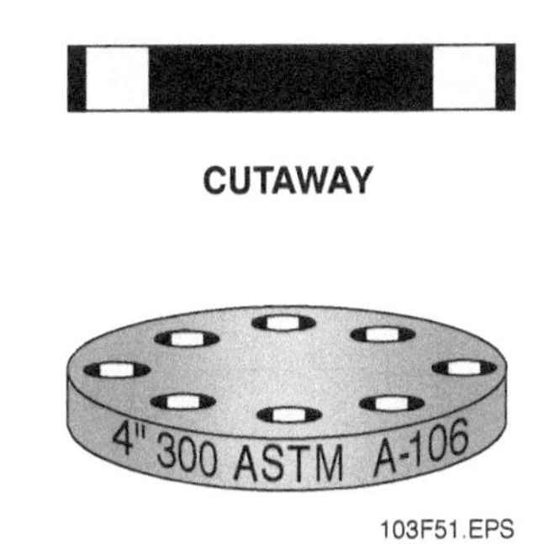

Figure 51 Blind flange.

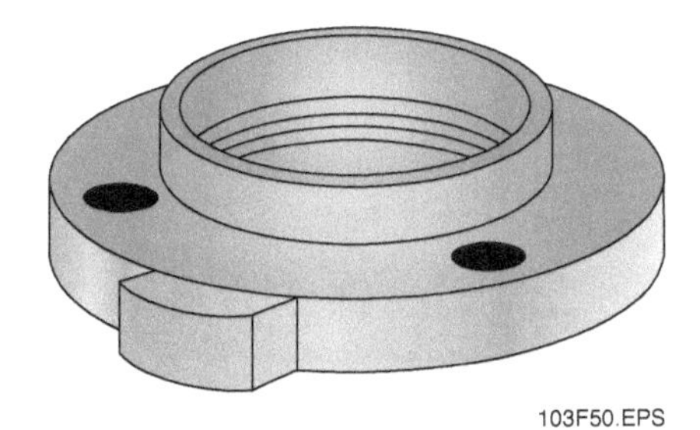

Figure 50 Companion flange.

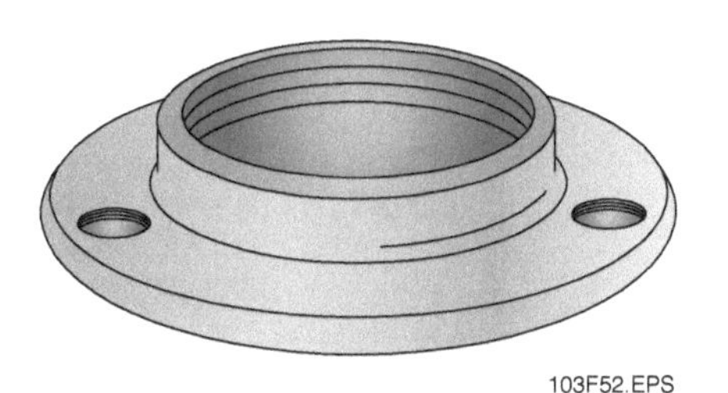

Figure 52 Floor flange.

Table 6 Reducing Flange Sizes

Size		Weight per 100 pcs. P.	Dimensions and Templates for Drilling							
Pipe Size	Flange OD		Over Hub R.	Outside Dia S.	Bolt Circle T.	Thickness Bolts	No. of Bolts	Dia. Holes	Dia. Length	Bolt
1	× 4 5/8	250	11/16	4 5/8	3 1/2	1/2	4	1/2	5/8	2
1 1/4	× 5	294	13/16	5	3 7/8	9/16	4	1/2	5/8	2
1	× 5	315	11/16	5	3 7/8	9/16	4	1/2	5/8	2
3/4	× 5	325	11/16	5	3 7/8	9/16	4	1/2	5/8	2
1 1/2	× 6	445	7/8	6	4 3/4	5/8	4	5/8	3/4	2 1/4
1 1/4	× 6	451	13/16	6	4 3/4	5/8	4	5/8	3/4	2 1/4
1	× 6	451	11/16	6	4 3/4	5/8	4	5/8	3/4	2 1/4
2	× 7	678	1	7	5 1/2	11/16	4	5/8	3/4	2 1/2
1 1/2	× 7	638	7/8	7	5 1/2	11/16	4	5/8	3/4	2 1/2
1 1/4	× 7	650	13/16	7	5 1/2	11/16	4	5/8	3/4	2 1/2
1	× 7	660	11/16	7	5 1/2	11/16	4	5/8	3/4	2 1/2
3/4	× 7	690	11/16	7	5 1/2	11/16	4	5/8	3/4	2 1/2
2 1/2	× 7 1/2	825	1 1/8	7 1/2	6	3/4	4	5/8	3/4	2 1/2
2	× 7 1/2	845	1	7 1/2	6	3/4	4	5/8	3/4	2 1/2
1 1/2	× 7 1/2	858	7/8	7 1/2	6	3/4	4	5/8	3/4	2 1/2
1 1/4	× 7 1/2	863	13/16	7 1/2	6	3/4	4	5/8	3/4	2 1/2
1	× 7 1/2	878	3/4	7 1/2	6	3/4	4	5/8	3/4	2 1/2
3	× 8 1/2	1060	1 3/16	8 1/2	7	13/16	8	5/8	3/4	2 3/4
2 1/2	× 8 1/2	1112	1 1/8	8 1/2	7	13/16	8	5/8	3/4	2 3/4
2	× 8 1/2	1130	1	8 1/2	7	13/16	8	5/8	3/4	2 3/4
1 1/2	× 8 1/2	1210	7/8	8 1/2	7	13/16	8	5/8	3/4	2 3/4
3 1/2	× 9	1345	1 1/4	9	7 1/2	15/16	8	5/8	3/4	3
3	× 9	1340	1 3/16	9	7 1/2	15/16	8	5/8	3/4	3
2 1/2	× 9	1415	1 1/8	9	7 1/2	15/16	8	5/8	3/4	3
2	× 9	1425	1	9	7 1/2	15/16	8	5/8	3/4	3
1 1/2	× 9	1140	15/16	9	7 1/2	15/16	8	5/8	3/4	3
1 1/4	× 9	1445	15/16	9	7 1/2	15/16	8	5/8	3/4	3
1	× 9	1453	15/16	9	7 1/2	15/16	8	5/8	3/4	3
4	× 10	1640	1 5/16	10	8 1/2	15/16	8	3/4	7/8	3
3 1/2	× 10	1710	1 1/4	10	8 1/2	15/16	8	3/4	7/8	3
3	× 10	1731	1 3/16	10	8 1/2	15/16	8	3/4	7/8	3
2 1/2	× 10	1765	1 1/8	10	8 1/2	15/16	8	3/4	7/8	3
2	× 10	1770	1	10	8 1/2	15/16	8	3/4	7/8	3
1 1/2	× 10	1780	15/16	10	8 1/2	15/16	8	3/4	7/8	3
5	× 11	1960	1 7/16	11	9 1/2	1	8	3/4	7/8	3 1/4
4	× 11	2080	1 5/16	11	9 1/2	1	8	3/4	7/8	3 1/4
3 1/2	× 11	2170	1 1/4	11	9 1/2	1	8	3/4	7/8	3 1/4
3	× 11	2140	1 3/16	11	9 1/2	1	8	3/4	7/8	3 1/4
2 1/2	× 11	2250	1 1/8	11	9 1/2	1	8	3/4	7/8	3 1/4
2	× 11	2230	1	11	9 1/2	1	8	3/4	7/8	3 1/4
1 1/2	× 11	2410	1	11	9 1/2	1	8	3/4	7/8	3 1/4
6	× 13 1/2	3240	1 3/16	13 1/2	11 3/4	1 1/8	8	3/4	7/8	3 1/2
5	× 13 1/2	3450	1 7/16	13 1/2	11 3/4	1 1/8	8	3/4	7/8	3 1/2
4	× 13 1/2	3640	1 5/16	13 1/2	11 3/4	1 1/8	8	3/4	7/8	3 1/2
3 1/2	× 13 1/2	3800	1 1/4	13 1/2	11 3/4	1 1/8	8	3/4	7/8	3 1/2
3	× 13 1/2	3720	1 3/16	13 1/2	11 3/4	1 1/8	8	3/4	7/8	3 1/2
2 1/2	× 13 1/2	4005	1 1/8	13 1/2	11 3/4	1 1/8	8	3/4	7/8	3 1/2
2	× 13 1/2	4100	1 1/8	13 1/2	11 3/4	1 1/8	8	3/4	7/8	3 1/2
1 1/2	× 13 1/2	4150	1 1/8	13 1/2	11 3/4	1 1/8	8	3/4	7/8	3 1/2
8	× 16	4800	1 3/4	16	14 1/4	13/16	12	7/8	1	3 3/4

NOTE: All dimensions in inches.

103T06.EPS

Table 7 Companion Flange Sizes

Size		Weight per 100 pcs.	Dimensions and Templates for Drilling							
Pipe Size	Flange OD		Over Hub P.	Outside Dia. R.	Bolt Circle S.	Thickness T.	No. of Bolts	Dia. Bolts	Dia. Holes	Bolt Length
1	× 4¼	187	11⁄16	4¼	3⅛	7⁄16	4	½	⅝	1¾
1¼	× 4⅝	235	13⁄16	4⅝	3½	½	4	½	⅝	2
1½	× 5	305	⅞	5	3⅞	9⁄16	4	½	⅝	2
2	× 6	442	1	6	4¾	⅝	4	⅝	¾	2¼
2½	× 7	653	1⅛	7	5½	11⁄16	4	⅝	¾	2½
3	× 7½	769	13⁄16	7½	6	¾	4	⅝	¾	2½
3½	× 8½	1010	1¼	8½	7	13⁄16	8	⅝	¾	2¾
4	× 9	1235	15⁄16	9	7½	15⁄16	8	⅝	¾	3
5	× 10	1465	17⁄16	10	8½	15⁄16	8	¾	⅞	3
6	× 11	1778	19⁄16	11	9½	1	8	¾	⅞	3¼
8	× 13½	2900	1¾	13½	11¾	1⅛	8	¾	⅞	3½
10	× 16	3850	115⁄16	16	14¼	13⁄16	12	⅞	1	3¾
12	× 19	5840	23⁄16	19	17	1¼	12	⅞	1	3¾
14	× 21	7600	2¼	21	18¾	1⅜	12	1	1⅛	4¼
16	× 23½	9800	2½	23½	21¼	17⁄16	16	1	1⅛	4½
18	× 25	11250	211⁄16	25	22¾	19⁄16	16	1⅛	1¼	4¾

NOTE: All dimensions in inches.

103T07.EPS

9.1.8 Extra-Heavy Flanges

Extra-heavy flanges are used for special applications where pressures are high. The gaskets used for these flanges are different from those used for standard flanges. Extra-heavy flanges have different bolt hole patterns and different bolt requirements than standard flanges. These flanges are thicker and are not compatible with standard flanges. Consult the manufacturer's instructions before installation.

9.2.0 Nuts, Bolts, and Gaskets

To ensure that the composition of various types of gaskets, gasket material, and some packing meet the required specifications, the American Society for Testing and Materials International (ASTM) devised a method of identifying different types of gasket material.

Gaskets are available in a variety of types and materials. The type of gasket used must be matched to the process characteristics and operating conditions to which it will be exposed. For example, different gasket materials are capable of withstanding different temperature and pressure ranges and certain gasket materials are compatible with different process fluids. In piping systems, gaskets are placed between two flanges to make the joint leakproof. Gaskets are normally sized by the thickness of the gasket material in fractions of an inch. Rubber gaskets, flat-ring gaskets, and full-face gaskets are used in flanged piping systems.

9.2.1 Gasket Materials

Gaskets are made of many different types of materials to meet the demands of the particular process system in which they are installed. The conditions affect the types of gaskets that can be used in the system.

> **NOTE**
> Bolts, nuts, and gaskets are usually combined in a package known as a flange-pack.

9.2.2 Rubber Gaskets

Rubber gaskets are available in a variety of pressure and temperature ratings. Standard black and red rubber gaskets come in thicknesses from 1⁄16" to 1". These gaskets are used primarily in low-pressure/low-temperature water, gas, air, and refrigerant systems. Rubber gaskets are used for saturated steam up to 100 psi. They have an approximate temperature range of 20°F to 170°F.

CAUTION

Some gasket materials are not compatible with CPVC, and some older gaskets may contain asbestos.

9.2.3 Types of Flange Gaskets

There are different types of gaskets that must be matched to different types of flanges. For example, flat ring gaskets are used with raised face flanges; full-face gaskets are used with flat-face flanges.

9.2.4 Flat-Ring Gaskets

Flat-ring gaskets (*Figure 53*) are used on flanges with raised faces. The outside diameter of the flat-ring gasket is slightly larger than the outside diameter of the raised face. The materials used to manufacture flat-ring gaskets may be metallic, nonmetallic, or a combination of both. The raised-face flange dimensions and class rating of the flange determine the size of flat-ring gasket needed.

9.2.5 Full-Face Gaskets

Full-face gaskets are used with flat-face flanges and extend all the way to the outer edge of the flange. These gaskets have bolt holes that match the holes of the flanges with which they are used. *Figure 54* shows a full-face gasket.

9.3.0 Flange Bolts

Two types of bolts are used on flange connections: machine bolts and stud bolts (*Figure 55*). Machine bolts have square or hexagonal heads. Stud bolts are threaded the entire length or are threaded on both ends. There are some instances in which cap screws are used instead of bolts. *Figure 55* shows types of flange bolts.

The following four characteristics must be considered when selecting bolts for flanges:

- Thread standards and series
- Thread class
- Thread designation
- Fastener grade

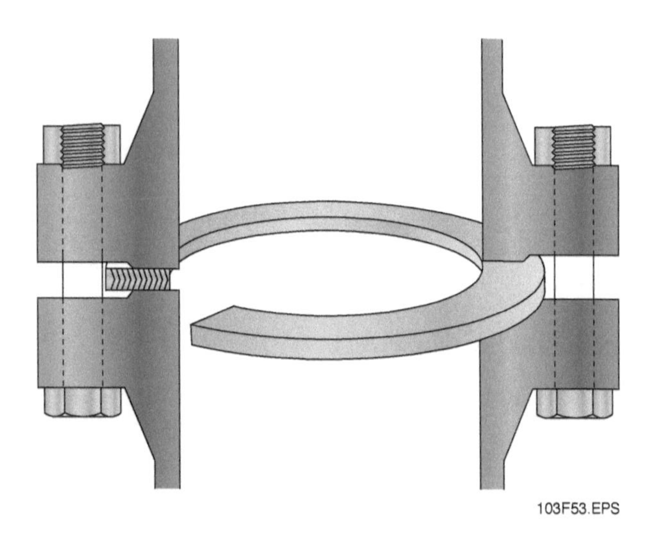

Figure 53 Flat-ring gasket.

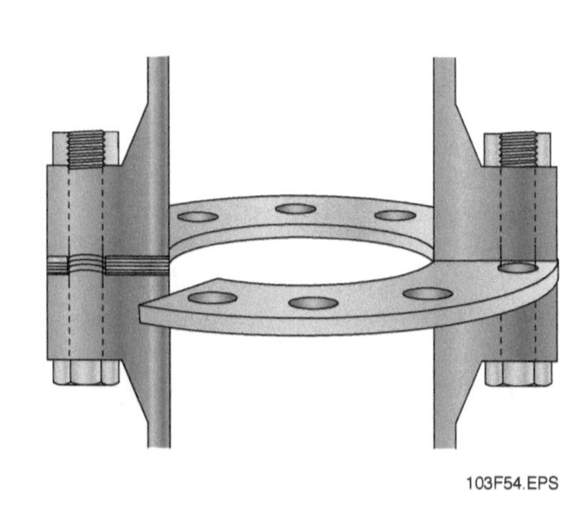

Figure 54 Full-face gasket.

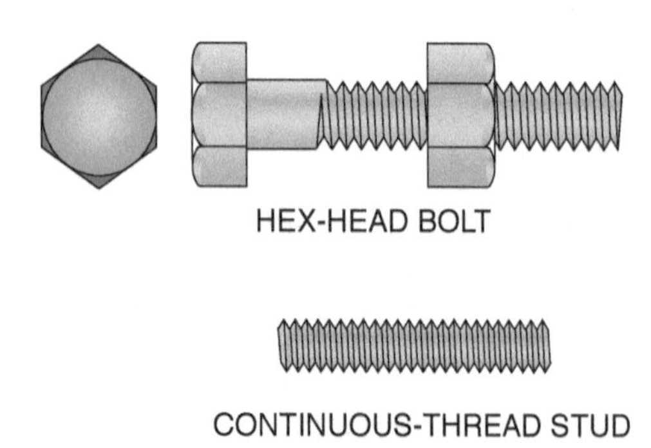

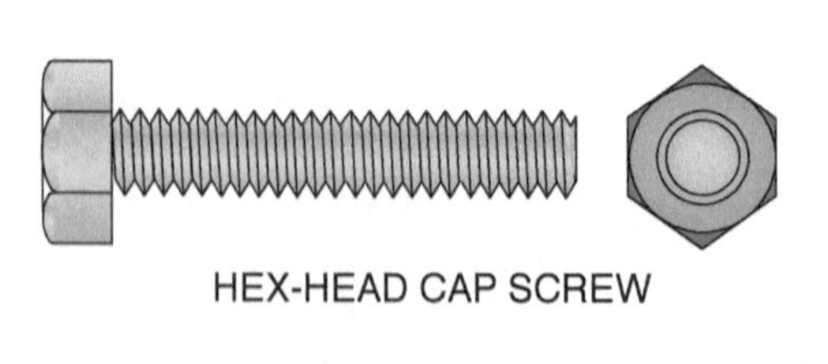

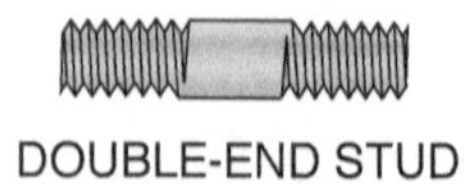

Figure 55 Flange bolts.

9.3.1 Thread Standards and Series

Fastener threads are made to established standards. The most common standard is the unified, or American, standard. Unified standards are established for three series of threads, depending on the number of threads per inch for a certain diameter fastener. These three series and their abbreviations are as follows:

- *Unified National Coarse Thread (UNC)* – Used for bolts, screws, nuts and other general uses. Fasteners with UNC threads are used for rapid assembly or disassembly of parts where corrosion or slight damage may occur.
- *Unified National Fine Thread (UNF)* – Used for bolts, screws, nuts and other uses where a finer thread than UNC is required.
- *Unified National Extra-Fine Thread (UNEF)* – Used for thin-walled tubes, nuts, ferrules, and couplings.

9.3.2 Thread Designation

Bolt and screw threads are designated by a standard method. The following are the standard designations for bolt and screw threads:

- *Nominal size* – Diameter.
- *Number of threads per inch* – Standard for different diameters.
- *Thread series symbol* – The unified standard thread used (UNC, UNF or UNEF).
- *Thread class symbol* – The closeness of fit between the bolt threads and nut threads.
- *Left-hand thread symbol* – Specified by the symbol LH. (Unless specified LH, all threads are right-hand.)

Figure 56 shows the thread designation for a common fastener.

9.3.3 Fastener Grade

The strength and quality of a fastener is easily determined by special grade markings on the head

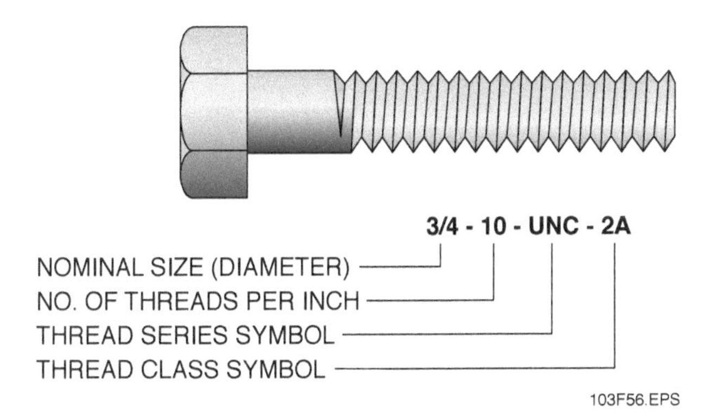

Figure 56 Thread designation.

of the fastener. These markings are standardized by the Society of Automotive Engineers (SAE) and by ASTM. Grade markings are sometimes called line markings. Stud bolts are stamped on one end only. When cutting stud bolts, be sure not to cut off the stamped end.

CAUTION

There are many counterfeit bolts on the market that do not have the ASTM grade markings and which may be inferior bolts.

9.4.0 Installing Pipe Flanges

ANSI standards require that flange bolts always straddle the horizontal and vertical center lines unless otherwise indicated. This straddling is known as the two-hole method. It is possible, though, for a job specification to require that the bolt holes line up with the vertical and horizontal lines. This method is known as the one-hole method. The general rule is to reference one center line only, depending on the type of drawing used. In elevation drawings, the usual reference is the vertical center line, and in plan drawings, the usual reference is the north-south center line. This rule ensures that all fabricators refer to the same center line.

The four steps in pipe flange installation are cleaning the parts, aligning the parts, installing the gaskets, and tightening the flange bolts.

9.4.1 Cleaning Parts

Like all pipe joints, flanged joints must be cleaned to ensure the best possible fit. Flanges from the factory and on other equipment often have a gum or other residue that must be scraped off the face. Flanges with grooved faces must be clean and free of dirt so that the gasket will fit properly. It is sometimes necessary to use solvents to remove the protective coating put on flanges at the factory. Always check the engineering specifications before applying any type of solvent to the flanges.

Gaskets should also be cleaned before installation. Flanged joints are designed so that the face of each flange matches up squarely with a gasket to form a tight seal. Any type of obstruction or debris that could cause the flanges to make up improperly must be removed.

9.4.2 Aligning Parts

Aligning flanges and component parts properly is extremely important to ensure a leak-free connection. First the pipe must be cleaned. After it is cleaned, the pipe must be put in place and properly supported. Hangers and support stands must then be adjusted until the flanges are properly aligned. A drift pin can be used to align the bolt holes. Never try to align the bolt holes with your fingers, as injury may result. After the alignment is complete you are ready to install the gasket. *Figure 57* shows good and bad alignment.

9.4.3 Installing Gaskets

After the parts are aligned, the gaskets should be selected based on the type of flange used. For training purposes, assume that a flat-face flange is used and a full-face gasket is installed. Follow these steps to install a full-face gasket between flat-face flanges:

Step 1 Check the specification to determine the type of gasket needed.

Step 2 Place the gasket between the flange faces.

Step 3 Hold the gasket in place, using a drift pin, and insert the bottom bolts into the flanges.

> **WARNING**
>
> Never use your fingers to hold the gasket in place. Use a drift pin to avoid pinching your fingers.

Step 4 Insert all the bolts through the flange bolt holes.

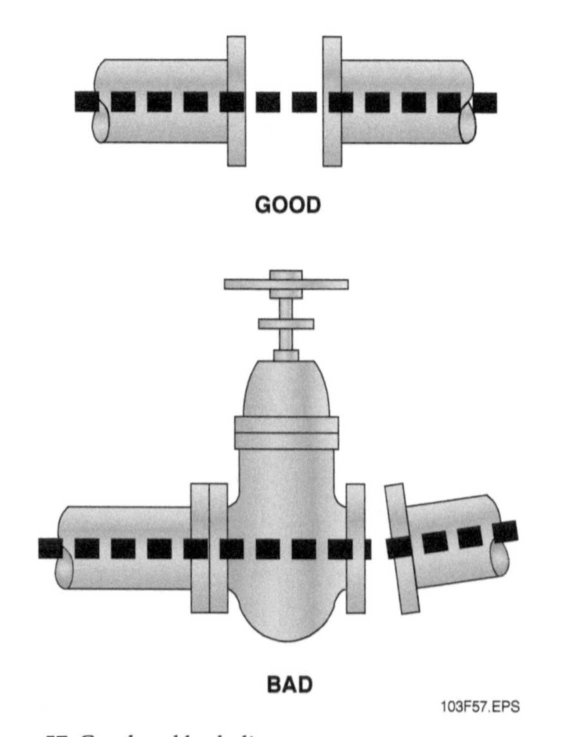

Figure 57 Good and bad alignment.

Step 5 Place the nuts on the bolts and tighten the nuts finger-tight.

If raised-face flanges are used, install a flat-ring gasket. Follow these steps to install a flat-ring gasket between raised-face flanges:

Step 1 Insert the bottom bolts into the flanges.

Step 2 Lubricate the gasket if necessary.

Step 3 Place the gasket between the flange faces.

Step 4 Insert all the bolts through the flange bolt holes.

Step 5 Place the nuts on the bolts and tighten them finger-tight.

9.4.4 Tightening Flange Bolts

Flange bolts must be tightened carefully to avoid warping the flange. If the bolts are not tightened correctly, the joint may leak or crack. The correct amount of pressure must be applied evenly and in the correct tightening sequence to make a good seal. *Figure 58* shows the proper tightening sequence for flange bolts.

10.0.0 Determining Pipe Lengths Between Fittings

To determine the length of pipe between two fittings, the sprinkler fitter must determine the takeout of the fittings. Takeout, also referred to as take-up or takeoff, refers to the dimensions of the fittings within a run of pipe. Knowing the takeout for the fittings is essential when figuring the lengths of straight pipe.

10.1.0 Takeouts

Typically, pipe that is dimensioned on drawings uses center-to-center dimensions. Pipe does not actually extend from center-to-center. If it did, the side outlet on a tee would be blocked. When cutting the pipe, something less than a center-

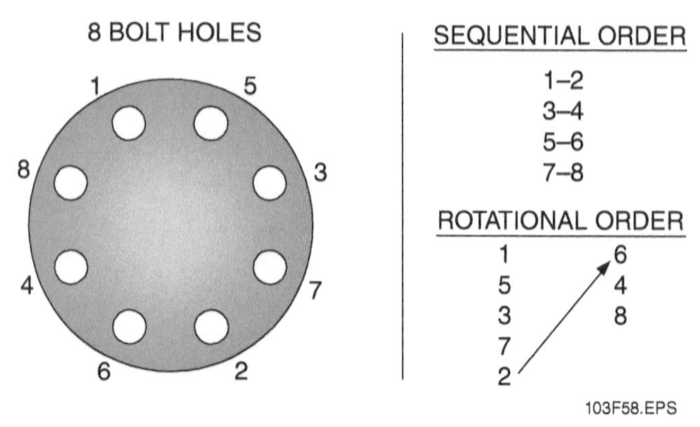

Figure 58 Proper tightening sequence.

to-center dimension is required. A takeout, also called thread allowance, must be applied to obtain the cut dimensions of the pipe. This is actually a fitting allowance. In sprinkler system work, the takeout is the dimension subtracted from the center-to-center dimension so that when pipe and fittings are joined, the center-to-center dimensions are correct.

Makeup is the term used to describe the amount of pipe that screws into a threaded fitting. The face-to-face or shoulder-to-shoulder dimension of a threaded fitting and the makeup are used to calculate the takeout. *Figure 59* shows takeout conventions.

Valves such as outside stem and yokes (OS&Ys), check valves, and alarm valves, like threaded fittings, take up room in a pipe run. For the system to fit properly, the dimensions of these valves must be included in the design. In creating a sprinkler drawing, detailing pipe cutting lengths is referred to as cutting pipe, even though the actual process is merely calculating a number. There are times in the field when a fitter must physically cut pipe based on the local job conditions. The fitter, in these circumstances, must calculate the takeouts.

The takeout numbers in this section are developed from manufacturer's literature. The data in the figures is used with the exercises and the module examination. Takeouts are useful for making field changes and for designing sprinkler layout. Different manufacturers use different designs and takeouts may differ.

10.2.0 Threaded Fittings

Takeouts for threaded fittings with 90-degree connections are applicable to tees and 90-degree elbows. Each manufacturer of threaded fittings has its own molds, so the dimensions of the molds vary slightly from manufacturer to manufacturer. This results in slight variations in the related takeouts. In addition, with a given manufacturer, there is a difference in takeouts with different reducing tees. With threaded fittings there are two takeouts that need to be added, one for each end of the pipe. *Table 8* shows a takeout chart for threaded fittings.

10.2.1 Tolerances

The takeouts in *Table 8* are dimensioned to $\frac{1}{16}$". Most pipe fabrication shops only dimension pipe to $\frac{1}{2}$", although some shops operate to $\frac{1}{4}$" accuracy. In reality, with a little care, pipe could be cut consistently to an accuracy of $\frac{1}{8}$". In a machine shop,

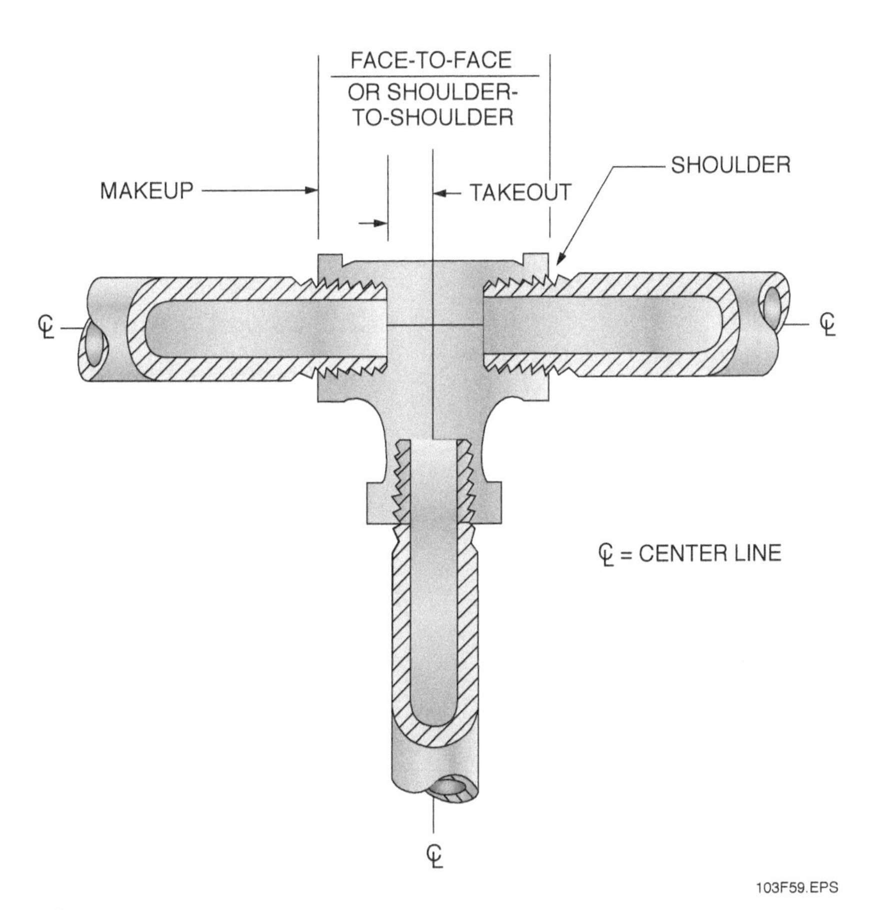

Figure 59 Takeout conventions.

Table 8 Takeout Chart for Threaded Fittings

Pipe @ Right Angle to Pipe Being Cut	Pipe Size on Run													45° ELL	Pipe Makeup
	½	¾	1	1¼	1½	2	2½	3	3½	4	5	6	8		
½	5/8	¾	9/16	5/8	¾	13/16	9/16	7/8	7/8	13/16	1 1/16	–	–	3/8	–
¾	¾	¾	¾	¾	13/16	15/16	13/16	7/8	7/8	15/16	1 1/16	1½	–	7/16	–
1	7/8	7/8	7/8	7/8	15/16	1 1/16	1	1 1/16	7/8	1½	1 1/16	1½	–	½	5/8
1¼	1	1 1/16	1 1/16	1 1/16	1 1/8	1 3/16	1 1/8	1 3/16	1	1¼	1 1/16	1½	2 1/16	5/8	11/16
1½	1 3/16	1 3/16	1 3/16	1 3/16	1¼	1 5/16	1¼	1 5/16	1 3/16	1 3/8	1½	1 5/8	2 1/16	¾	11/16
2	1 3/8	1 7/16	1 3/8	1 7/16	1½	1 9/16	1 7/16	1 9/16	1 3/8	1 5/8	1 11/16	1¾	2 1/16	1	11/16
2½	1 15/16	1 11/16	1 11/16	1 11/16	1 13/16	1 15/16	1¾	1 13/16	1 5/8	1 15/16	2	2 1/8	2 5/16	1	15/16
3	2 1/8	2 1/16	2 1/16	2 1/16	2 1/8	2¼	2 1/8	2 1/16	1 15/16	2 3/16	2¼	2 3/8	2 5/8	1 3/16	1
3½	2¾	2 11/16	2 5/8	2 9/16	2 9/16	2 11/16	2 3/8	2¼	2 3/8	2¼	2½	2 5/8	3 1/8	1 1/16	1 1/16
4	2 5/8	2 9/16	2 9/16	2 9/16	2 5/8	2¾	2 9/16	2 5/8	2½	2 11/16	2¾	2 7/8	3 1/8	1½	1½
5	3½	3 7/16	3 3/8	3 5/16	3¼	3 7/16	3 3/16	3¼	3¼	3 5/16	3¼	3 3/8	3 5/8	1 13/16	1½
6	–	3 7/8	3 13/16	3¾	3 13/16	3 15/16	3¾	3¾	3 11/16	3 13/16	3¾	3 7/8	4 3/16	2 3/16	1½
8	–	–	–	5 1/8	5 1/8	5 1/8	5 1/16	5 1/16	5 1/8	5 1/16	5	5 1/8	5 3/16	2 7/8	1 3/8

NOTE: All dimensions in inches.

103T08.EPS

the dimensions could be cut to a tolerance of 0.001"; however, the more accurate the cutting, the more expensive the pipe.

Be aware that field measurements generally are only accurate to ±½". In the construction industry, a ±½" tolerance is usually sufficient. Exceptions occur when cutting drops and when centering heads in ceiling tiles. Accuracy is also required for gridded systems and pump rooms. Your fabrication shop may routinely use a ¼" tolerance. The following principles, although based on ±½", can be used to realize your tolerance goal if it is less than ½".

On Site

Piping Installation Direction

When installing piping for any system, start the piping at the source and work toward the unit being installed. If piping drawings and specifications are supplied for the job, they must be followed when installing the piping runs. Measure and cut the pipe as the installation proceeds to ensure a neat appearance. If the unit is not yet in place, pipe to the approximate location of the unit and plug or cap the pipe. The final piping can be accomplished when the unit is in place. If a material specification sheet exists for the job, check the sheet for the unit rough-in location information and pipe to that location. If the unit is roughed-in on the wrong side, most units will have a provision to allow piping to be installed from two or more sides.

In design, when there is a takeout of a value such as 2¼", it can be rounded up or rounded down to reach a ½" increment. (If your tolerance goal is ±¼", the same reasoning applies for a value of 2⅛".) For the first such dimension, it does not matter which way you round; however, the second takeout of the same value must be rounded in the opposite direction. That is, 2¼"+ 2¼" = 4½". The final result can be reached as: 2" + 2½" = 4½" or 2½" + 2" = 4½". If we use the process as 2½" + 2½", the result of 5" is ½" too much. In rounding to ½" increments, keep a record of what is left over from the rounding off process, balancing the pluses and the minuses. This prevents tolerance build-up.

10.2.2 Using Takeout Table

Figure 60 shows a typical sprinkler system design using 3", Schedule 40 steel pipe. Use the information in *Table 8* to determine the pipe cut lengths required in *Figure 60*.

The center-to-center distance between branch lines is 12'-6", but the cutting length is found by subtracting the takeouts from the center-to-center distance. The amount of takeout is specifically related to the pipe at right angles.

Read across the top of *Table 8* until you reach 3". Then move down the column until you find the 1½" row. The value of the takeout for our combination of pipe and fittings is 1 5/16". That is the value of the takeout for one side of that one fitting, with 3" pipe on the run and 1½" pipe at right angles.

When cutting the pipe between two such connections, 1 5/16" are taken out at A and 1 5/16" taken out at B. Thus, the total takeout for this piece of pipe is

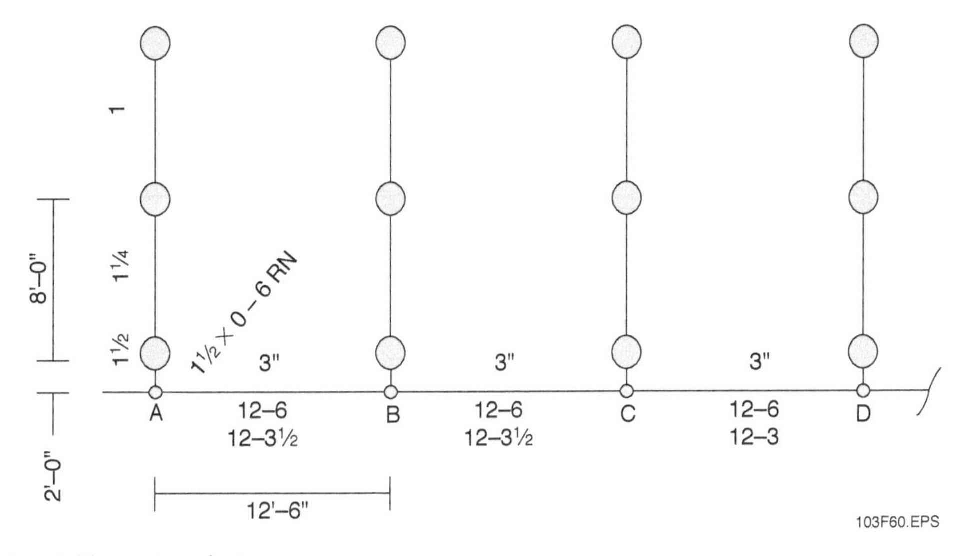

Figure 60 Typical sprinkler system design.

1⁵⁄₁₆" + 1⁵⁄₁₆" = 2⅝". The nearest ½" increment is 2½". Therefore, use a takeout of 2½", but keep track of the fact that this piece of pipe is cut ⅛" too long.

To maintain center-to-center dimensions accurately, we must reduce the pipe by ⅛". Continuing with the line segment from B to C, we have two 1⁵⁄₁₆" takeouts. The total takeout now needed is 2⅝" + ⅛" (left over) = 2¾". If we round to 2½", the cut is ¼" too long. If we round to 3", the cut is ¼" too short. For purposes of this module, use 2½" for the takeout.

With the line segment from C to D, the takeout again is 2⅝". If we use 2½" for the takeout, the cut is ⅜" too long. In this case, we take the excess of ¼" and add it to the takeout of 2⅝" to obtain a total needed takeout of 2⅞". Rounding to the nearest ½" increment, we use a takeout of 3". Thus, the cut is ⅛" too short. In all cases, this approach ensures that the identified cutting dimensions are never more than ¼" off the correct value. To maintain our ½" accuracy, this means that the installer must physically cut the pipe to an accuracy of ±¼". Notice that the main in *Figure 60* has been dimensioned with the nominal pipe diameter on the top of the line. Below the line, we first note the center-to-center dimension, and below that the planned cutting length.

10.3.0 Welded Outlets

The two most common types of welded outlets are the threaded and grooved models. The threaded outlets are used particularly for small threaded connections (branch lines, riser nipples, or sprinkler outlets) to light-wall pipe. The grooved outlet is used when the outlet connection is to Schedule 10 pipe. It is important for the sprinkler fitter to understand takeouts for welded outlets. *Table 9* shows takeouts for welded outlets.

10.4.0 Flanged Fittings

Flanged fittings have a ⅛" gasket between the two flanges. In a pump room, for example, there might be three or four flanged connections in series. The build-up (4 × ⅛" = ½"), if not considered, can result in the material not fitting. This is an area where ¼" accuracy must be considered, which requires pipe to be dimensioned and cut to an accuracy of ⅛". *Table 10* lists the takeouts for specific flanged fittings. Notice that these dimensions are face-to-face or flange-to-flange. The gaskets must be added when cutting the pipe. For example, when using a Mueller 4" swing check, 12⅞" must be allowed, flange-to-flange, when laying out the pipe. With a gasket on each side, a total of 13⅛" must be allowed for this installation.

10.5.0 Flange-Thread Fittings

Flange-thread fittings have one side made for a threaded connection. The other side is made for a flanged connection. In the past these were used in fire department connections to a riser because they have a short turn radius that conserves space. This is an area where the use of ¼" dimensioning might also be indicated. Takeouts for these fittings are shown in *Table 10.*

10.6.0 Gasketed Grooved Couplings

Gasketed, grooved fittings are considered to require no takeout allowance, but when they are pressurized, gasketed, grooved couplings can expand by ⅛" to ¼" per joint. The results of this expansion can create a serious problem in a combined standpipe/sprinkler riser in a high-rise building. In a building 50 stories high, ⅛" × 50 = 6½", or 50 × ¼" = 1'-1". To avoid this problem, use

Table 9 Takeouts for Welded Outlets

Pipe Size on Run

Welded Female Thread Takeouts

Outlet Size	1	1¼	1½	2	2½	3	3½	4	5	6	8
½	1 5/16	1½	1 13/32	1 21/32	1 29/32	2 7/32	2 15/32	2 23/32	3¼	3 25/32	4 25/32
¾	1 5/32	1 21/64	1 29/64	1 11/16	1 15/16	2¼	2½	2¾	3 9/32	3 13/16	4 13/16
1	1 3/16	1 23/64	1 31/64	1 23/32	2	2 9/32	2 17/32	2 25/32	3 5/16	3 27/32	4 27/32
1¼	1 3/16	1 31/64	1 39/64	1 27/32	2⅛	2 13/32	2 21/32	2 29/32	3 7/16	4	4 15/16
1½	1 13/32	1 19/32	1 45/64	1 15/16	2 3/16	2½	2¾	3	3 17/32	4 5/64	5 1/16
2	1 11/32	1 25/32	1 53/64	2 1/16	2 5/16	2⅝	2⅞	3¼	3 21/32	4 3/16	5⅛
2½	1 51/64	1 25/32	2 5/64	2 5/16	2 9/16	2⅞	2⅛	3⅜	3 15/16	4 7/16	5 7/16

Chart Dimension

Outlet Size	1	1¼	1½	2	2½	3	3½	4	5	6	8
1	3 21/32	3 53/64	3 61/64	4 3/16	4 7/16	4¾	5	5¼	5 25/32	6 5/16	7 5/16
1¼	–	3 53/64	3 61/64	4 3/16	4 7/16	4¾	5	5¼	5 25/32	6 5/16	7 5/16
1½	–	–	3 61/64	4 3/16	4 7/16	4¾	5	5¼	5 25/32	6 5/16	7 5/16
2	–	–	–	4 3/16	4 7/16	4¾	5	5¼	5 25/32	6 5/16	7 5/16
2½	–	–	–	–	4 7/16	4¾	5	5¼	5 25/32	6 5/16	7 5/16
3	–	–	–	–	–	4¾	5	5¼	5 25/32	6 5/16	7 5/16
4	–	–	–	–	–	–	–	6¼	6 25/32	7 5/16	8 5/16
6	–	–	–	–	–	–	–	–	–	7 5/16	8 5/16
8	–	–	–	–	–	–	–	–	–	–	8 5/16

Chart Dimension

Welded Groove Outlet Takeouts

NOTE: All dimensions in inches.

103T09.EPS

Table 10 Takeouts for Flange-Thread, Flanged, and Other Fittings

Size		½	¾	1	1¼	1½	2	2½	3	3½	4	5	6	8
F.T. ELL ℄ to F.	90°	–	–	–	–	–	–	2 11/16	3 1/16	3 7/16	3¾	4½	5⅛	6 9/16
	45°	–	–	–	–	–	–	1 15/16	2 3/16	2⅛	2⅝	3 1/16	3 7/16	4¼
Flanged ELLs & Tees ℄ to F.	90°	–	–	–	–	–	–	5	5½	6	6½	7½	8	9
	45°	–	–	–	–	–	–	3	3	3½	4	4½	5	5½
C.I. ELLs	90°	⅝	¾	⅞	1 1/16	1¼	1 9/16	1¾	2 1/16	2⅜	2 11/16	3¼	3⅞	5 3/16
	45°	⅜	7/16	½	⅝	¾	1	1	1 3/16	1 1/16	1½	1 13/16	2 3/16	2⅞
Swing Checks F. to F.	M&H & Kennedy	–	–	–	–	–	10	10¼	–	13	15	16	19	–
	Mueller	–	–	–	–	–	–	9⅛	10⅜	11¾	12⅞	14⅛	16½	19½
	Stockham	–	–	–	–	–	–	10	11	–	13⅛	15	16	20
Screwed Check VA, ℄ to End		–	1⅛	1⅜	1 13/16	1⅞	2 5/16	–	–	–	–	–	–	–
Flanged OS&Y Gate Valve	F. to F.	–	–	–	–	–	–	7½	8	8½	9	10	10½	11½
	℄ of VA to End of Stem	–	–	–	–	–	–	15⅞	16⅞	19	20¾	26	30	39¾
Screwed OS&Y, VA. ℄ to End		–	¾	⅞	15/16	1 1/16	1 3/16	–	–	–	–	–	–	–
Gate Valve ℄ to End		–	1 1/16	1¼	1½	1 11/16	2 1/16	–	–	–	–	–	–	–
Standard Companion Flange	FLG. DIA.	–	–	4½	4⅝	5	6	7	7½	8½	9	10	11	13½
	No. Bolts	–	–	4	4	4	4	4	4	8	8	8	8	8
	B&N	–	–	½ × 1½	½ × 2	½ × 2	⅝ × 2¼	⅝ × 2¼	⅝ × 2½	⅝ × 2¾	⅝ × 3	¾ × 3	¾ × 3¼	¾ × 3½
T.O. for G.J. Union		13/16	⅞	13/16	1⅛	1¼	1 3/16	–	–	–	–	–	–	–

NOTE: All dimensions in inches.

103T10.EPS

riser clamps at the top and bottom of the floor penetrations and use couplings that do not flex.

If the flexibility of a grooved joint is not needed, use a rigid coupling. If the flexibility of a normal gasketed coupling is required, consider the series effect of a number of them. The stretching of mains, for instance, could result in overspacing of branch lines. In a long run of pipe, hanger or fitting interference can result.

10.7.0 Grooved Fittings

There are numerous grooved fittings available for use in sprinkler systems. Some have the same takeouts, but some new products may differ. For example, a Fire-Lok® grooved fitting is a product that has a shorter center-to-center dimension than standard grooved fittings and has special installation requirements. They are to be used only in conjunction with Fire-Lok® couplings. The use of other products with Fire-Lok® products may result in bolt pad interference.

Fire-Lok® grooved fittings are used in valve connections, fire mains, and long straight runs because of their rigidity. Check manufacturer's data for takeouts and installation requirements. *Table 11* shows takeouts for standard grooved fittings.

10.8.0 Plain-End Fittings

Plain-end fittings come in various sizes and reducing combinations. Check the manufacturer's data sheet for takeout dimensions.

10.9.0 Devices

Dimensions are given face-to-face on flanged devices. On threaded fittings, values given are center line to end. With threaded fittings, the pipe makeup must be subtracted from the center line to end dimension to identify the actual takeout required. The final authority on these takeouts is the manufacturer. As new products come on the market, takeout information may require updating. No data is shown for alarm, dry pipe, or deluge valves because they vary from manufacturer to manufacturer and style to style. Takeouts for valves are obtained directly from the manufacturers' data.

NOTE

There are two types of grooved fittings that vary from industry standard because they have special 90-degree fittings with special takeouts. Check manufacturer's instructions for specific information.

Table 11 Takeouts for Standard Grooved Fittings

Nominal Pipe Size	1	1¼	1½	2	2½	3	3½	4	5	6	8
STD. 90° ELL	2¼	2¾	2¾	2¼	3¾	4¼	4½	5	5½	6½	7¾
STD. 45° ELL	1¾	1¾	1¾	2	2¼	2½	2¾	3	3¼	3½	4¼
STD. 22° ELL	3¼	1¾	1¾	3¾	4	4½	3½	2⅞	2⅞	6¼	7¾
STD. 11° ELL	1⅜	1⅜	1⅜	1⅜	1½	1½	1¾	1¾	2	2	2
STD. Tee & Cross	2¼	2¾	2¾	3¼	3¾	4¼	4½	5	5½	6½	7¾

NOTE: All dimensions in inches.

103T11.EPS

Summary

Steel pipe is the type most often used in sprinkler fitter work. Steel pipe can be joined with threaded fittings, grooved joint fittings, flanges, or plain-end couplings.

Threaded pipe is a versatile, widely used material for joining pipe that is relatively simple to install. All pipe is threaded according to national standards. The tapered pipe thread, or NPT, is the most common thread used on pipe and fittings for industrial and commercial low-pressure applications. The threaded pipe system requires a thread compound and proper assembly to ensure a tight, leakproof joint.

The ability to install steel pipe in accordance with the job requirements is a skill that every sprinkler fitter must develop. This module has introduced you to the basics of installing steel pipe. All you have to do is put these basics to work for you to become a skilled worker in the sprinkler fitting trade.

Review Questions

1. Galvanized pipe is pipe that has been _____.
   a. painted with an aluminum coating
   b. sprayed with Galvan
   c. dipped in a hot zinc mixture
   d. threaded using a hand threading tool

2. One advantage of stainless steel is that it is _____.
   a. inexpensive
   b. easy to thread
   c. resistant to corrosion
   d. difficult to make leakproof

3. The wall thickness of Schedule 40 and Schedule 80 pipe is the same; it is the outside diameter that is different.
   a. True
   b. False

4. The type of pipe required by FM Global in dry systems is _____.
   a. galvanized
   b. black iron
   c. copper
   d. CPVC

5. Which of the following is a correct statement about thin-wall pipe?
   a. It can be used in sprinkler work as long as it is designated thin-wall pipe by *NFPA 13*.
   b. It must be Schedule 40 or thicker, regardless of the use.
   c. It can only be used in single-story buildings.
   d. It can be used if is specifically Listed for the application.

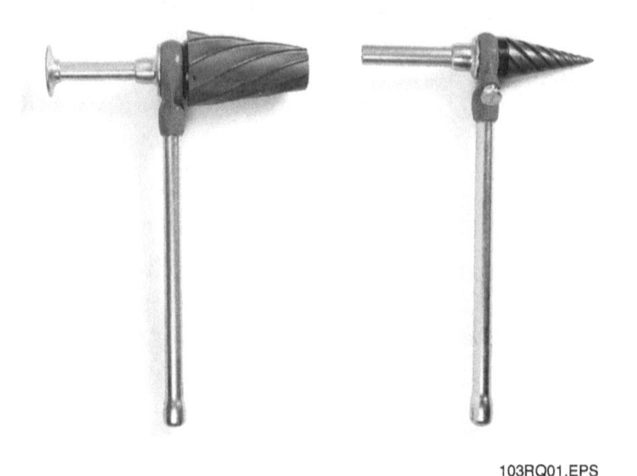

103RQ01.EPS

Figure 1

6. The tools shown in *Figure 1* are used to _____.
   a. cut pipe
   b. thread pipe
   c. ream pipe
   d. clean a die head

7. In the thread designation 3⁄8–18–NPT, the 18 represents the number of threads per inch.
   a. True
   b. False

8. Given a properly threaded pipe, a threaded fitting should be able to be hand-threaded _____.
   a. 1 to 2 full rotations
   b. 2 to 3 full rotations
   c. 3 to 4 full rotations
   d. 4 to 5 full rotations

9. Any high-quality pipe thread compound can be used on any sprinkler system piping.
   a. True
   b. False

10. The most common method of grooving pipe used in sprinkler fitting work is the cut groove method.
    a. True
    b. False

103RQ02.EPS

Figure 2

11. The fitting shown in *Figure 2* is a _____.
    a. cap
    b. coupling
    c. nipple
    d. bushing

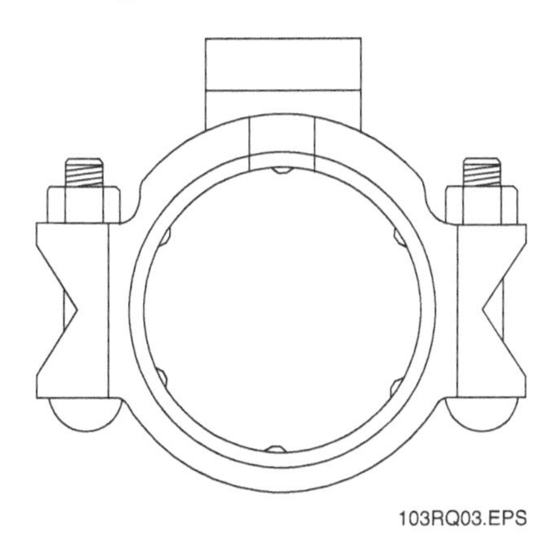

103RQ03.EPS

Figure 3

12. The component shown in *Figure 3* is a(n) _____.
    a. eccentric reducer
    b. reducing coupling
    c. bullhead tee
    d. outlet coupling

13. Flanged joints are most common on sprinkler systems with connections smaller than 2" inside diameter.
    a. True
    b. False

14. The term *takeout* refers to the _____.
    a. amount of pipe that screws into a threaded fitting
    b. space between the makeup of the fitting and the center of the fitting
    c. distance between the threaded ends of the pipes connected to the straight-through run of the tee
    d. shoulder-to-shoulder dimension of the fitting

15. The takeout dimension is the same for all grooved fittings.
    a. True
    b. False

Trade Terms Quiz

Fill in the blank with the correct trade term that you learned from your study of this module.

1. An elbow that has one male thread and one female thread is a(n) _________.
2. _________ are formed by pressing the metal around the pipe.
3. The U.S. standard for pipe threads is _________.
4. A pipe thread compound applied to the male threads on a pipe is _________.
5. A fitting with a flanged connection on one end and a threaded connection on the other end is called a(n) _________.
6. _________ makes an angle between adjacent pipes.
7. A fitting with four branches all at right angles to each other is a(n) _________.
8. A piece of pipe about two times the length of a standard thread and threaded on both ends is a(n) _________.
9. A(n) _________ is a pipe fitting that is a hollow plug with internal and external threads.
10. _________ is wrapped around the male threads of a pipe before that pipe is screwed into a fitting.
11. A horizontal or vertical supply main running through buildings and having hose valves on each floor is a(n) _________.
12. A pipe that has no groove on the end and has all imperfections removed is called _________.
13. _________ is a metallic fitting material generally used for air and water applications.
14. A(n) _________ is used to seal a joint between metal surfaces to prevent gas or liquid from entering or escaping.
15. _________ refers to pipe thicker than standard or to valves suitable for higher working pressures.
16. A fitting that displaces to one side the center line of the smaller of two pipes being joined is a(n) _________.
17. A(n) _________ is used to connect a fitting or flange directly to threaded pipe.
18. A(n) _________ is generally used in water applications and has a classification of either 125 or 250.
19. A fitting that has one side outlet 90 degrees to the run is a(n) _________.
20. A(n) _________ is a piece of pipe threaded on both ends and less than 12 inches long.
21. _________ is a term used to describe the amount of pipe that screws into a threaded fitting.
22. _________ refers to a type of cast iron in which magnesium is added to the molten gray iron to reduce brittleness.
23. A term used to describe the face-to-face dimension of a fitting is _________.
24. A plate of metal machined to hold other parts in place, to afford a bearing surface, or to provide a means for overcoming looseness is called a(n) _________.

25. A(n) _________ is formed around pipe by cutting away the pipe material.

26. Pipe whose ends have been prepared by the manufacturer or fabrication shop is referred to as _________ pipe.

27. A reducing coupling that maintains the same center line between the two pipes that it joins is a(n) _________.

Trade Terms

Bushing
Cast-iron fitting
Close nipple
Companion flange
Concentric reducing coupling
Cross
Cut groove
Ductile
Eccentric reducing coupling
Elbow
Extra heavy
Flange-thread
Gasket
Gib
Liquid Teflon®
Makeup
Malleable iron
Nipple
National Pipe Thread (NPT)
Plain-end
Prepared-end
Rolled grooves
Shoulder-to-shoulder
Standpipe
Street elbow
Tee
Teflon® tape

Cornerstone of Craftsmanship

Martha Graizer

Advanced Fire Protection
System Designer

Martha won the first AFSA Apprentice Competition in 1994. Since then, she has worked with the same company, advancing from apprentice to installer to system designer.

How did you choose a career in the sprinkler fitting field?
My college had a training program where you could work with the fire department. I wanted to be a firefighter or paramedic and joined the program. My eyesight was not good enough to pass the physical without surgery. But I still wanted to be part of the industry. One of the guys in the program suggested that I look for a job in fire protection systems. I applied for a job at Advanced Fire Protection, and they hired me. Now I am helping to keep firefighters out of buildings by building sprinkler systems.

What types of training have you been through?
Most of my training has been on the job. We worked during the day learning to cut and thread pipe. We went to class one night a week. After 10,000 hours in the field and four years of classes, I was qualified as a journeyman level sprinkler fitter by the state of Washington.

When I started working in the office, I learned the design aspects of sprinkler systems: AutoCAD, fire codes, and how to submit plans. I am always learning new things. The codes are constantly changing, so you have to stay up-to-date.

What kinds of work have you done in your career?
I worked for 11 years in the field as an installer. As an installer, I mostly worked on residential projects like custom homes and apartment complexes. I did some commercial work, like small steel systems for parking garages and warehouses.

Now I am doing design work for all types of systems. I design systems, do the calculations, determine the pipe sizes, and lay out the project. I also do project management, materials takeoff, scheduling, and coordination.

What do you like about your job?
When I first started out, I enjoyed the physical and mental demands of the job. I don't like sitting still, but I also wanted to do something where I had to think.

Every day I get new challenges. I am constantly learning new things. I enjoy working on a variety of different projects. It is satisfying to see a project installed the way I designed it and know that the system went in smoothly for the fitters. Overall, it is very satisfying to know that I am protecting lives.

What factors have contributed most to your success?
Primarily, it is the willingness of the company I work for to provide good training combined with my own willingness to learn. In our company, most of us have worked together for a long time. Learning to work as part of a team is very important. I get along well with the other guys in the company.

What advice would you give to those new to the sprinkler fitter field?
Have a backup plan. After 10-15 years of climbing up and down ladders, your knees go on you. My nephew dropped out of high school to work as a carpenter because he was making good money. After a few years, he realized that if he got hurt he couldn't support himself. Bodies wear out, and you need to have other skills. Now he is going back to school to learn construction management, so he can own his own company one day.

Trade Terms Introduced in This Module

Bushing: A pipe fitting that connects a pipe with a fitting of a larger nominal size. A bushing is a hollow plug with internal and external threads to suit the different diameters.

Cast-iron fitting: A cast-iron fitting is generally used in water applications and has a classification of either 125 or 250.

Close nipple: A nipple that is about twice the length of a standard thread that is threaded from each end with no shoulder.

Companion flange: A flange that is used to connect a fitting or flange directly to threaded pipe.

Concentric reducing coupling: A reducing coupling that maintains the same center line between the two pipes that it joins.

Cross: A pipe fitting with four branches that are all at right angles to each other.

Cut groove: A groove that is formed around pipe by cutting away the pipe material.

Ductile: In reference to ductile iron, a type of cast iron in which magnesium is added to the molten gray iron to reduce brittleness.

Eccentric reducing coupling: A reducing coupling that displaces the center line of the smaller of the two joining pipes to one side.

Elbow: A fitting that makes an angle between adjacent pipes. An elbow is always 90 degrees unless another angle is stated on the fitting or in the drawing specifications. Also known as an ell.

Extra heavy: When applied to pipe, extra heavy means pipe thicker than standard pipe. When applied to fittings and valves, extra heavy indicates units suitable for higher working pressures.

Flange-thread: A fitting that provides for a threaded connection on one end and a flanged connection on the other.

Flexible drop: Flexible fitting developed to span the distance from the branch line to the sprinkler.

Gasket: A flat sheet or ring of rubber or other material used to seal a joint between metal surfaces to prevent gas or liquid from entering or escaping.

Gib: A plate of metal or other material machined to hold other parts in place, to afford a bearing surface, or to provide a means for overcoming looseness.

Liquid Teflon®: A pipe thread compound that is applied to the male threads on a pipe to serve as a lubricant and sealant.

Makeup: A term used to describe the amount of pipe that screws into a threaded fitting.

Malleable iron: A metallic fitting material that is generally used for air and water applications and has a classification of either 150 or 300.

Nipple: A piece of pipe that is threaded on both ends and is less than 12" in length. Any pipe over 12" is referred to as cut pipe.

National Pipe Thread (NPT): The U.S. standard for pipe threads. NPT has a 1⁄16" taper per inch from back to front.

Plain-end: Pipe that has no groove on the end and has all imperfections removed.

Prepared-end: Pipe whose ends have been prepared by the manufacturer or fabrication shop.

Rolled grooves: Grooves that are formed by pressing the metal around the pipe.

Shoulder-to-shoulder: A term used to describe the face-to-face dimension of a fitting.

Standpipe: A horizontal or vertical supply main running through buildings and having hose valves on each floor.

Street elbow: An elbow that has one male thread and one female thread.

Tee: A fitting that has one side outlet 90 degrees to the run.

Teflon® tape: Tape that is made of Teflon® that is wrapped around the male threads of a pipe before the pipe is screwed into a fitting. Teflon® tape serves as both a lubricant and a sealant.

Additional Resources

This module presents thorough resources for task training. The following resource material is suggested for further study.

FM Global Approval Guide, Latest Edition. Norwood MA: Factory Mutual.

NFPA 13, Standard for the Installation of Sprinkler Systems, Latest Edition. Quincy, MA: National Fire Protection Association.

Standard USAS (ASME) B1.20.1. New York, NY: American National Standards Institute Inc.

The Handbook of Steel Pipe, Latest Edition. Washington, DC: American Iron and Steel Institute.

The Pipefitters Blue Book, 2002. W.V. Graves. Webster, TX: Graves Publishing Company.

Underwriters Laboratories Fire Protection Equipment Directory. Latest Edition. Northbrook, IL: Underwriters Laboratories.

Figure Credits

FlexHead Industries, Inc., Module opener, 103F29

National Fire Protection Association, 103T01

Reprinted with permission from *NFPA 13-2013, Installation of Sprinkler Systems,* Copyright © 2012, National Fire Protection Association, Quincy, MA 02169. The reprinted material is not the complete and official position of the NFPA on the referenced material, which is represented only by the standard in its entirety.

Topaz Publications, Inc., 103SA01, 103F14, 103F15, 103SA02, 103SA03

Ridge Tool Co./Ridgid®, 103F03-103F06, 103F07 (photo), 103F08, 103F33 (portable in-air groover)

Coleman Cable, Inc., 103F09

Koike Aronson, Inc., 103F10 (table roller)

Sumner Manufacturing Company, Inc., 103F10 (roller head jack, V-head jack, floor stand roller)

Mueller/B&K Industries, Inc., 103F17, 103F19, 103F22, 103F28

Anvil International, Inc., 103F27, 103F39-103F41, 103F43, 103T08

Victaulic Company, 103F30,103F33 (power roll groover), 103F34-103F38, 103F42, 103F44, 103F59, 103F60, 103T11

Tyco Fire and Building Products, 103T05

Stockham Valves & Fittings, 103T07, 103F52

MODULE 18103-13 — ANSWERS TO REVIEW QUESTIONS

Answer		Section
1.	c	2.0.0
2.	c	2.0.0
3.	b	2.2.1
4.	a	2.3.4
5.	d	2.6.0
6.	c	3.1.3; Figure 5
7.	a	4.1.0
8.	c	4.2.0
9.	b	5.3.0
10.	b	8.1.1
11.	b	8.4.2; Figure 36
12.	d	8.4.6; Figure 40
13.	b	9.1.0
14.	b	10.1.0
15.	b	10.7.0

MODULE 18103-13 — ANSWERS TO TRADE TERMS QUIZ

1. Street elbow
2. Rolled grooves
3. NPT
4. Liquid Teflon®
5. Flange-thread
6. Elbow
7. Cross
8. Close nipple
9. Bushing
10. Teflon® tape
11. Standpipe
12. Plain-end
13. Malleable iron
14. Gasket
15. Extra heavy
16. Eccentric reducing coupling
17. Companion flange
18. Cast-iron fitting
19. Tee
20. Nipple
21. Makeup
22. Ductile
23. Shoulder-to-shoulder
24. Gib
25. Cut groove
26. Prepared end
27. Concentric reducing coupling

NCCER CURRICULA — USER UPDATE

NCCER makes every effort to keep its textbooks up-to-date and free of technical errors. We appreciate your help in this process. If you find an error, a typographical mistake, or an inaccuracy in NCCER's curricula, please fill out this form (or a photocopy), or complete the online form at **www.nccer.org/olf**. Be sure to include the exact module ID number, page number, a detailed description, and your recommended correction. Your input will be brought to the attention of the Authoring Team. Thank you for your assistance.

Instructors – If you have an idea for improving this textbook, or have found that additional materials were necessary to teach this module effectively, please let us know so that we may present your suggestions to the Authoring Team.

NCCER Product Development and Revision
13614 Progress Blvd., Alachua, FL 32615
Email: curriculum@nccer.org
Online: www.nccer.org/olf

❑ Trainee Guide ❑ AIG ❑ Exam ❑ PowerPoints Other ______________________________

Craft / Level: Copyright Date:

Module ID Number / Title:

Section Number(s):

Description:

Recommended Correction:

Your Name:

Address:

Email: Phone:

CPVC Pipe and Fittings

18104-13

SPRINKLER FITTING LEVEL ONE

18106-13 Underground Pipe

18105-13 Copper Tube Systems

18104-13 CPVC Pipe and Fittings

18103-13 Steel Pipe

18102-13 Introduction to Components and Systems

18101-13 Orientation to the Trade

Core Curriculum: Introductory Craft Skills

This course map shows all of the modules in *Sprinkler Fitting Level One*. The suggested training order begins at the bottom and proceeds up. Skill levels increase as you advance on the course map. The local Training Program Sponsor may adjust the training order.

Objectives

When you have completed this module, you will be able to do the following:

1. Follow basic safety precautions for preparing and installing CPVC pipe.
2. Recognize chemical compatibility issues when joining CPVC to other materials.
3. Identify approved types of CPVC pipe and fittings.
4. Recognize tools for cutting and chamfering CPVC pipe.
5. Calculate takeouts.
6. Set up equipment.
7. Cut, chamfer, and clean CPVC pipe.
8. Properly prepare pipe ends.
9. Join and cure CPVC pipe.

Trade Terms

Ambient temperature
Chamfering
Chlorinated polyvinyl chloride (CPVC)
CPVC compound
CPVC resin
Escutcheon
Iron pipe size (IPS)
Standard dimensional ratio (SDR)
Takeout
Thermoplastic
Transverse

Prerequisites

Before you begin this module, it is recommended that you successfully complete *Core Curriculum* and *Sprinkler Fitting Level One*, Modules 18101-13 through 18103-13.

NOTE

Unless otherwise specified, references in parentheses following figure and table numbers refer to *NFPA 13*.

Special Caution

CPVC can be damaged by contact with chemicals found in some construction products. Reasonable care needs to be taken to ensure that products such as cutting oil, pipe dope, and plumbing tape coming into contact with CPVC systems are chemically compatible. Refer to the *Appendix* for general notes on CPVC chemical compatibility, and for a list of products that must not be used with CPVC pipe and fittings.

There are several different manufacturers of CPVC pipe and fittings. Although materials from various manufacturers may be dimensionally equivalent, listings and manufacturer's installation instructions may vary. Care should be exercised before mixing different manufacturer's materials.

Contents

Topics to be presented in this module include:

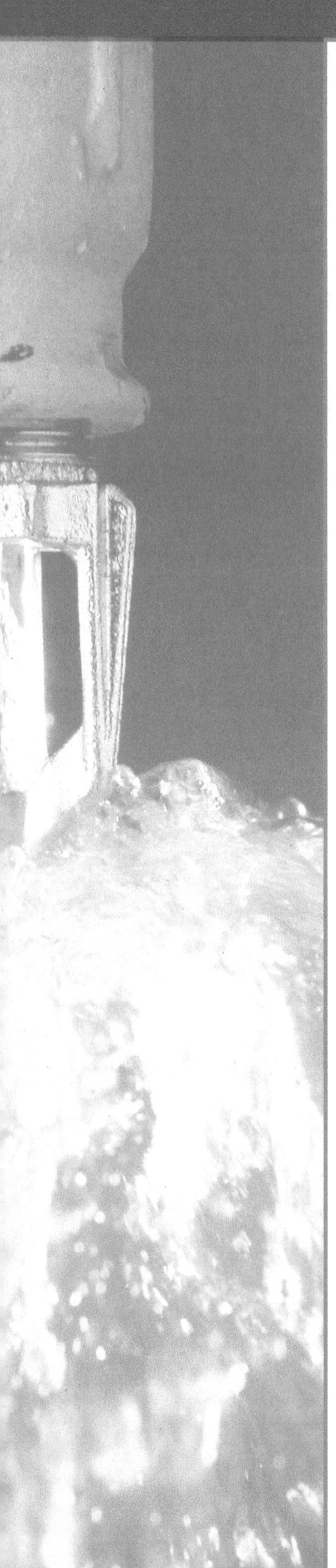

Figures and Tables

1.0.0 Introduction

Chlorinated polyvinyl chloride (CPVC) pipe is light, flexible, tough, and corrosion-resistant. With its ability to withstand higher temperatures than PVC applications, CPVC can be found in fire sprinkler systems, hot water systems, hot acid distribution, and waste systems.

CPVC systems are easier and quicker to install than metal systems. Because no torch is needed to join pipe and fittings, there is no fire hazard like there is in sweating or brazing copper. Although CPVC systems have many advantages, they do require careful attention to installation techniques and close adherence to NFPA code requirements and guidelines.

2.0.0 CPVC Pipe

CPVC is known chemically as post-chlorinated polyvinyl chloride. In 1960, the B.F. Goodrich Chemical Corporation (now known as Lubrizol Advanced Materials) introduced a family of specialty thermoplastics. After careful study of fire sprinkler engineering requirements, a special type of CPVC was developed and Listed specifically for fire sprinkler systems. CPVC has been widely accepted for its cleanliness and lightweight handling characteristics, compatibility with potable water systems, economic installation costs, superior flow characteristics, and proven solvent-cement joining method.

CAUTION

CPVC pipe, fittings, and cement must be Listed for use in fire sprinkler systems.

2.1.0 CPVC Characteristics

CPVC resin is produced from white PVC resin. Then, ingredients are added to the white CPVC resin to produce an orange CPVC compound. CPVC compounds come in two forms (pellets and powder). Pellets are used in the process of injection-molding fittings; powder is extruded into pipe.

CPVC pipe for sprinkler systems is manufactured in a standard dimensional ratio (SDR) of 13.5 for all diameters in accordance with *ASTM Standard F-442*. SDR means the pipe wall thickness is directly proportional to the outside diameter. For SDR 13.5 pipe, it means that the outside diameter of the pipe is 13.5 times the wall thickness. Therefore, all diameters of SDR 13.5 pipe have the same pressure-bearing capability (175 psi at 150°F).

SDR 13.5 CPVC pipe has a larger inside diameter than Schedule 40 steel pipe. *Figure 1* shows a comparison of the inside diameters of Schedule 40 steel pipe and SDR 13.5 CPVC pipe.

CPVC typically has less hydraulic friction loss than steel pipe. Because of the larger inside diameter and the inside smoothness, it may be possible to use a smaller diameter CPVC pipe than steel pipe for the same job. CPVC pipe is considered a rigid pipe. The pipe weighs approximately one-sixth that of steel and approximately one-half that of copper. *Table 1* gives a weight comparison of CPVC pipe versus metal pipe.

CPVC pipe and fittings are rated for a system design operating pressure of 175 psi at a temperature of 150°F. In accordance with the listing requirement for a system to have a safety factor of five, CPVC pipe and fittings have a required

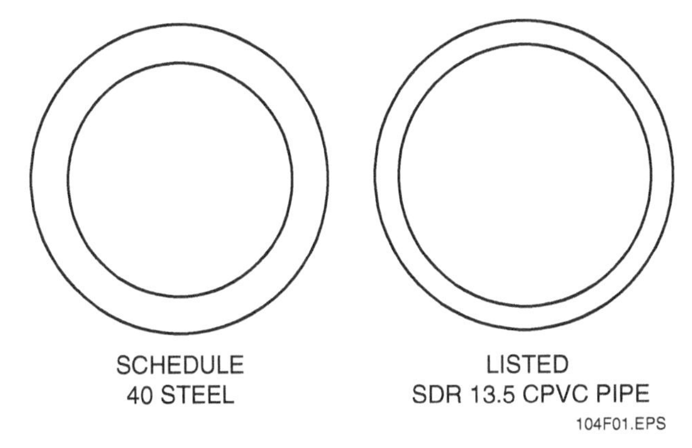

Figure 1 ID comparison.

Did You Know?

The city of Chicago was the only major metropolitan area of the United States that did not allow CPVC for sprinklers. In 2005, that changed. The city of Chicago can now use CPVC fire protection systems in its high-rise retrofits. This marks the first time in Chicago's history that commercial buildings can have installed alternatives to metal fire sprinkler pipe.

A previous section in the *Chicago Building Code* required that only metal pipe be used in fire sprinkler installations. In 2005, the city council approved the code allowing pipe, including CPVC, which meets or exceeds the requirements of *NFPA 13*. The many advantages of CPVC systems are being recognized.

Table 1 Weight Comparison of CPVC Versus Metal Pipe

Nominal Pipe Size	Pounds per Foot		
	CPVC SDR 13.5	Steel Pipe Schedule 40	Copper Tube Type M
¾"	0.17	NA	0.33
1"	0.26	1.68	0.47
1¼"	0.42	2.27	0.68
1½"	0.55	2.72	0.94
2"	0.86	3.65	1.46
2½"	1.26	5.79	2.03
2"	1.87	7.58	2.68

104T01.EPS

minimum quick burst strength of 5 times the rated pressure or 875 psi. Since the design operating pressure of CPVC pipe is 175 psi, it might appear that testing at pressures above 175 psi is not allowed. This is not the case. Test pressures of 200 psi, or a system test that requires a test pressure of 50 pounds greater than operating pressure (225 psi), are not restricted by the 175 psi limitation. Such system tests may be conducted without concern.

2.2.0 Listings for CPVC Pipe

CPVC pipe and fittings are intended for use in sprinkler systems in various types of occupancies. The following are listings for CPVC pipe and fittings based on occupancy.

- Light hazard occupancies as defined in *NFPA 13, Standard for the Installation of Sprinkler Systems*
- Residential occupancies as defined in *NFPA 13R, Standard for the Installation of Sprinkler Systems in Residential Occupancies up to and Including Four Stories in Height*
- Residential occupancies as defined in *NFPA 13D, Standard for the Installation of Sprinkler Systems in One- and Two-Family Dwellings and Manufactured Homes*

Did You Know?

The first automatic fire extinguishing system on record was patented in England in 1723 and consisted of a cask of water, a chamber of gunpowder, and a system of fuses. In about 1852, the perforated-pipe system represented the first form of a sprinkler system used in the United States. In 1874, Henry S. Parmelee of New Haven, Connecticut, patented the first practical automatic sprinkler.

- Return air plenum installations as described in *NFPA 90A, Standard for the Installation of Air Conditioning and Ventilating Systems*
- Underground installations using pipe ranging in size from ¾" to 3" installed in accordance with *ASTM D2774* and *NFPA 13*
- CPVC sprinkler pipe and fittings certified by NSF (The Public Health and Safety Company) for potable water usage

FM Global approvals also permit the use of CPVC pipe and fittings. Specific approvals vary among pipe and fitting manufacturers. Always follow the pipe and/or fitting manufacturer's installation instructions.

CPVC sprinkler pipe and fittings from different manufacturers should not be combined unless their combination has been evaluated by their listing. Consult the pipe and/or fitting manufacturer's installation instructions for requirements.

NOTE

If you run out of material on a job, do not go to the local store for replacements. Check with your supervisor for the specific job requirements.

2.3.0 Handling and Storage of CPVC

Most CPVC fittings (*Figure* 2) and pipe are packaged for ease of handling and storage, minimizing the potential damage to pipe and fittings during transit and storage. CPVC pipe and fittings have specific handling and storage requirements.

2.3.1 Handling of CPVC

CPVC piping products have a lower impact strength than metal piping products. Pipe fit-

104F02.EPS

Figure 2 CPVC fittings.

tings, cartoned or loose, must never be tossed or thrown to the ground. Pipe must never be dropped or dragged on the ground (such as when unloaded from a truck) and must remain boxed until ready to use. Heavy or sharp objects must not be thrown into or against CPVC pipe or fittings. When handling CPVC pipe, ensure that the pipe is well supported so that sagging is minimized. Very cold weather makes CPVC pipe and fittings brittle. Use extra care during handling to prevent damage.

Pipe and fittings must always be inspected for damage before actual installation. Cracks, splits, or scratches can weaken or damage the pipe and fittings. Pipe and fittings with cuts, gouges, scratches, splits, or other signs of damage from improper handling or storage must not be used. Damaged sections on lengths of pipe can easily be cut out using proper techniques for cutting CPVC pipe.

2.3.2 Storage of CPVC

CPVC pipe must be covered with a non-transparent material when stored outside for extended periods of time. Brief exposure to direct sunlight on the job site may result in color fade, but will not affect physical properties. Store CPVC pipe and fittings in their original containers to keep them free from dirt and to reduce the possibility of damage.

When storing CPVC pipe and fittings inside, keep them in a well-ventilated area, away from steam lines or other types of heat sources. To prevent damage, always store CPVC pipe and fittings in their original packaging until needed.

Store pipe on a clean, flat surface that provides an even support for the entire length of the pipe. When palletized pipe is stored, ensure that the wooden pallet bracings are in full contact with each other. Store loose pipe in packaging away from previously used CPVC pipe. Racks used for storing pipe must have continuous or close support-arms to prevent the pipe from sagging. Store CPVC pipe fittings in their original cartons on pallets. Wrap cartons with thin plastic sheeting to prevent moisture from causing them to collapse. Never mix pipe fittings in storage bins with metal fittings. Take special care to avoid contamination of CPVC pipe and fittings with petroleum-based products, such as cutting or packing oils that may be present on metallic system components.

2.4.0 CPVC Protection

CPVC pipe is approved for use in wet systems. A wet pipe system contains water and is connected to a water supply so that the water discharges immediately when the sprinkler is opened. CPVC pipe and fittings may be installed in both concealed and exposed sprinkler systems.

All piping systems should be tested for leaks with water following installation. Pipe and fitting manufacturers do not recommend air testing. They cannot be held liable for any injuries occurring during the air testing of their product, as noted in the manufacturer's literature.

WARNING

Never use air or compressed gas for pressure testing unless specifically permitted by the pipe and the fitting manufacturer. Because air is compressible, it can store much more energy than water. That stored energy can lead to an explosion, particularly if too much air pressure is used.

2.4.1 Concealed Installations

Underwriters Laboratories® requires that, for concealed installations, the minimum protection between the protected area and the piping must consist of one layer of ½" plywood soffit, ⅜" gypsum wallboard, or a suspended (membrane) ceiling with lay-in panels or tiles having a weight of not less than 0.35 pound per square foot when installed with metal support grids. Most commercial ceiling tiles weigh at least 0.70 pound per square foot. For residential occupancies, as defined by *NFPA 13R* and *NFPA 13D*, protection may also be ½" plywood.

2.4.2 Exposed Installations

UL has also Listed CPVC pipe for use in exposed installations in residential occupancies as defined in *NFPA Standards 13D* and *13R*, and in light hazard occupancies as defined in *NFPA Standard 13*. The following restrictions apply for an exposed installation:

- The system must be installed beneath a smooth, flat, horizontal ceiling.
- The system must use Listed quick-response sprinklers with deflectors installed within 8" from the ceiling, or Listed residential sprinklers installed according to their listing, using a maximum spacing between sprinklers not to exceed 15'.
- The system must use Listed quick-response horizontal sidewall sprinklers that have deflectors installed within 6" of the ceiling and within 4" of the sidewall, or Listed residential horizontal sidewall sprinklers installed according to their listing, using a maximum spacing between sprinklers not to exceed 14'.

- Exposed pipe employing Listed quick-response upright heads must be installed not more than 7½" from the ceiling to the center line of the pipe. Upright heads must be installed with the deflectors within 4" of the ceiling, and a maximum distance between sprinklers not to exceed 15'. For upright heads, the maximum distance from the center line of a sprinkler to a hanger is 3".
- Some manufacturers have additional listings for exposed installations. For example, BlazeMaster® pipe and fittings are Listed for use exposed in unfinished basements using solid wood joist construction, in exposed *13R* and *13D* risers, and with extended coverage sprinklers. Refer to the manufacturer's installation for listings and limitations.

In general, CPVC sprinkler pipe and fittings are not approved for installation in combustible concealed spaces requiring sprinklers, as referenced in *NFPA 13*. However, there are sprinklers that have been tested and are Specially Listed for use with CPVC sprinkler pipe and fittings in combustible concealed spaces requiring sprinklers, such as attics and open-joist wood construction.

CAUTION

The listings do not apply to CPVC pipe and fittings from all manufacturers. Consult the sprinkler installation instructions for details and limitations.

NFPA 13R and *13D* also permit the omission of sprinklers from combustible concealed spaces, and CPVC pipe and fittings can be installed in these areas when installing sprinklers in residential occupancies according to specific standards. CPVC pipe and fittings may be suitable for use in areas where ambient temperatures are above freezing, and do not exceed 150°F. CPVC pipe can be installed in areas where the temperatures will exceed 150°F only if ventilation is provided, or insulation is used around the pipe to maintain a pipe temperature of 150°F or lower. Protection from freezing must also be provided.

CAUTION

The only allowed antifreeze solution is glycerin. It must be installed in accordance with *NFPA 13*.

In attic spaces that require sprinkler systems, CPVC piping must be protected. The Authority Having Jurisdiction (AHJ) must be consulted prior to any installation of CPVC in attic spaces requiring sprinklers. Protection methods and requirements may vary by jurisdiction and are subject to interpretation.

CPVC in return-air plenums can be installed flush with the edge of the opening, but must not be positioned directly over open ventilation grills.

CPVC will not burn without a significant flame source, and once that flame source is removed, CPVC will not continue to burn. Fire sprinkler pipe tested and Listed in accordance with *UL 1887, Fire Test of Plastic Sprinkler Pipe for Flame and Smoke Characteristics*, meets the requirements of *NFPA 90A* for installation in return-air plenums. Consult the manufacturer's installation instructions for requirements.

Before penetrating fire walls and partitions, consult building codes and the AHJ in your area. CPVC systems must be designed and installed so that the piping is not exposed to excessive temperatures from specific heat-producing sources such as light fixtures, ballasts, transformers for fluorescent lights, and steam lines.

NOTE

There is no exact minimum distance that CPVC pipe and fittings must be installed away from heat sources. Minimum distances are a function of the specific heat producing source, the maximum ambient temperature, heat shielding, if any, and the proximity of CPVC piping systems to the heat sources. Please consult the manufacturer for questions regarding specific heat sources and recommended CPVC location.

During periods of remodeling and renovation, appropriate steps must be taken to protect the piping from fire exposure if the ceiling is temporarily removed. Consult the AHJ.

CPVC pipe and other Listed fittings can be protected from freezing using glycerin antifreeze solutions only, as outlined by *NFPA 13*. Consult the local AHJ before using any antifreeze solutions in fire sprinkler applications.

NOTE

An antifreeze solution must be prepared with a freezing point below the expected minimum temperature for the locality. The antifreeze solution must be checked each year prior to the start of freezing weather. The use of glycol-based antifreeze solutions in CPVC fire sprinkler systems is specifically prohibited.

2.5.0 CPVC Fittings

CPVC pipe and fittings are manufactured in iron pipe sizes (IPS) only. Fittings in diameters of 3/4" to 1 1/4" are produced to Schedule 40 dimensions (*ASTM F-438*). In diameters of 1 1/2", 2", 2 1/2", and 3", the fittings are produced to Schedule 80 dimensions (*ASTM 437*).

2.6.0 CPVC Takeouts

The takeout is the distance from one pipe stop in a fitting to the center line of the fitting. Takeouts may vary with different manufacturers. Check your source of materials before cutting pipe. *Table 2* shows typical takeouts for CPVC fittings.

> **CAUTION**
>
> The values shown in *Table 2* are to the nearest 1/16". Verify takeout dimensions prior to cutting pipe.

3.0.0 Cutting, Chamfering, and Cleaning CPVC Pipe

CPVC pipe must be properly cut, chamfered, and cleaned prior to being joined. The specific functions in joining CPVC pipe are:

- Cut
- Chamfer
- Clean
- Check dry fit

The following sections describe the tools and methods used to cut, chamfer, clean, and join CPVC pipe. *Figure 3* shows typical tools and supplies used when cutting, chamfering, and cleaning CPVC pipe.

Table 2 Typical Takeouts for CPVC Fittings

	3/4	1	1 1/4	1 1/2	2	2 1/2	3
Tee	1 1/8	1 7/8	1 13/16	2 1/16	2 1/2	3 1/8	3 3/4
90° ELL	9/16	3/4	15/16	1	1 1/4	1 9/16	1 7/8
45° ELL	5/16	5/16	3/8	7/16	11/16	3/4	13/16
Coupling	1/16	1/16	1/16	1/16	1/16	3/16	3/16
Cap	3/8	7/16	1/2	5/8	1 1/2	1 3/4	1 7/8
Flange	1/16	1/16	1/16	1/8	3/16	3/16	3/16
Male Adapter	9/16	11/16	3/4	3/4	13/16	–	–
Sprinkler Head Adapter (Socket)	9/16	9/16	–	–	–	–	–
Sprinkler Head Adapter (Spigot)	9/16	9/16	–	–	–	–	–
Sprinkler Head Adapter Tee	1	1	–	–	–	–	–
Reducing Tee	–	1 7/16	1 13/16	–	–	–	–
Sprinkler Head Adapter 90° ELL	1/2	1/2	–	–	–	–	–

NOTE: All dimensions in inches.

104T02.EPS

3.1.0 Cutting CPVC

CPVC pipe is cut with a variety of tools (*Figure 4*). Regardless of which tool is used, it is extremely important that a square cut be made. This is critical to making a good solvent-cement connection. If, after making the cut, there is any indication of

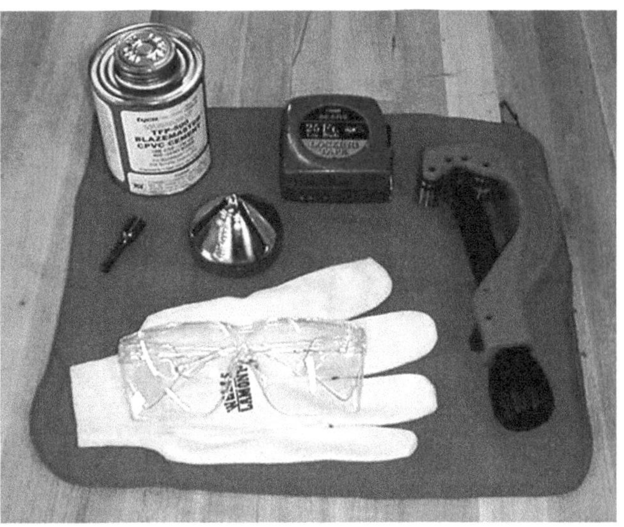

104F03.EPS

Figure 3 CPVC tools and supplies.

> **Did You Know?**
>
> In addition to the standard text, annexes, and formal interpretations, part one of the *Automatic Sprinkler Handbook* includes commentary that provides the history and other background information for specific paragraphs in the *NFPA 13* standard. This insightful commentary takes the reader behind the scenes, into the reasons underlying the requirements.

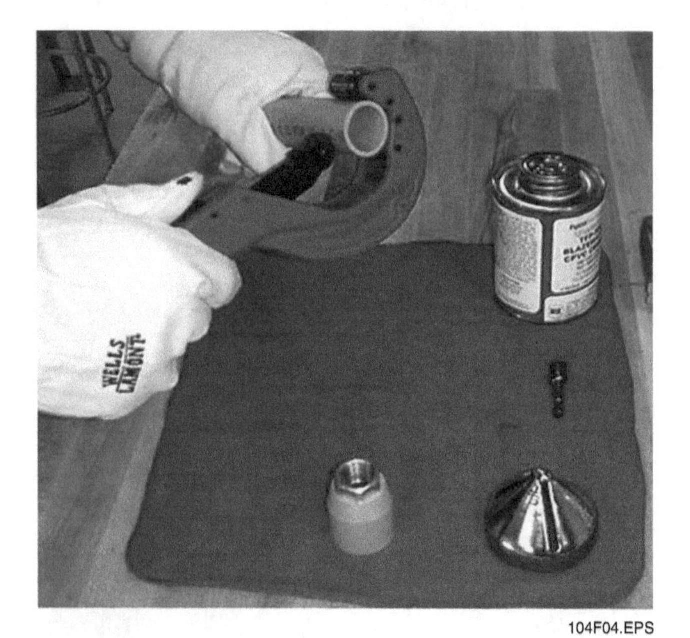

104F04.EPS

Figure 4 Cutting CPVC pipe.

damage or cracking at the pipe end, cut off at least 2" beyond any visible crack. The following sections describe tools that are used for cutting CPVC pipe.

> **NOTE**
>
> The correct cutting wheel for CPVC pipe must be used.

3.1.1 Tubing Cutters

Some tubing cutters have rollers that are tightened to force the pipe against the cutting wheel. Other models have a sliding cutting wheel, or a cutting blade that is forced against the pipe. All tubing cutters must be rotated completely around the pipe to make the cut. Tubing cutters are sized according to pipe capacity. *Figure 5* shows tubing cutters.

3.1.2 Ratchet Shears

Ratchet shears (*Figure 5*) are designed for smooth, one-hand ratchet cutting action. It is important to keep the shear blades sharp, and never use ratchet shears with chipped or broken blades. Faulty shears may weaken or damage the pipe. Generally, ratchet shears are used on pipe sizes of $^{3}/_{4}$" and 1" when temperatures are above 65°F.

3.1.3 CPVC Pipe Saw

CPVC pipe saws work on the same principle as all other hand saws. The tooth design on the CPVC pipe saw allows cutting CPVC pipe evenly in sizes as large as 4" pipe. When using a CPVC pipe saw, use a miter box to ensure a square cut. *Figure 6* shows CPVC pipe saws.

3.1.4 Chop Saw

Chop saws may also be used for cutting CPVC pipe. Chop saws are electrically driven saws in a pull-down frame, available from many manufac-

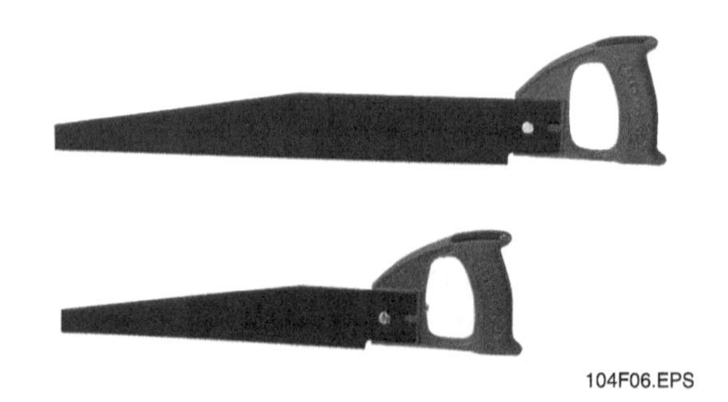

104F06.EPS

Figure 6 CPVC pipe saws.

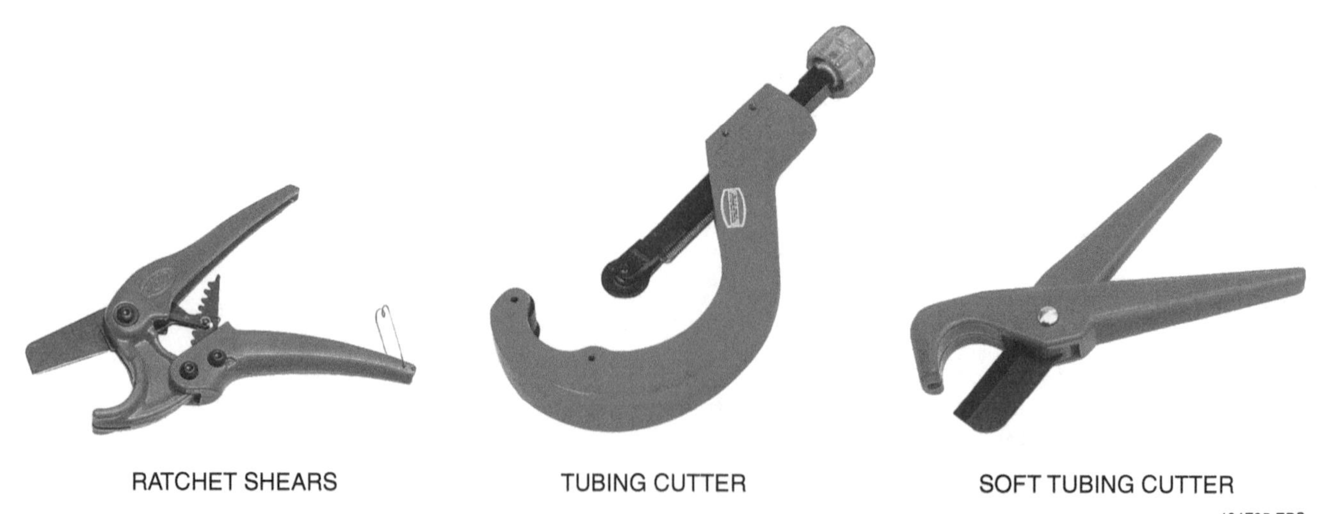

104F05.EPS

Figure 5 Tubing cutters.

turers. The saw blade and speed must be matched to the pipe being cut. Proper safety precautions must be taken to avoid personal injury. *Figure 7* shows a chop saw.

3.2.0 Chamfering CPVC

CPVC pipe must be deburred and chamfered after it has been cut (*Figure 8*). Burrs can prevent proper contact between the pipe and fitting during assembly. Particles in the solvent-cement may result in leaking. Chamfering allows the pipe to slide into the fitting easier, and prevents the pipe edge from wiping the solvent-cement from the fitting. *Figure 9* shows tools that deburr and chamfer at the same time.

3.3.0 Cleaning CPVC

Clean pipe and fittings are important for a proper seal. The deburred and chamfered pipe must be wiped clean of all foreign material. A clean, dry rag should be used to wipe loose material and moisture from the fitting socket and the pipe end. After cleaning, you are ready to begin the joining process. Because the solvent-cement joining process is a chemical reaction, care must be taken to avoid any contamination. Faulty joints caused by contamination may not show up for weeks.

Figure 7 Chop saw.

After the pipe end and fitting socket have been cleaned, check the dry fit. The pipe should enter the fitting socket easily ¼ to ¾ of the way. If the pipe bottoms (slides in) with little interference, another fitting must be used.

4.0.0 Joining CPVC Pipe

CPVC pipe is joined by a one-step solvent-cement process. Plan ahead by estimating the amount of solvent-cement needed for the job. Count the

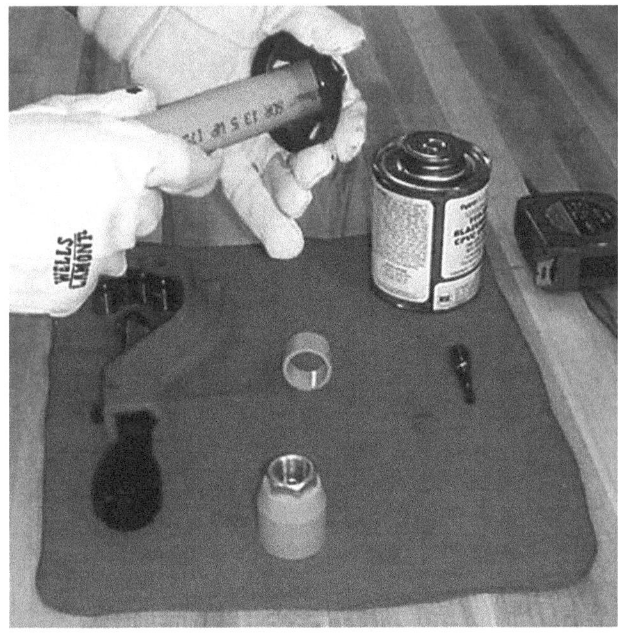

Figure 8 Deburring CPVC pipe.

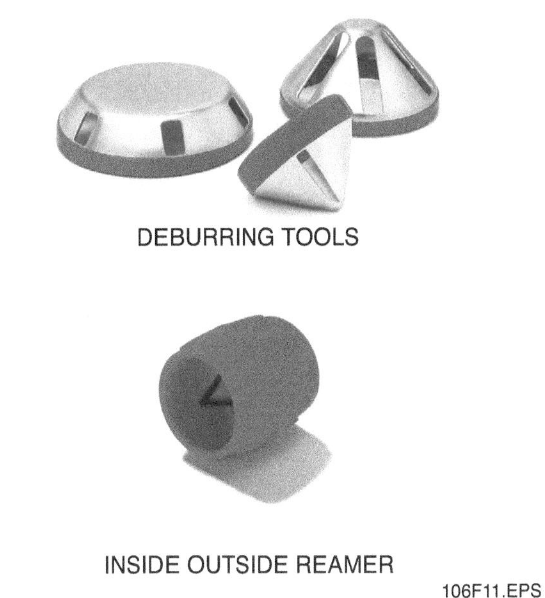

Figure 9 Deburring tools.

number of joints to be made, and divide the total by the joints-per-quart figures in the manufacturer's data for the solvent-cement in use. This gives the number of quarts of solvent-cement needed for the job. *Table 3* shows the number of joints-per-quart of solvent-cement for various fittings.

4.1.0 Solvent-Cement Safety Precautions

Prior to using CPVC solvent-cements, read the directions and take precautions found on the container labels, material safety data sheets (MSDSs), and *ASTM F402, Standard Practice for Safe Handling of Solvent-Cements, Primers, and Cleaners Used for Joining Thermoplastic Pipe and Fittings*. Solvent-cements contain volatile solvents that evaporate rapidly and create fumes. Always provide proper ventilation and avoid breathing the vapors. If necessary, use a fan to keep the work area clear of fumes. Avoid skin or eye contact with the solvent-cement and keep the container closed when not in use. If the solvent-cement thickens beyond its original consistency, discard it. Do not attempt to dilute it with primer or thinner, as this may change the characteristics of the solvent-cement and make it ineffective.

CAUTION

CPVC solvent-cement has a limited shelf life of approximately one to two years. Do not use it beyond the period recommended by the manufacturer.

Before applying solvent-cement, observe all appropriate safety precautions:

- Review the MSDS for the solvent-cement.
- Store the solvent-cement between 40°F and 100°F and keep it away from direct sunlight.
- Eliminate all ignition sources and do not smoke when using solvent-cement.
- Avoid breathing vapors.
- Use only with adequate ventilation. Explosion-proof general mechanical ventilation or local exhaust is recommended to maintain vapor concentrations below the recommended exposure limits.

On Site

Chamfering

Some people think that a bevel and a chamfer are the same thing. They are not. When a pipe is chamfered, the end is beveled as well as squared to ensure a good joint for welding or seating in a fitting.

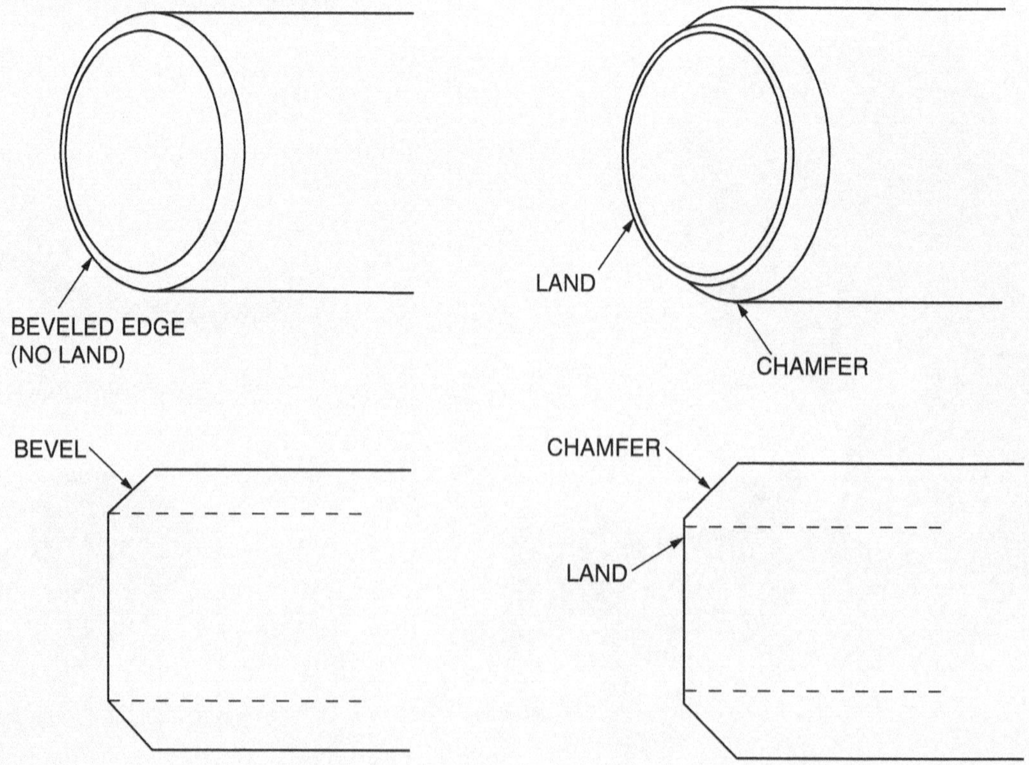

Table 3 Joints-Per-Quart of Solvent-Cement

Listed Pipe SDR 13.5 (*ASTM F442*) Estimating One Step Cement Requirements	
Fitting Size (inches)	**Solvent Cement (joints/quart)**
¾	270.0
1	180.0
1¼	130.0
1½	100.0
2	70.0
2½	50.0
3	40.0

104T03.EPS

- In confined or partially enclosed areas, use a National Institute for Occupational Safety and Health (NIOSH)-approved organic vapor cartridge respirator with a full face piece.
- Keep containers of solvent-cement tightly closed when not in use, and cover them as much as possible when they are in use.
- Avoid frequent contact with skin. Wear appropriate protective clothing and an impervious apron.
- Avoid any contact with eyes. Wear splash-proof chemical goggles. For further information, refer to the MSDS for the solvent-cement in use.

WARNING

Solvent-cement contains volatile solvents that evaporate rapidly and create fumes. Provide proper ventilation and avoid breathing the vapors. Avoid skin or eye contact with the solvent-cement and keep the container closed when not in use.

4.2.0 Solvent-Cement Application

The solvent-cement used in the one-step method is manufactured to be used without primer. Always refer to the manufacturer's instructions before using solvent-cement. Joining surfaces have to be penetrated and softened (see *Figure 10*). Apply a heavy, even coat of solvent-cement to the outside of the pipe end, and a medium coat of solvent-cement to the fitting socket. A second application on the pipe end is required for pipe sizes 1¼" and above.

The proper sequence is to apply solvent-cement on the pipe end, then the fitting socket, and then on the pipe again. It is important that you ensure sufficient penetration of the solvent-cement into the pipe and the fitting surfaces. Do this by wiping the solvent-cement with a dauber until the pipe markings have been removed (*Figure 11*). Usually three to five rotations around the pipe with the dauber are sufficient to achieve proper softening.

WARNING

Before you use CPVC solvent-cements, review and follow all precautions found on the container labels, material safety data sheets, and *ASTM F402*.

4.3.0 Assembly

Insert the pipe into the fitting socket immediately, while rotating the pipe one-quarter turn (*Figure 12*). Align the fitting for the installation at this time. The pipe must bottom to the stop. To ensure the initial set, hold the assembly without movement for 10 to 15 seconds. A continuous bead of cement should be evident around the pipe and fitting juncture. If there is any break in the bead

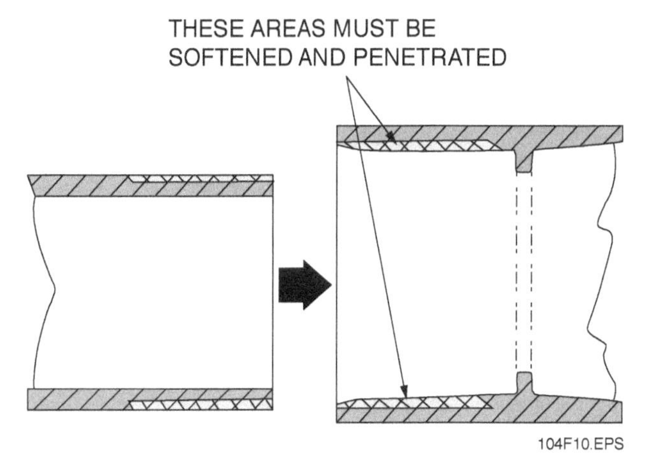

104F10.EPS

Figure 10 Area of coverage (bonding area).

104F11.EPS

Figure 11 Applying solvent cement.

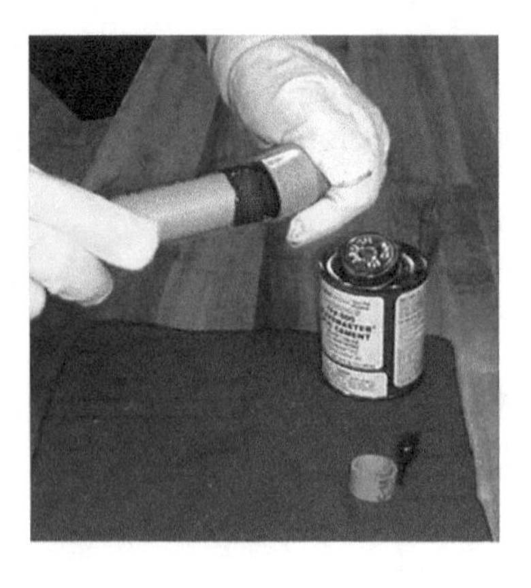

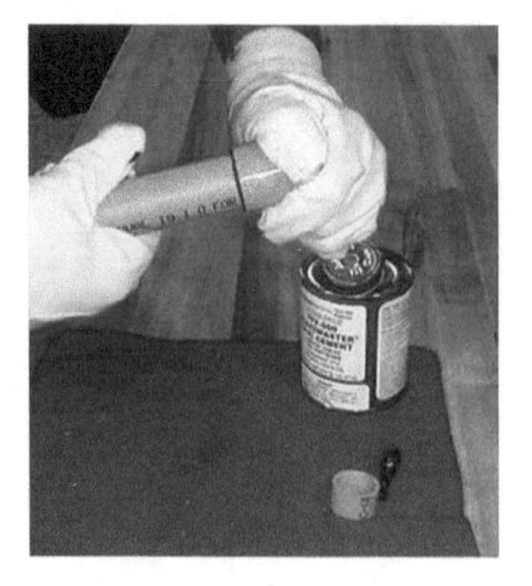

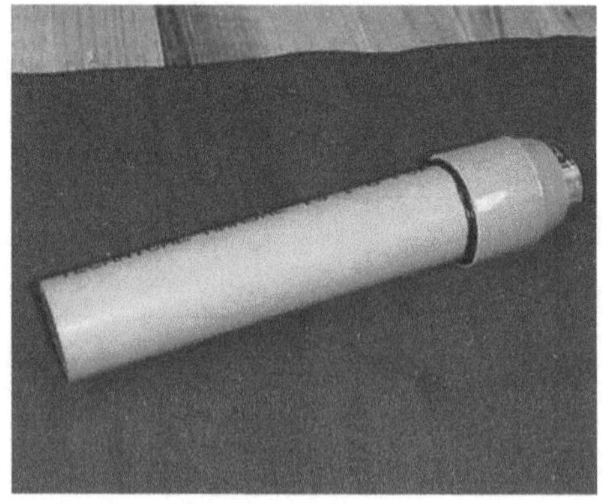

104F12.EPS

Figure 12 Assembling cemented pipe and fitting.

around the juncture, it may indicate that insufficient solvent-cement was applied. If this occurs, the fitting must be discarded and the process repeated. If there is solvent-cement in excess of the bead, it can be wiped off with a rag.

NOTE

Rotating the pipe eliminates tracks made from inserting the pipe into the fitting. Most of the joint strength is located close to the pipe stop. This is why it is important to make a square cut. The holding power is in the last ¼" of the tapered fitting.

Before installing sprinklers, let the sprinkler fittings cure for a minimum of 30 minutes. Prior to installing the sprinkler, the pipe drop must be held or anchored securely to avoid rotating the pipe in the previously cemented connections.

CAUTION

Do not install sprinklers on the head adapter fittings, and then solvent-cement the head adapter fittings to the drop. This may allow the cement to run into the sprinkler orifice and plug the sprinkler.

4.4.0 Set and Cure Times

Solvent-cement set and cure times depend on pipe size, temperature, relative humidity, tightness of fit, and the type of solvent-cement used. Drying time is faster for dry environments than wet environments. It is also faster for smaller pipe sizes, higher temperatures, and tighter fits. In the initial set, the assembly must be allowed to set, without any stress on the joint, for the recommended amount of time, depending on pipe size and temperature. The assembly can then be handled carefully, provided significant stresses to the joint are avoided. *Table 4* shows the Listed minimum cure times, prior to pressure testing, for Listed CPVC pipe and fittings using one-step solvent-cement.

CAUTION

Table 4 shows one example. Check with specific manufacturer's installation instructions. Cure times may vary.

Sprinkler fittings must be visually inspected and probed with a wooden dowel to ensure that the pipe and threads are clear of any excess solvent-cement. Once the installation is complete and cured for the specified amount of time, the system must be hydrostatically tested.

4.5.0 Joining CPVC in Cold Weather

Methods for joining CPVC in hot and cold weather are the same, but you need to consider your equipment and materials for various weather situations. The bonding of CPVC pipe and fittings is a function of temperature and time; therefore, very cold weather requires proper care to be taken. CPVC solvent-cement is approved for cold weather use down to 0°F. Very cold weather makes CPVC pipe and fittings brittle, so take extra care to prevent damage during handling.

Follow the installation instructions when working in cold weather, taking special note that solvents penetrate and soften the surfaces more slowly than in warm weather. Installation temperatures below 0°F invalidate the listing of the CPVC and solvent-cement. Surfaces below this temperature become more resistant to solvent attack.

Table 4 Cure Times

Cure Times with One Step Solvent Cement 225 psi (Maximum) Test Pressure			
Pipe Size (inches)	Ambient Temperature During Cure Period		
	60°F to 120°F	40°F to 59°F	0°F to 39°F
¾	1.0 hr.	4.0 hr.	48.0 hr.
1	1.5 hr.	4.0 hr.	48.0 hr.
1¼	3.0 hr.	32.0 hr.	10.0 days
1½	3.0 hr.	32.0 hr.	10.0 days
2	8.0 hr.	48.0 hr.	See Note 1
2½	24.0 hr.	96.0 hr.	See Note 1
3	24.0 hr.	96.0 hr.	See Note 1

NOTE 1: For these sizes, the solvent cement can be applied at temperatures below 32°F; however, the sprinkler system temperature must be raised to a temperature of 32° or above and allowed to cure per the above recommendations prior to pressure testing.

Cure Times with One Step Solvent Cement 200 psi (Maximum) Test Pressure			
Pipe Size (inches)	Ambient Temperature During Cure Period		
	60°F to 120°F	40°F to 59°F	0°F to 39°F
¾	45.0 min.	1.5 hr.	24.0 hr.
1	45.0 min.	1.5 hr.	24.0 hr.
1¼	1.5 hr.	16.0 hr.	120.0 hr.
1½	1.5 hr.	16.0 hr.	120.0 hr.
2	8.0 hr.	36.0 hr.	See Note 1
2½	8.0 hr.	72.0 hr.	See Note 1
3	8.0 hr.	72.0 hr.	See Note 1

NOTE 1: For these sizes, the solvent cement can be applied at temperatures below 32°F; however, the sprinkler system temperature must be raised to a temperature of 32° or above and allowed to cure per the above recommendations prior to pressure testing.

Cure Times with One Step Solvent Cement 100 psi (Maximum) Test Pressure			
Pipe Size (inches)	Ambient Temperature During Cure Period		
	60°F to 120°F	40°F to 59°F	0°F to 39°F
¾	15.0 min.	15.0 min.	30.0 min.
¾ CTS	15.0 min.	15.0 min.	30.0 min.
1	15.0 min.	30.0 min.	30.0 min.
1¼	15.0 min.	30.0 min.	2.0 hr.

104T04.EPS

Colder temperatures result in greater cure time due to the slower penetration of the solvent-cement. The following precautions should be observed when cementing during cold weather:

- Carefully read and follow all directions before installation.
- Prefabricate as much of the system as possible in a heated work area.
- Store the solvent-cement in a warm area when it is not in use and make sure it remains fluid.
- Take special care to remove moisture, including ice and snow.
- Allow the manufacturer's recommended cure period before the system is used.

4.6.0 Joining CPVC in Hot Weather

CPVC solvent-cements contain volatile solvents. Higher temperatures and wind accelerate evaporation. Pipe stored in direct sunlight may have surface temperatures of 20°F to 30°F above air temperatures. Solvents attack these hot surfaces more deeply and it is very important to avoid puddling the solvent-cement inside the fitting socket. Always ensure that the excess solvent-cement is wiped from the outside of the joint. Follow the standard installation instructions and observe the following precautions:

- Refer to the pipe manufacturer's information or the appropriate temperature-related expansion and contraction information.
- Store solvent-cements in a cool, shaded area prior to use.
- If possible, store pipe and fittings, or at least the ends to be joined, in a shady area before cementing.
- Make sure that both surfaces to be joined are still wet with solvent-cement when putting them together. With larger sizes of pipe, more people may be required in order to complete the application successfully.
- Carefully read and follow all directions before installation.

4.7.0 Repair and Modification to Existing CPVC Systems

The risk of extreme water damage exists when working on active wet systems. Follow the manufacturer's cut-in procedures and consult with your company's service department before doing this type of work.

Follow these cut-in procedures for system modifications or repairs to existing systems in existing buildings:

- Be sure to review the manufacturer's solvent cementing procedures and follow them to ensure that proper joining techniques are used. Always cut pipe square and to the correct length. Ensure that the pipe has been deburred and beveled and is dry. Always use the correct tools for the task.
- To obtain faster cure times before filling the system with water, begin by making the cut-in on the smallest diameter pipe, as close as possible to the area to be repaired. Completely drain all existing lines prior to applying solvent cement, as moisture increases the curing time and results in weaker joints.
- Measure and cut pipe to the proper length and dry-fit the components to ensure proper fit. Follow the manufacturer's instructions for assembling the cut-in tee, as a one-quarter turn is required when inserting pipe into the fitting, particularly on pipe of 1½" diameter and larger. You may have to assemble some components using socket unions, flanges, or grooved coupling adapters to allow the quarter-turn.
- Before applying the solvent cement, clean moisture and dirt out of fitting sockets and pipe ends with a clean, dry rag. Always use a fresh, unexpired can of solvent cement for cut-in connections.
- When you have completed the cut-in work, allow the joints to cure completely before filling the system with water. Refer to the manufacturer's tables for the curing times as determined by pipe size and ambient temperature.

> **CAUTION**
>
> One of the most frequent types of insurance claims by fire sprinkler contractors is failure of CPVC cut-ins and modifications to existing CPVC systems. These claims can be very large if the failure occurs in the upper floors of a rise building.

4.8.0 Connecting CPVC to Other Materials

CPVC can be connected to other materials using male threaded adapters, female threaded adapters, flanges, or grooved coupling adapters. Teflon® or TFE thread tape is the recommended thread compound and should be used with all threaded connections. The fitting manufacturer must be consulted before any other thread compound is used. Never use both TFE thread tape and a thread paste on the same connection. When used together, these two products create a poor connection.

> **CAUTION**
>
> When using thread sealants, lubricants, and fire stops, CPVC compatibility issues exist. Always make sure that all products coming in contact with CPVC are chemically compatible. Consult the CPVC manufacturer for guidance. See the *Appendix* for additional information.

5.0.0 Rules For Using Hangers on CPVC

Protect CPVC from abrasion by sharp edges of concrete or steel. This includes metal studs and plates. Most hangers and supports that are suitable for metal pipe systems are also acceptable for CPVC pipe assemblies. Regardless of what type of hanger is used, it must be free of burrs and rough or sharp edges. Residential occupancies can use plumbing-type hangers under *NFPA 13D*. When plumbing-type hangers are used, they must comply with local plumbing codes.

It is very important to properly place hangers and supports to prevent vertical thrust when a sprinkler discharges. Thermal expansion also must be taken into consideration. The pipe hanger or support must be located immediately adjacent to the sprinkler fitting, or the tee to the sprinkler drop. If the location of the sprinkler prevents installing a hanger or clamp adjacent to the tee, change the location of the sprinkler or provide a support for the hanger or clamp. Vertical restraints are required within 6" to 9" of the sprinkler. Consult the manufacturer's installation instructions.

5.1.0 Support Spacing for CPVC Pipe

Support spacing for CPVC pipe is different than for other piping materials because of the flexibility of the pipe itself. Different types of CPVC pipe have different requirements for support spacing.

5.1.1 Thermal Expansion

All materials expand as the temperature increases, and contract as the temperature decreases. This dimensional change causes pipe to change its length as the temperature changes. The change with CPVC is greater than with steel pipe. As a result, supporting and/or restraining

CPVC pipe is necessary with all sprinkler systems to prevent sprinklers from moving. For details on thermal expansion, including the use of expansion loops, see the manufacturer's installation guide.

5.1.2 Spacing Limitations for CPVC Pipe

Spacing between pipe hangers is determined by the pipe size. *Table 5* gives one manufacturer's recommended maximum support spacing for CPVC pipe.

When using a two-hole strap near a fitting, care should be taken not to stress the pipe. UL allows for upsizing the hanger by one size over the pipe in order to not stress the pipe because of the larger diameter fitting.

5.1.3 Restraints, Anchors, and Guides

In addition to hangers and supports, the use of restraints, anchors, and guides is required for CPVC pipe. These are used on long runs of pipe and whenever there are directional changes in the pipe. They must also be used wherever expansion joints are used.

Anchors direct the movement of the pipe within a defined reference frame. An anchor prevents vertical movement and transverse movement at the anchoring point. Guides prevent transverse movement of the pipe, but allow longitudinal movement. Restraints prevent vertical movement of the pipe.

Other materials that are commonly used for restraints, anchors, and guides are special fittings, plastic poly strapping, and U-bolt clamps. In all cases, the manufacturer's recommendations must be consulted. There are two strap hangers specifically engineered for CPVC piping. These are combination hanger-restraint devices. Both of these strap fasteners have flared edges to prevent pipe abrasion. They are made of galvanized carbon-steel sheet metal and meet the requirements of *NFPA 13, 13R,* and *13D. Figure 13* shows a two-hole strap.

When a sprinkler activates, a significant reactive force is exerted on the pipe, especially at system pressures greater than 100 psi. This reactive force will cause the pipe to lift vertically if it is not properly secured, especially if the sprinkler drop is from a small diameter pipe.

When a sprinkler drop is from a ¾", 1", or 1¼" pipe, the closest hanger should brace the pipe against the vertical lift-up. A number of techniques can be used to brace the pipe, such as using a standard band hanger to position the threaded support rod to 1⁄16" above the pipe, or using a split ring or wrap-around hanger.

When piping is suspended from a deck, hangers are required to suspend the pipe and provide vertical lift restraint. One support can serve as both. Drop locations between supports are acceptable in any location as long as support spacing requirements are followed.

When the piping is supported by wood joists or trusses, the structure provides the support, especially when the joists are close together. The only requirement with this type of construction is to provide vertical restraint. When supporting CPVC pipe below the deck and when the supporting members are spaced far apart, it is important to brace for upward restraint. Drop location supports are acceptable in any location as long as support spacing requirements are followed.

Table 5 Hanger Spacing for CPVC Pipe

Nominal Pipe Size		Maximum Support Spacing	
inches	mm	feet	meters
¾	20	5½	1.7
1	25	6	1.8
1¼	32	6½	2.0
1½	40	7	2.1
2	50	8	2.4
2½	65	9	2.7
3	80	10	3.0

104T05.EPS

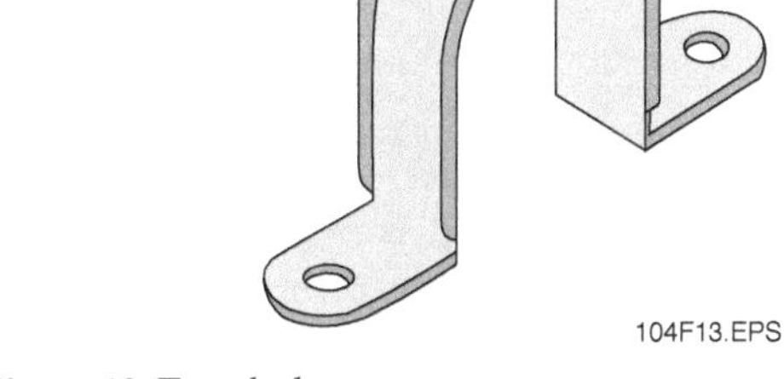

Figure 13 Two-hole strap.

5.1.4 Recommended Practices and Precautions

The following DOs and DON'Ts are general instructions typically pertaining to CPVC products.

DO:

- Install the product according to the manufacturer's installation instructions.
- Follow recommended safe work practices.
- Make certain that thread sealants, gasket lubricants, or fire stop materials are compatible with Listed CPVC.
- Use only latex-based paints if painting is desired.
- Keep pipe and fittings in original packaging until needed.
- Cover pipe and fittings with an opaque tarp if stored outdoors.
- Follow proper handling procedures.
- Use tools specifically designed for use with plastic pipe and fittings.
- Use proper solvent cement and follow application instructions.
- Use a drop cloth to protect interior finishes.
- Cut the pipe ends square.
- Deburr and bevel the pipe end before solvent cementing.
- When solvent cementing, rotate the pipe one-quarter turn when bottoming pipe in fitting socket.
- Avoid puddling of solvent cement in fittings and pipe.
- Make certain that solvent cement does not run and plug the sprinkler orifice.
- Follow the manufacturer's recommended cure times prior to pressure testing.
- Fill lines slowly and bleed the air from the system prior to pressure testing.
- Support sprinkler properly to prevent lift up of the head through the ceiling when activated.
- Keep threaded rod within 1⁄16" of the pipe.
- Install Listed CPVC pipe and fittings in wet systems only or specially Listed dry systems.
- Use only insulation and/or glycerin and water solutions for freeze protection.
- Allow for movement due to expansion and contraction.
- Renew your Listed CPVC pipe and fittings installation training every two years.

DON'T:

- Do not use edible oils such as Crisco® as a gasket lubricant.
- Do not use petroleum or solvent-based paints, sealants, lubricants or fire stop materials.
- Do not use any glycol-based solutions as an antifreeze.
- Do not mix glycerin and water solution in contaminated containers.
- Do not use both Teflon® tape and thread sealants simultaneously.
- Do not use solvent cement that exceeds its shelf life or has become discolored or gelled.
- Do not allow solvent cement to plug the sprinkler orifice.
- Do not connect rigid metal couplers to Listed CPVC grooved adapters.
- Do not thread or groove Listed CPVC pipe.
- Do not use solvent cement near sources of heat, open flame, or when smoking.
- Do not pressure test until recommended cure times are met.
- Do not use dull or broken cutting tool blades when cutting pipe.
- Do not use Listed CPVC pipe that has been stored outdoors, unprotected and is faded in color.
- Do not allow threaded rod to come in contact with the pipe.
- Do not install Listed CPVC pipe in cold weather without allowing for expansion.
- Do not install Listed CPVC pipe and fittings in dry systems, unless specially Listed for such use.

SUMMARY

CPVC pipe is becoming increasingly popular for residential and light hazard applications. Because it is lightweight, it is easier to handle than ferrous pipe material.

CPVC pipe is manufactured in a Standard Dimensional Ratio (SDR) of 13.5 for all diameters per *ASTM F-442*. While CPVC pipe does have advantages, it also has special requirements that must be followed, particularly for handling and storage. CPVC pipe is approved for wet systems and for both concealed and exposed sprinkler systems, but consult the sprinkler installation instructions for details and limitations.

CPVC has special cutting, chamfering, and cleaning requirements. The methods for joining CPVC in hot and cold weather are the same, but you need to consider your equipment and materials for various weather situations.

The manufacturer's instructions must be followed closely when using CPVC and its unique solvent-cement joining method. Proper pipe preparation, and allowing the proper amount of time for the joint to cure, are the keys to making strong, leakproof joints. Detailed installation instructions are available from the manufacturer, or from any CPVC resource material.

Review Questions

1. All diameters of SDR 13.5 pipe have pressure-bearing capacity that is _____.
   a. unequal
   b. the same
   c. standard
   d. graduated

2. CPVC pipe and fittings are rated for a system design operating pressure of 175 psi at a temperature of 150°F. Therefore, a listing requirement is a quick-burst strength of 5 times the rated pressure or _____.
   a. 525 psi
   b. 700 psi
   c. 875 psi
   d. 1,050 psi

3. Testing CPVC pipe at pressures above the CPVC pipe design operating pressure of 175 psi is *not* permitted.
   a. True
   b. False

4. Compared to metal piping, CPVC piping products have _____. This results in specific handling and storage requirements for CPVC piping.
   a. greater impact strength
   b. lower impact strength
   c. about the same impact strength
   d. significantly greater impact strength

5. When storing CPVC material, you should _____.
   a. keep it outside with or without cover
   b. use two supports for pipe, one at each end
   c. store loose pipe separately from boxed pipe
   d. store pipe and fittings in their original packaging until needed

6. CPVC pipe and fittings may be suitable for use in areas where ambient temperatures are above freezing and do not exceed _____.
   a. 100°F
   b. 150°F
   c. 200°F
   d. 250°F

7. CPVC fittings in diameters of ¾" to 1¼" are produced to _____
   a. Schedule 40 dimensions
   b. Schedule 50 dimensions
   c. Schedule 60 dimensions
   d. Schedule 80 dimensions

8. When cutting CPVC pipe, it is extremely important to make a _____.
   a. slightly beveled cut
   b. quick cut
   c. fully beveled cut
   d. square cut

9. Because the solvent-cement joining process is a chemical reaction, thorough cleaning is not necessary.
   a. True
   b. False

10. When dry-fitting CPVC pipe, the pipe should enter the socket easily ¼ to ¾ of the way. But if the pipe bottoms with little interference, which of the following may be required when making the joint?
    a. Shims
    b. A different type of solvent-cement
    c. Less solvent-cement
    d. A new fitting

11. The standard practice for handling CPVC solvent-cements is found in _____.
    a. *NFPA 13*
    b. *ASTM F204*
    c. *ASTM F402*
    d. *NFPA 13R*

12. Before applying solvent-cement, _____.
    a. observe all appropriate safety precautions
    b. install all sprinklers
    c. pressure test the system
    d. determine that the solvent-cement is not more than seven years old

13. The maximum support spacing for 1½" CPVC pipe is _____.

    a. 5'
    b. 6'
    c. 7'
    d. 8'

14. An anchor prevents _____.

    a. diagonal movement of the pipe
    b. horizontal movement of the pipe
    c. vertical movement of the pipe
    d. oblique movement of the pipe

15. If CPVC pipe is to be painted, a latex-based paint should be used.

    a. True
    b. False

Trade Terms Quiz

Fill in the blank with the correct trade term that you learned from your study of this module.

1. The distance from one pipe stop in a fitting to the centerline of the fitting is called __________.
2. __________ is a white thermoplastic material that accounts for 85 percent of the weight of finished CPVC products.
3. A common thermoplastic pipe used in the fire sprinkler industry is __________.
4. When pipe has shifted laterally, it has done a(n) __________ shift.
5. The nominal pipe size of iron pipe is referred to as __________.
6. __________ is the name for the surrounding air temperature.
7. __________ refers to plastic piping that is soft and pliable when heated, but hard when cooled.
8. __________ is an orange thermoplastic material used to produce fire sprinkler pipe and fittings.
9. Beveling and squaring the end of plastic pipe to make a clean end for a fitting is called __________.
10. The protective faceplate for sprinklers is called a(n) __________.
11. __________ is the pipe wall thickness in proportion to its outside diameter.

Trade Terms

Ambient temperature
Chamfering
Chlorinated polyvinyl chloride (CPVC)
CPVC compound
CPVC resin
Escutcheon
Iron pipe size (IPS)
Standard dimensional ratio (SDR)
Takeout
Thermoplastic
Transverse

Cornerstone of Craftsmanship

Richard Conley

Regional Service Manager
HFP Fire Protection Services, Inc.
Westfield, Massachusetts

Richard competed in the first AFSA National Apprenticeship Competition in 1994 and won the Region 7 championship. The annual convention was in Fort Lauderdale, Florida, and Richard placed second nationwide. Before he completed his apprenticeship, he was running jobs for his company. One of his long-term goals is to complete a service and repair manual for service and inspection techs that will help them to troubleshoot emergencies.

How did you choose a career in the sprinkler fitting field?
I had a background as a carpenter and construction worker/heavy machine operator; however, I was looking for more year-round work with a good benefit package as I was starting a family.

What types of training have you been through?
I started with the ASFA Wheels of Learning apprenticeship program that coincides with four years of on-the-job Sprinkler Fitter Apprentice training. I have also attended numerous seminars addressing management skills, construction foremanship skills, NFPA code changes and upgrades, residential sprinklers, fire pumps, construction equipment, and lift safety.

What kinds of work have you done in your career?
In my 15 years at HFP, I have held many different titles: apprentice, licensed journeyman, foreman, special projects foreman, service technician, field superintendent, and service manager.

I have installed mostly water-based sprinkler systems, from small drugstores to large warehouse grid systems. I've also had the opportunity to install many types of fire pumps, including horizontal split case and vertical split case, and vertical turbine pumps, both electric and diesel driven. For a couple of years, I performed emergency service work, including troubleshooting older systems with both alarm check valves and dry pipe valves.

For four years I was the superintendent of our Connecticut contracting division, overseeing a crew of 25 to 30 people. This job involved coordinating each job from start to finish with regards to manpower, tools, and quality control.

For the last five years, I have been the Connecticut service manager for our company. My duties include pricing and scheduling of deficiency repairs and emergency service calls, as well as being responsible for tenant fit-out work, residential sprinkler systems, and other small contracting jobs. My day is split between going on the road to look at work; working at the computer to do pricing, quoting, and scheduling; and overseeing of our warehouse and service department in general.

What do you like about your job?
The things I like most about being service manager is that I am directly responsible for the profitability and success of my department. It is also very satisfying to know that both my opinions and knowledge are respected and appreciated.

What factors have contributed most to your success?
Some of the factors that have contributed to my success are my willingness to perform to completion whatever task has been asked of me, my interest in knowing and studying as much about the sprinkler trade as possible, a take-charge attitude, and the luck of picking one of the fastest-growing industries in the country today.

What advice would you give to those new to the sprinkler fitter field?
Study your apprenticeship books and learn as much as you can about the trade. Be reliable and willing to work hard; you will be appreciated. Set goals that you want to attain. And the most important advice is to work safely.

CHEMICAL COMPATABILITY

CPVC domestic, water, fire sprinkler, and industrial piping systems have been used successfully for more than 45 years in new construction, re-pipe, and repair. CPVC products are ideally suited for these applications due to their outstanding corrosion resistance. Occasionally, however, CPVC and PVC can be damaged by contact with chemicals found in some construction products (and site preparations). Reasonable care needs to be taken to ensure that products coming into contact with CPVC systems are chemically compatible. Chemical compatibility with CPVC should be confirmed with the manufacturer of the product in contact with CPVC piping systems. If chemical compatibility with CPVC is in question, it is recommended to isolate the suspect product from contact with CPVC pipe or fittings.

Ongoing research conducted by industry experts has identified various products to be incompatible with CPVC systems. As products are continually tested for compatibility, new information is frequently released. Always check with the CPVC manufacturer for the latest information to determine its compatibility with products it may contact.

Some CPVC manufacturers and their contact information are given:

- Harvel Plastics; www.harvelsprinklerpipe.com.
- Lubrizol Advanced Materials; www.system-compatible.com.
- Spears Manufacturing; www.spearsmfg.com.

One CPVC manufacturer, Lubrizol, has identified the following products to be *unacceptable* when used in direct contact with Lubrizol CPVC systems (updated Feb. 11, 2010):

Caulk

- Polyseamseal® Tub & Tile Adhesive Caulk (OSI Sealants)
- Polyseamseal® All Purpose Adhesive Caulk (OSI Sealants)
- ProSeries PC-158 Caulk (OSI Sealants)
- Grabber® Acoustical Sealant GSCS (John Wagner Associates)
- AC-20® Acrylic Latex Caulk & Silicone (Pecora)
- Sikaflex™ Self-Leveling Sealant (Sika Corp.)
- SHEETROCK® Brand Acoustical Sealant (United States Gypsum)
- 3006 All Purpose Adhesive Caulk (White Lightning)

Fire Stopping Systems

- Fire Barrier 2003 Silicone Fire Barrier CP25WB+ (3M™)
- Flame Stop V (Flame Stop)
- Proseal Plug, black or red (ProSet)

Leak Detector

- Gasoila Leak Tech (Federal Process Co.)
- Masters Leak Detector (G.F. Thompson Co., Ltd.)
- Multitec Leak Detecting Spray (Unipak A/S)
- RectorSeek™ Low-Temp (RectorSeal™)

Miscellaneous

- Shockwave (Fiberlock Technologies)
- WD-40 lubricant
- Silicone pipe lubricant
- Peppermint oil
- Roofing tar
- Vaseline®
- Vegetable oils

Pipe Clamp

- Acousto®-Clamp, Acousto®-Plumb System Incompatible information is based on testing of products manufactured prior to Oct. 2007. (LSP Specialty Products)
- Naylon vinyl-coated wire pipe hangers (Naylon Products)

Pipe Tape

- Pipe Wrap Tape (Christy's)
- Pipe Wrap Tape, black (Pro Pak, Inc.)
- All Weather PVC Pipe Wrap (PASCO)
- No. 413 Pipe Wrap Tape (Wonder)

Thread Sealants

- Super Dope (Allied Rubber & Gasket Co. [ARGCO])
- TFE Paste (Anti-Seize Technology)
- SuperLock® Hi-Strength, Stud Lock Grade 2271 (Devcon)
- GS-600 (General Sealant)
- Masters® Pro-Dope® with Teflon® (G.F. Thompson Co., Ltd.)
- Brush-on/Blue Block™ (Hercules)
- Powerseal #932 (Hernon Mfg. Inc.)
- White Seal (IPS)
- Seal Unyte Thread & Gasket Sealer (JC Whitlam Mfg. Co.)
- Jet Lube V-2® (Jet Lube, Inc.)
- Tighter-than-Tite™ (Jomar)
- Threadlocker 242 (Locktite)
- Proseal (Lyn-Car Products Ltd.)
- Permabond LH-050 and LH-054 (National Starch & Chemical Permabond Division)
- Permatex® 14H (Permatex Co., Inc.)
- High Performance Teflon Thread Sealing Compound (Rule)
- Saf-T-Lok TPS Anaerobic Adhesive/Sealant, Industrial Grade TPS (Saf-T-Lok Chemical)
- SWAK® (Swagelock Co.)

Other products also raise compatibility concerns when in contact with CPVC systems. You should alert your supervisor immediately if you suspect CPVC material has come in contact with these or other questionable products:

- Acetone in primers, cleaners and solvent cements
- Antifreeze: glycerin from biodiesel
- Antimicrobial coatings
- Cleaning CPVC pipe (dishwashing liquids)
- Flexible wiring and cable (containing plasticizers)
- Fragrances – perfumes
- Fungicides and mold inhibitors
- Grease and cooking oils
- Leak detectors
- Molten solder and solder flux
- Paint (oil or solvent based)
- Polyurethane (spray-on) foams
- Residual oils (including cutting oils) with steel pipe
- Residual oils with HVAC applications
- Rubber and flexible materials containing plasticizers
- Sleeving material
- Spray-on coatings
- Steel piping with antimicrobial coating
- Teflon® tape
- Termiticides and insecticides

Again, it is imperative to contact your CPVC manufacturer to obtain the latest compatibility information, as changes are made as more tests are conducted. Always be mindful of these concerns while on the jobsite. If you suspect product contact that could compromise the integrity of CPVC systems, notify your supervisor at once.

Trade Terms Introduced in This Module

Ambient temperature: The surrounding air temperature.

Chamfering: Beveling and squaring the end of CPVC pipe to make a clean end for a fitting.

Chlorinated polyvinyl chloride (CPVC): A common thermoplastic pipe used in the fire sprinkler industry.

CPVC compound: An orange thermoplastic material used to produce fire sprinkler pipe and fittings.

CPVC resin: A white thermoplastic material that accounts for 85 percent of the weight of finished CPVC products.

Escutcheon: A protective faceplate for sprinklers.

Iron pipe size (IPS): The nominal pipe size of iron pipe.

Standard dimensional ratio (SDR): The pipe wall thickness in proportion to its outside diameter.

Takeout: The distance from one pipe stop in a fitting to the center line of the fitting.

Thermoplastic: Refers to plastic piping that is soft and pliable when heated, but hard when cooled.

Transverse: The lateral shifting of pipe.

Additional Resources

This module presents thorough resources for task training. The following resource material is suggested for further study.

BlazeMaster® Online Training Program, www.blazemastertraining.com.

Multimedia Apprenticeship Training Supplement for Fire Sprinkler Fitters (CD set: Level 1 through Level 4). American Fire Sprinkler Association. www.sprinklernet.org.

NFPA 13, Standard for the Installation of Sprinkler Systems, Latest Edition. Quincy, MA: National Fire Protection Association.

NFPA 13D, Standard for the Installation of Sprinkler Systems in One- and Two-Family Dwellings and Manufactured Homes, Latest Edition. Quincy, MA: National Fire Protection Association.

NFPA 13R, Standard for the Installation of Sprinkler Systems in Residential Occupancies up to and Including Four Stories in Height, Latest Edition. Quincy, MA: National Fire Protection Association.

Standard 203, Standard for Pipe Hanger Equipment for Fire Protection Service, Latest Edition. Northbrook, IL: Underwriters Laboratories.

Figure Credits

Tyco Fire and Building Products, Module opener, 104F02-104F04, 104F08, 104F11, 104F12

Reed Manufacturing Company, 104F05, 104F06

Milwaukee Electric Tool Corporation, 104F07

Ridge Tool Co./Ridgid®, 104F09

Lubrizol Advanced Materials, 104T03-104T05, Appendix

NIBCO INC., 104F13

ANSWERS TO REVIEW QUESTIONS

Answer		Section Reference
1.	b	2.1.0
2.	c	2.2.0
3.	b	2.1.0
4.	b	2.3.1
5.	d	2.3.2
6.	b	2.4.2
7.	a	2.5.0
8.	d	3.1.0
9.	b	3.3.0
10.	d	3.3.0
11.	c	4.1.0
12.	a	4.1.0
13.	c	5.1.2; Table 5
14.	c	5.1.3
15.	a	5.1.4

MODULE 18104-13 — ANSWERS TO TRADE TERMS QUIZ

1. Takeout
2. CPVC resin
3. Chlorinated polyvinyl chloride (CPVC)
4. Transverse
5. Iron pipe size (IPS)
6. Ambient temperature
7. Thermoplastic
8. CPVC compound
9. Chamfering
10. Escutcheon
11. Standard dimensional ratio (SDR)

NCCER CURRICULA — USER UPDATE

NCCER makes every effort to keep its textbooks up-to-date and free of technical errors. We appreciate your help in this process. If you find an error, a typographical mistake, or an inaccuracy in NCCER's curricula, please fill out this form (or a photocopy), or complete the online form at **www.nccer.org/olf**. Be sure to include the exact module ID number, page number, a detailed description, and your recommended correction. Your input will be brought to the attention of the Authoring Team. Thank you for your assistance.

Instructors – If you have an idea for improving this textbook, or have found that additional materials were necessary to teach this module effectively, please let us know so that we may present your suggestions to the Authoring Team.

NCCER Product Development and Revision

13614 Progress Blvd., Alachua, FL 32615

Email: curriculum@nccer.org
Online: www.nccer.org/olf

❑ Trainee Guide ❑ AIG ❑ Exam ❑ PowerPoints Other ______________________________

Craft / Level: Copyright Date:

Module ID Number / Title:

Section Number(s):

Description:

Recommended Correction:

Your Name:

Address:

Email: Phone:

Copper Tube Systems

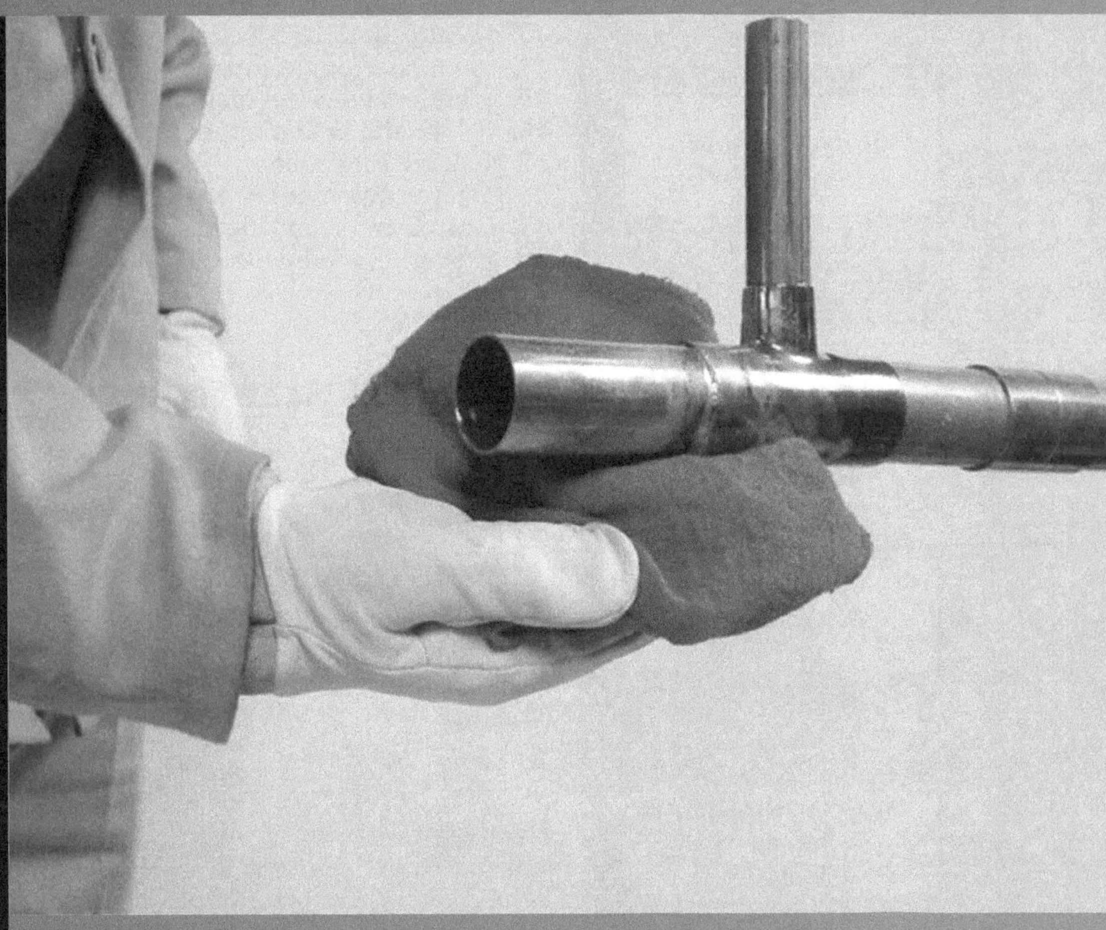

18105-13

SPRINKLER FITTING LEVEL ONE

18106-13 Underground Pipe

18105-13 Copper Tube Systems

18104-13 CPVC Pipe and Fittings

18103-13 Steel Pipe

18102-13 Introduction to Components and Systems

18101-13 Orientation to the Trade

Core Curriculum: Introductory Craft Skills

This course map shows all of the modules in *Sprinkler Fitting Level One*. The suggested training order begins at the bottom and proceeds up. Skill levels increase as you advance on the course map. The local Training Program Sponsor may adjust the training order.

Objectives

When you have completed this module, you will be able to do the following:

1. Follow basic safety precautions for preparing and installing copper tube.
2. Identify approved types of copper tube and fittings.
3. Identify and describe cast bronze fittings.
4. Identify wrought fittings.
5. Identify and select dielectric fittings.
6. Solder and braze copper tubing joints.
7. Calculate takeouts.
8. Set up equipment.
9. Cut, chamfer, and clean copper tube.
10. Properly prepare tube ends.

Trade Terms

Alloy
Borates
Brazing
Capillary action
Dielectric
Dimple
Flux
Galvanic
Liquidus
Oxidation
Oxyacetylene
Potable
Purge
Solder
Soldering
Solidus
Sweat joint
Wick
Wrought fittings

Prerequisites

Before you begin this module, it is recommended that you successfully complete *Core Curriculum* and *Sprinkler Fitting Level One*, Modules 18101-13 through 18104-13.

NOTE

Unless otherwise specified, references in parentheses following figure and table numbers refer to *NFPA 13*.

Contents

Topics to be presented in this module include:

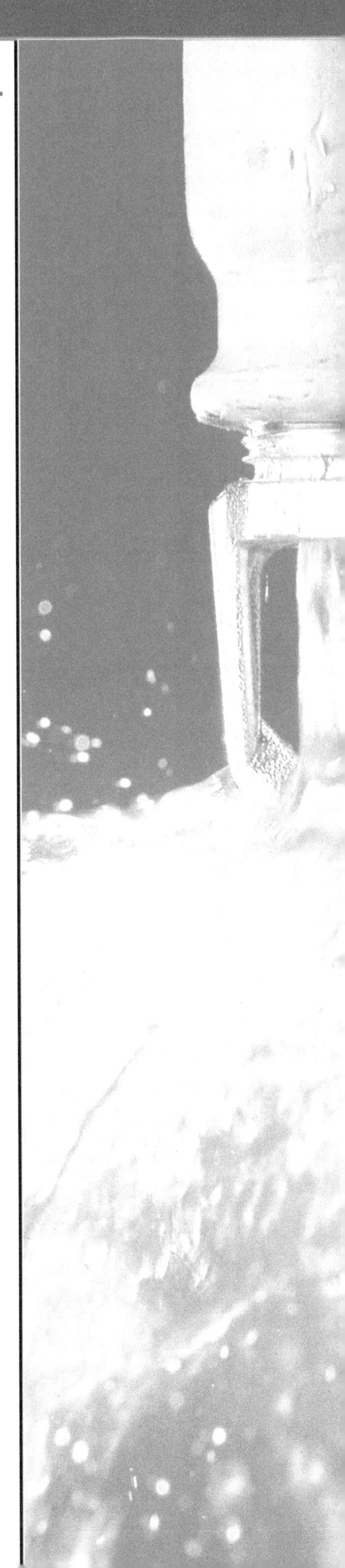

Figures and Tables

1.0.0 INTRODUCTION

A major advantage of copper tube is that it resists corrosion much better than steel pipe. It is useful in potable water supplies. Copper also has better flow characteristics than steel pipe, but is not as resistant to internal buildup of water-hardness products as plastic pipe. Copper has a high thermal conductivity and tends to freeze more readily than steel. Because of its positive flow characteristics, copper tube can be smaller than steel pipe in a given situation. For example, if an application required a minimum 1" diameter steel pipe, a ¾" copper tube could be used instead. The use of copper tube in fire sprinkler systems requires different joining and support methods than those used for steel or plastic pipe.

2.0.0 COPPER TUBING AND FITTINGS

Types K, L, and M copper tubing are allowed by *NFPA 13, 13R,* and *13D* for use in sprinkler systems. Types K and L are very ductile and come in straight lengths, as well as annealed or soft rolls. Type K has the thickest tube wall and Type M has the thinnest. Type M is hard-drawn copper and is available in straight sections only. Hard-drawn copper tube can be bent without kinking when bent with tools specifically designed to bend copper tube. Care must be taken to maintain the quality of the tubing during installation. Copper is expensive; take care to not waste this material. The cost of copper tube is directly related to wall thickness and the thinnest wall permitted is normally used. The actual internal diameter of copper tube is very close to the nominal tube size. Copper specifications are often referred to as CTS, meaning copper tube size, as differentiated from IPS, originally standing for iron pipe size. *Table 1* shows the differences in wall thickness.

There are other types of copper tubing available, such as drain, waste, and vent (DWV) and air conditioning and refrigeration (ACR). These types are not acceptable under *NFPA 13, 13R,* and *13D* for use in fire sprinkler systems.

Did You Know?

U.S. industries use close to 3 million tons of copper each year. About two-thirds of the copper is mined; the rest comes from recycled copper. Discarded objects, dismantled buildings, and worn-out machinery are sources of scrap copper. This scrap copper can be melted down and reused. Many recycling stations pay for scrap copper.

Table 1 Differences in Wall Thickness (Table A.6.3.5)

Nominal Tube Size		Outside Diameter		Type K Inside Diameter		Type K Wall Thickness		Type L Inside Diameter		Type L Wall Thickness		Type M Inside Diameter		Type M Wall Thickness	
inch	(mm)	inch	(mm)	inch	(mm)	inch	(mm)	inch	(mm)	inch	(mm)	inch	(mm)	inch	(mm)
¾	(20)	0.875	(22.2)	0.745	(18.9)	0.065	(1.7)	0.785	(19.9)	0.045	(1.1)	0.811	(20.6)	0.032	(0.8)
1	(25)	1.125	(28.6)	0.995	(25.3)	0.065	(1.7)	1.025	(26.0)	0.050	(1.3)	1.055	(26.8)	0.035	(0.9)
1¼	(32)	1.375	(34.9)	1.245	(31.6)	0.065	(1.7)	1.265	(32.1)	0.055	(1.4)	1.291	(32.8)	0.042	(1.1)
1½	(40)	1.625	(41.3)	1.481	(37.6)	0.072	(1.8)	1.505	(38.2)	0.060	(1.5)	1.527	(38.8)	0.049	(1.2)
2	(50)	2.125	(54.0)	1.959	(49.8)	0.083	(2.1)	1.985	(50.4)	0.070	(1.8)	2.009	(51.0)	0.058	(1.5)
2½	(65)	2.625	(66.7)	2.435	(61.8)	0.095	(2.4)	2.465	(62.6)	0.080	(2.0)	2.495	(63.4)	0.065	(1.7)
3	(80)	3.125	(79.4)	2.907	(73.8)	0.109	(2.8)	2.945	(74.8)	0.090	(2.3)	2.981	(75.7)	0.072	(1.8)
3½	(90)	3.625	(92.1)	3.385	(86.0)	0.120	(3.0)	3.425	(87.0)	0.100	(2.5)	3.459	(87.9)	0.083	(2.1)
4	(100)	4.125	(104.8)	3.857	(98.0)	0.134	(3.4)	3.905	(99.2)	0.110	(2.8)	3.935	(99.9)	0.095	(2.4)
5	(125)	5.125	(130.2)	4.805	(122.0)	0.160	(4.1)	4.875	(123.8)	0.125	(3.2)	4.907	(124.6)	0.109	(2.8)
6	(150)	6.125	(155.6)	5.741	(145.8)	0.192	(4.9)	5.845	(148.5)	0.140	(3.6)	5.881	(149.4)	0.122	(3.1)
8	(200)	8.125	(206.4)	7.583	(192.6)	0.271	(6.9)	7.725	(196.2)	0.200	(5.1)	7.785	(197.7)	0.170	(4.3)
10	(250)	10.130	(257.3)	9.449	(240.0)	0.338	(8.6)	9.625	(244.5)	0.250	(6.4)	9.701	(246.4)	0.212	(5.4)

105T01.EPS

2.1.0 Wrought Fittings for Copper Tubing

Wrought fittings for copper tubing are specified in *ASME B16.22, Wrought Copper and Copper Alloy Solder Joint Pressure Fittings*. Under *NFPA 13, 13D,* and *13R*, any solder meeting *ASTM B32*, having a melting point above 400°F, and not containing greater than 0.2 percent lead (Pb) would be accepted. Dry systems must be brazed with BCuP3 or BCuP4 **brazing** filler metal; 50 percent tin/50 percent lead solder is no longer allowed in *NFPA 13D* systems. In addition to the fittings shown in *Figure 1*, there are plugs, caps, and transition fittings.

2.2.0 Cast Bronze Fittings

Cast bronze fittings can be less convenient to use due to differences in handling of the cast fittings when brazing. Cast fittings require the use of flux when brazing. When brazing wrought fittings, the use of flux is optional. Cast bronze fittings for copper tubing are specified in *ASTM B16.18, Cast Copper Alloy Solder Joint Pressure Fittings.*

2.3.0 Dielectric Fittings

When a connection is made between dissimilar metals such as steel pipe and copper tubing, a galvanic cell can be formed. This cell is similar to a battery in that it can generate electric current. The flowing water in the tubing provides an electrolyte that allows the current to flow between the two different metals. Metals are ranked on the Standard Galvanic Series according to their relative corrosion potential. In the case of copper and steel, copper will become the cathode, while the steel will become the anode. The galvanic action produced can cause corrosion of the steel pipe and fittings and early failure of the system if corrective action is not taken. There is a specific requirement in *NFPA 13* for a dielectric fitting in closed loop systems that have these dissimilar (steel pipe and copper tube) metal connections. An insulating washer or some other non-conductive material is used to prevent the electric currents from passing between the dissimilar metals. In addition, a brass nipple is sometimes used as an insulating fitting. *Figure 2* shows a dielectric fitting.

Did You Know?

Copper is one of the most plentiful metals. It has been in use for thousands of years. In fact, a piece of copper pipe used by the Egyptians more than 5,000 years ago is still in good condition. When the first Europeans arrived in the New World, they found the Native Americans using copper for jewelry and decoration. Much of this copper was from the region around Lake Superior, which separates northern Michigan and Canada.

One of the first copper mines in North America was established in 1664 in Massachusetts. During the mid-1800s, new deposits of copper were found in Michigan. Later, when miners went west in search of gold, they uncovered some of the richest veins of copper in the United States.

NOTE

Dielectric unions are not needed for CPVC and copper.

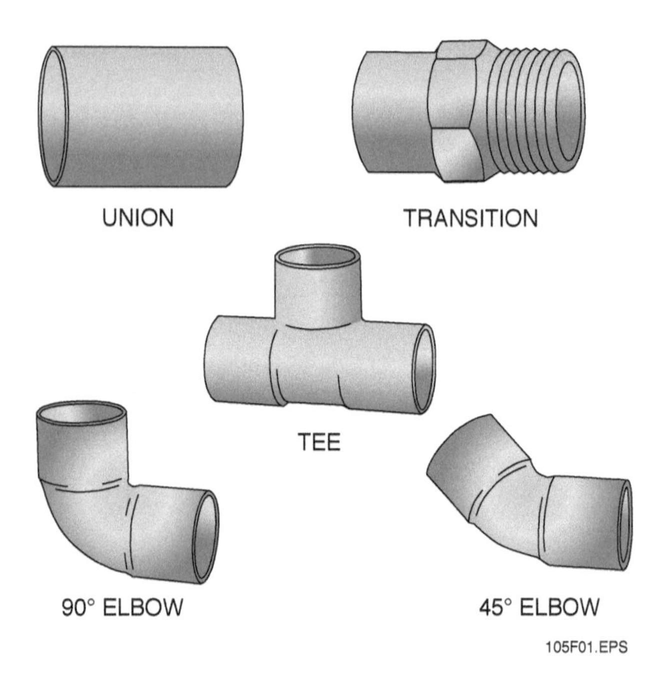

Figure 1 Sweat fittings.

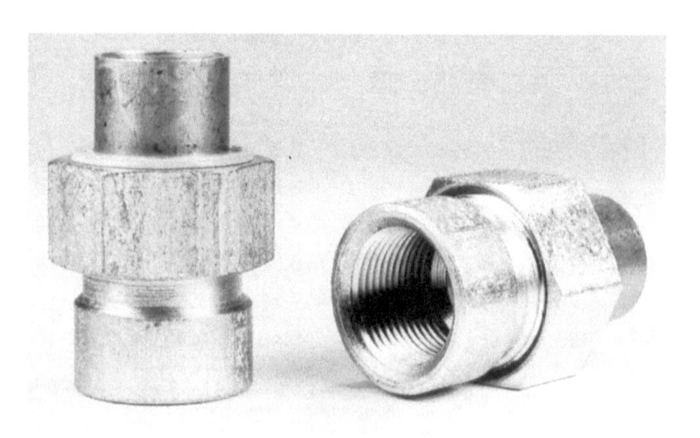

Figure 2 Dielectric fitting.

2.4.0 Branching without Fittings

A T-drill can be used to create branches without fittings. A T-drill drills a hole in the tube on the run. Once the T-drill has penetrated the tube, collaring pins are extended and the head extracts itself. This leaves an extruded outlet on the main. *Figure 3* shows the use of a T-drill.

After using a T-drill to cut the hole, a tube end notcher is used to notch the outlet tube to avoid obstruction of the waterway and, at the same time, it dimples the tube so that it can only penetrate the collar enough to provide the required overlap. This process produces a UL-approved assembly when brazed. The use of the notcher is extremely important because it controls the depth that the branch penetrates into the main. This method of branching is economical because it eliminates a fitting and two brazed joints. The outlet point can be readily tailored to the specifics of the installation. *Figure 4* shows a notched tube end.

2.5.0 Takeouts

Takeouts are required for copper tube in a manner similar to steel pipe. However, these takeouts are unique to the copper tube manufacturer.

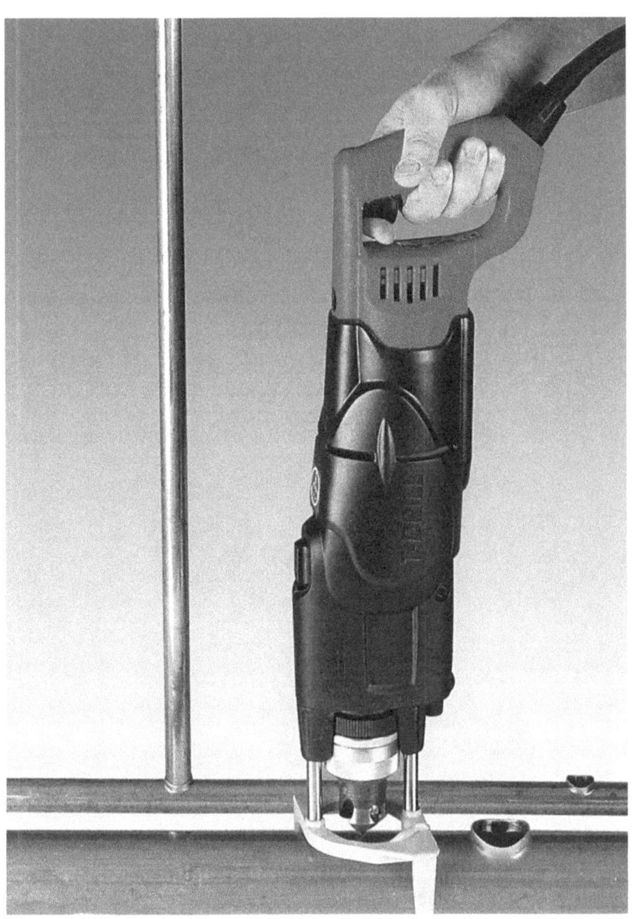

105F03.EPS

Figure 3 T-drill.

105F04.EPS

Figure 4 Mechanical tee connection with notched tube end.

The takeouts for copper tube fittings should be checked on the job for accurate dimensions. *Table 2* shows a takeout table developed for copper tube and wrought fittings from the data of one manufacturer.

3.0.0 Cutting Copper Tube

Copper tube must be cut square to the run of the tube. Thin-wall tube from ⅛" to 4" in diameter can be cut with tube-cutting tools, such as those shown in *Figure 5A*. Larger-diameter tube can be cut with pipe cutters (*Figure 5B*).

When using manual tube and pipe cutters, clamp the tool onto the tube and rotate it, tightening the adjusting screw as you go. Once the tube has been cut, it must be reamed to remove burrs that are left by the cutting process. Some tube cutters have built-in deburring blades for this purpose (*Figure 6*).

Power saws, specifically the portable band saw and the chop saw (*Figure 7*), are often used to cut tube. If you are using a power bandsaw, the tube must be held in place with a vise. The chop saw, which uses an abrasive blade, has a built-in vise.

> **WARNING**
>
> Because of their high speed and extremely sharp blades, power saws are inherently dangerous. Do not attempt to use one of these saws unless you have been specifically trained in its use.

Hacksaws can also be used for cutting copper tube, but it is more difficult to get a square cut this way. Cutting with a hacksaw also tends to produce a lot of burrs on the pipe. The hacksaw blade must have 24 or 32 teeth per inch. A coarser blade will tear and strip the thin copper tube wall.

Table 2 Pipe Takeouts

		Pipe Size Being Cut										
		3/4	1	1 1/4	1 1/2	2	2 1/2	3	3 1/2	4	5	6
Pipe Size at Right Angles	3/4	29/64	15/32	37/64	45/64	37/64	15/16	15/64	–	37/64	–	–
	1	37/64	9/16	37/64	53/64	37/64	1 5/64	1 5/64	–	37/64	–	–
	1 1/4	45/64	11/16	23/32	45/64	37/32	1 5/64	1 5/64	–	1 1/16	–	–
	1 1/2	1 5/8	1 21/32	27/32	53/64	53/64	1 5/64	1 5/64	–	1 1/16	–	–
	2	1 21/32	1 9/32	2 1/16	1 1/16	1 5/64	1 5/64	1 5/64	–	1 1/16	3 11/16	1 1/2
	2 1/2	2 55/64	2 25/32	2 1/2	1 25/32	1 11/16	1 5/16	1 5/64	–	1 37/64	–	1 1/2
	3	3 5/32	2 15/16	2 29/32	2 3/4	1 25/32	2 15/16	1 19/32	–	1 23/32	3 7/16	1 3/4
	3 1/2	–	–	–	–	–	–	1 41/64	1 55/64	–	–	–
	4	2 23/64	29/16	3 17/32	3 11/64	2 29/64	3 13/64	2 13/64	–	2 11/64	3 3/32	2 9/32
	5	–	–	–	–	–	–	3 7/16	–	–	2 21/32	3 1/8
	6	–	–	–	–	4 59/64	4 25/32	3 47/64	–	3 1/4	3 5/8	3 5/8
Std. 90° ELL	–	45/64	29/32	1 1/16	1 7/32	1 9/16	1 13/16	2 1/16	2 13/32	2 23/32	3 1/8	4 13/64
L.R. 90° ELL	–	1 13/16	1 1/2	1 7/8	2 1/4	2 7/8	3 11/16	–	–	–	–	–
45° ELL	–	23/64	15/32	25/32	15/16	1 1/4	1 9/16	1 21/32	1 3/4	1 15/16	1 7/16	1 5/8

NOTE: All dimensions in inches.

105T02.EPS

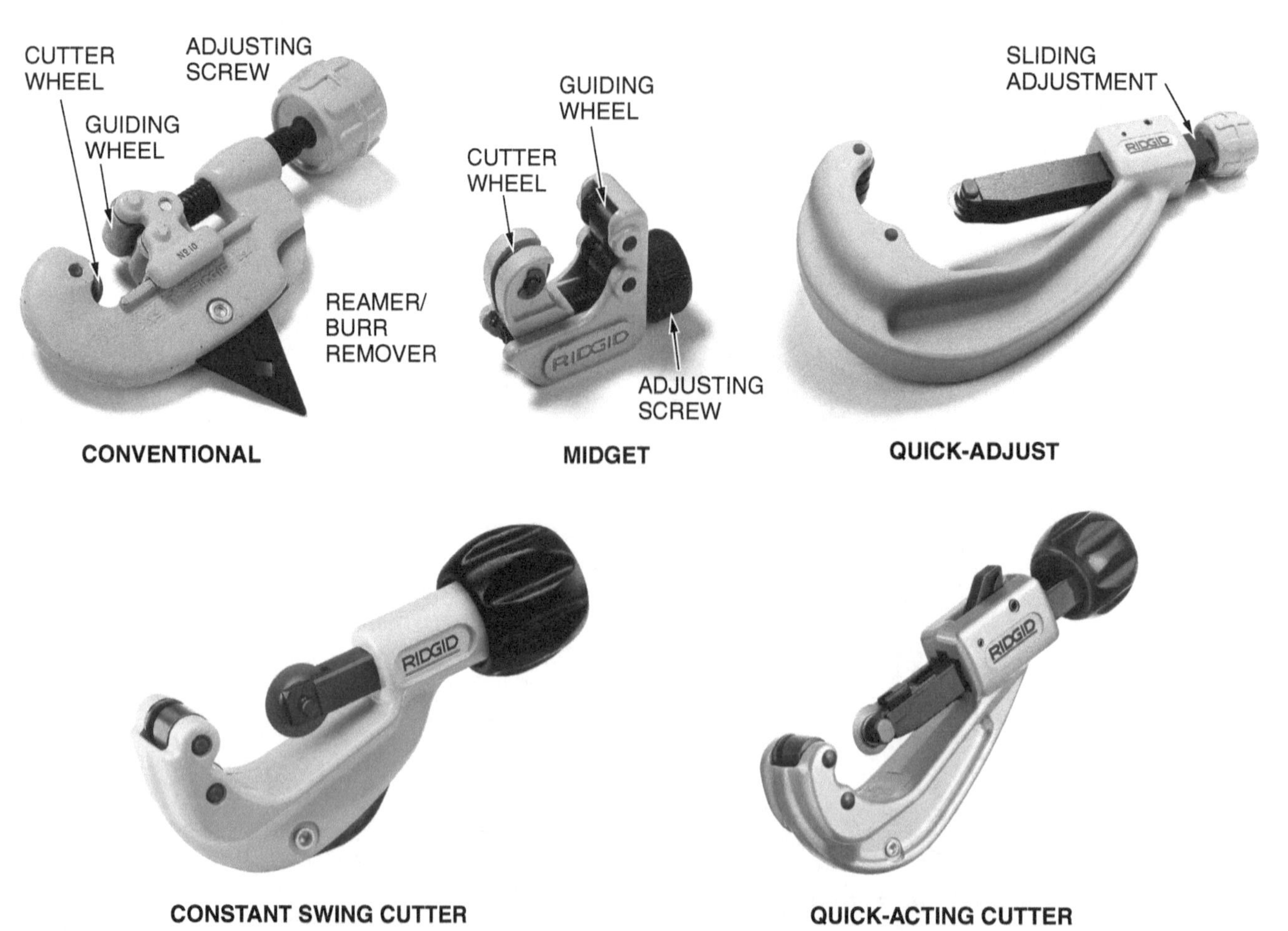

Figure 5 Tube cutters.

On Site

Chop Saw Blades

The abrasive cutting blades for chop saws have maximum rated speeds according to the abrasive blades they will accept. Make sure you choose a blade that is rated for the speed of the saw. For example, the typical maximum safe speed for a 12-inch blade is 5,000 revolutions per minute (rpm).

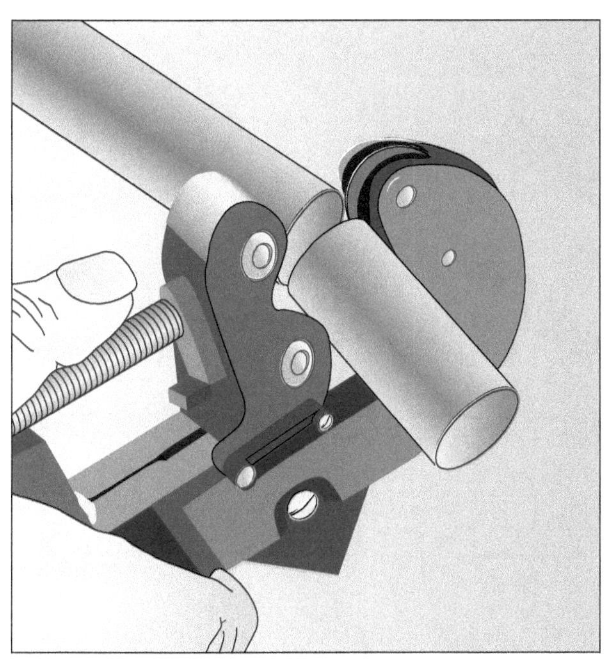

MEASURING, MARKING, AND CUTTING

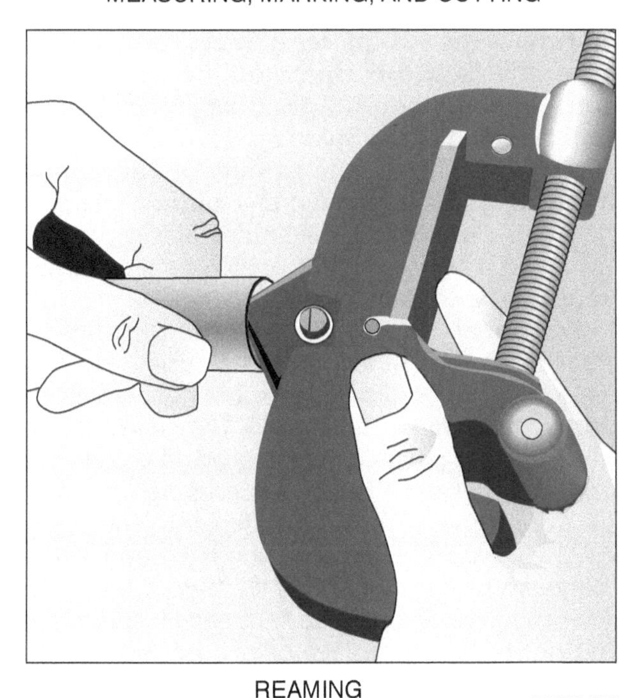

REAMING

105F06.EPS

Figure 6 Using a handheld tube cutter.

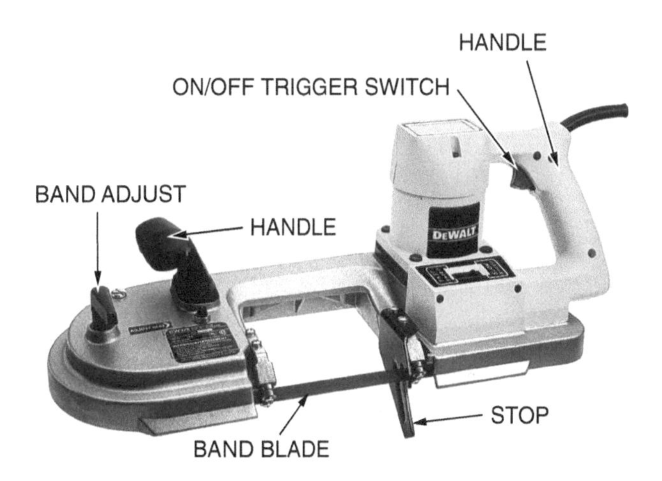

PORTABLE BAND SAW

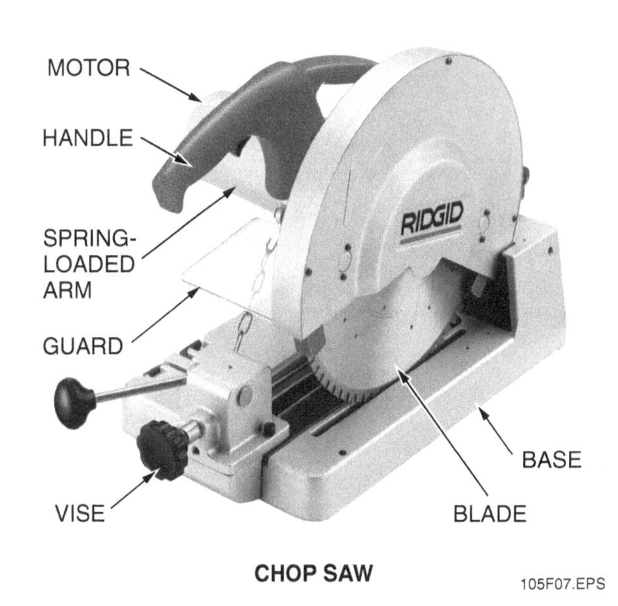

CHOP SAW

105F07.EPS

Figure 7 Power saws used to cut tube.

4.0.0 Bending Copper Tube

Unlike other types of rigid metal pipe, such as cast iron or steel, copper tube can be bent. In fact, tests have shown that the bursting strength of bent pipe is greater than that of regular tubing. Bending reduces the number of joints and fittings in a plumbing system. This can reduce leaks, as well as installation time. When you are using soft tubing, it is best to bend rather than cut it.

You can easily bend small-diameter tubing by hand. Take care not to flatten the tube with a bend that is too sharp. Tube-bending springs can be used on the outside of the tube to prevent it from collapsing while you are bending it by hand (see *Figure 8*). You can also use tube-bending equipment to make accurate and reliable bends in both soft and hard tubing. This equipment has various sizes of forming attachments for use in bending tubing with diameters up to ⅞" at any angle up to 180 degrees (see *Figure 9*).

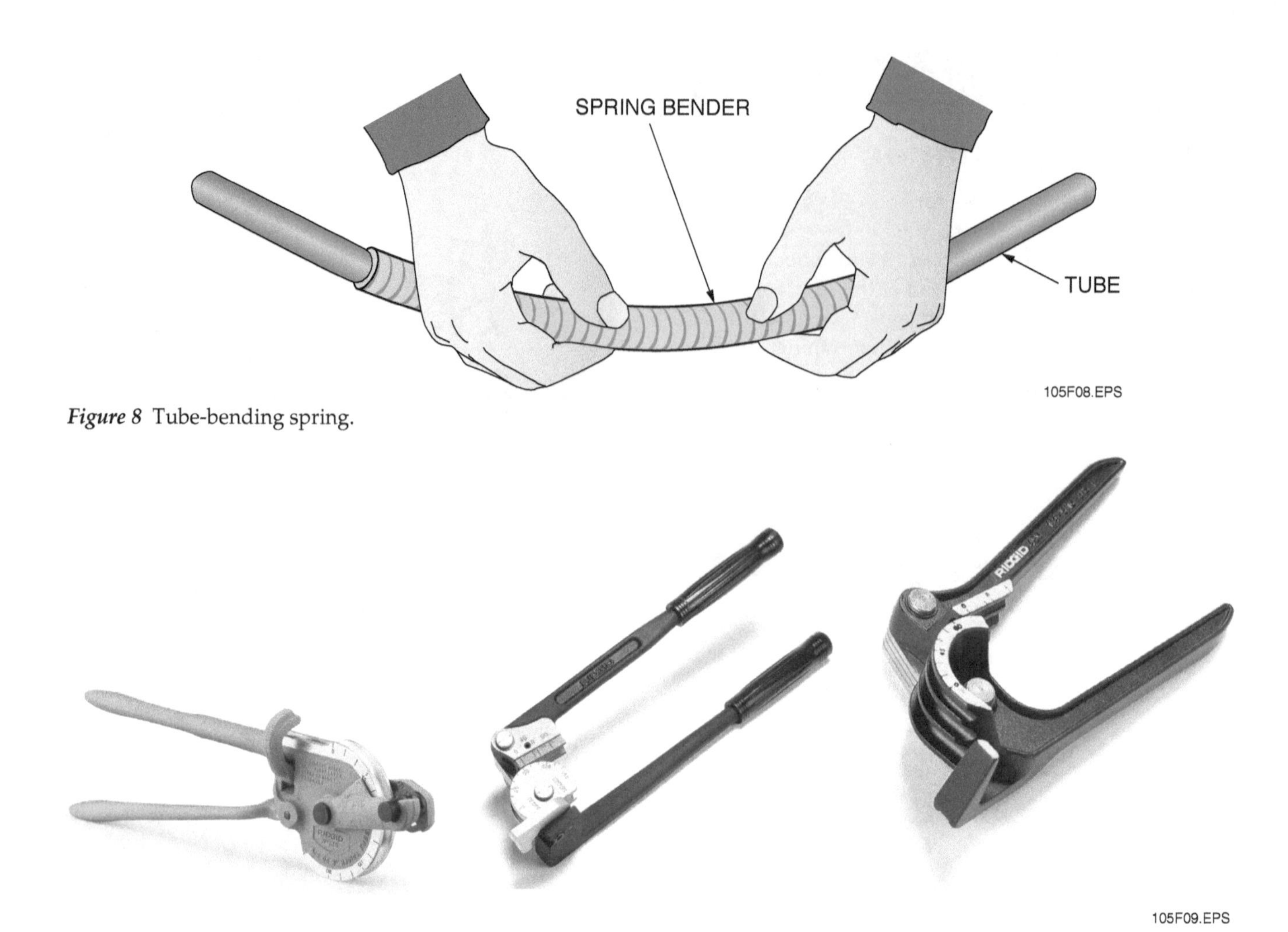

Figure 8 Tube-bending spring.

105F09.EPS

Figure 9 Tube-bending equipment.

5.0.0 Introduction to Soldering

Soldering is a simple and economical method of creating a joint with copper tube. Soldering is the process of joining (bonding) copper tube and fittings with a dissimilar metal or an alloy of two or more dissimilar metals that has a melting temperature (liquidus temperature) of 840°F or lower.

Soldering is distinguished from brazing and welding by the temperature at which joining takes place. To be a soldering operation, the temperature of liquification (liquidus temperature) of the solder alloy must be 840°F or below. A solder joint is also referred to as a sweat joint.

In soldering, molecules of the solder are attached to the molecules in the material being soldered. The molecular bond can take place only if the surface of the material is clean. This requires mechanical and chemical cleaning of the joint pieces. The use of a flux cleans and prevents oxidation while the material is being heated to the liquidus temperature of the solder. When the liquidus temperature is reached, the solder metal is drawn (wicked) into the joint by capillary action producing a pressure-tight seal when it cools (*Figure 10*). Maintaining the temperature of the two pieces of metal at or slightly above the liquidus point of the solder long enough for the bonding to take place is sometimes referred to as sweating.

Solder can be elastic in nature and should not be depended upon for mechanical strength. The main purpose of the solder is to provide a seal against leakage. The principal mechanical strength of a soldered joint depends on the tight fit and overlap between the tube and fitting. The integrity of the solder joint is a function of the character of the solder itself, how carefully the joint parts have been cleaned, the clearance between the two parts being joined, and the amount of overlap between the two pieces being joined. The tube must be fully inserted into the fitting and the clearance between the tube and fitting must be minimal. Generally, the tighter the fit, the better the joint. Due to the softness of solder, cold flow or creep takes place over a period of time. This can destroy the seal under some load conditions.

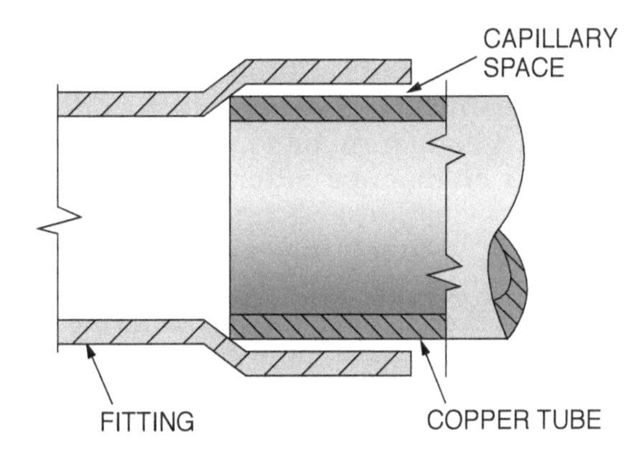

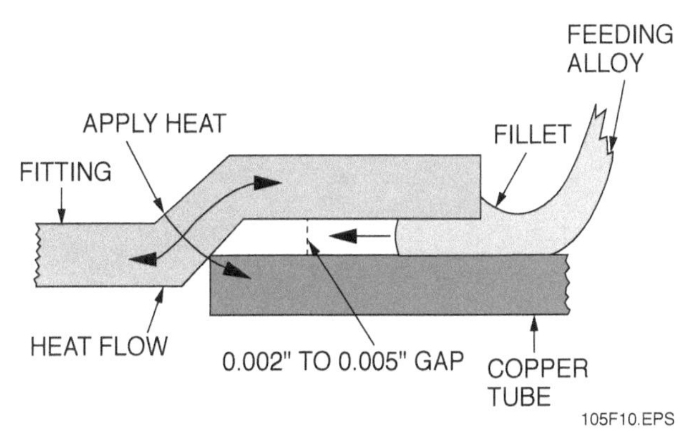

Figure 10 Capillary action.

5.1.0 Code Restrictions on Soldering

The use of solder in fire protection is restricted to use in wet systems. The solder must conform with *ASTM B32*, melt above 400°F, and contain less than 0.2 percent lead (Pb). The solder permitted in wet-pipe fire sprinkler systems is the 95/5 (tin-antimony) alloy. See *NFPA 13R* for details.

5.2.0 Solder Metal

Solder is an alloy of metals selected to produce specific melting, bonding, and physical characteristics. It has a melting point of 840°F or below. Solders approved for fire sprinkler systems are also used for potable (safe for human consumption) water connections because they contain no lead. A characteristic of many alloys used for solder is that they are formulated to have a temperature range between the solid state and the liquid state. This temperature range is usually called the working range or the pasty range. The true liquidus point is reached when the solder wicks. At that point, the joining area has reached the proper temperature for the solder to flow into the joint by capillary action. The joining area should not be heated beyond what is needed to melt the solder (liquidus temperature).

WARNING

Always solder in a well-ventilated area. Fumes from the flux can irritate your eyes, nose, throat, and lungs.

5.3.0 The Joining Process

Although the soldering operation is inherently simple, the deletion or misapplication of a single part of the process may mean the difference between a good joint and a failure. *ASTM B828, Standard Practice for Making Capillary Joints by Soldering of Copper and Copper Alloy Tube and Fittings* was created to standardize the process of preparing, heating, and applying alloy in soldered joints. The following joining process steps outline the basic requirements for consistently making a high-quality soldered joint. Each of these steps is discussed in the paragraphs that follow.

- Measuring and cutting
- Reaming
- Cleaning
- Fluxing
- Assembly and support
- Heating
- Applying the filler metal
- Cooling and cleaning

Figure 11 shows tools and materials commonly used in the soldering process.

5.3.1 Measuring and Cutting

Accurately measure the length of each tube segment. Inaccuracy can compromise joint quality. If the tube is too short, it will not reach all the way into the cup of the fitting and a proper joint cannot be made. If the tube segment is too long, system strain may be introduced, which could affect service life.

Cut the tube to the measured length. As previously discussed, cutting can be accomplished in a number of different ways to produce a satisfactory squared end. The tube can be cut with a disc-type tube cutter, a hacksaw, an abrasive wheel, or a stationary or portable bandsaw. Care must be taken to ensure the tube is not deformed while being cut. Regardless of the method, the cut must be square to the run of the tube so the tube will seat properly in the fitting cup.

5.3.2 Reaming

Ream all cut tube ends to the full inside diameter (ID) of the tube to remove the small burr created by the cutting operation. If this rough, inside edge is not removed by reaming, erosion/corrosion may occur due to local turbulence and increased local flow velocity in the tube. A properly reamed piece of tube provides an undisturbed surface for smooth flow.

TUBE CUTTER

SOLDERING TORCH

WIRE SOLDER

FITTING BRUSH

CLEANING MATERIALS

FLUX BRUSH AND SOLDER PASTE

105F11.EPS

Figure 11 Tools for soldering copper tubing to a fitting.

Remove any burrs on the outside of the tube ends created by the cutting operation to ensure proper assembly of the tube into the fitting cup.

Tools used to ream tube ends include the reaming blade on the tube cutter, half-round or round files, a pocketknife, and a suitable deburring tool (*Figure 12*). With soft tube, care must be taken not to deform the tube end by applying too much pressure. Soft-temper tube, if deformed, can be brought back to roundness with a sizing tool consisting of a plug and sizing ring.

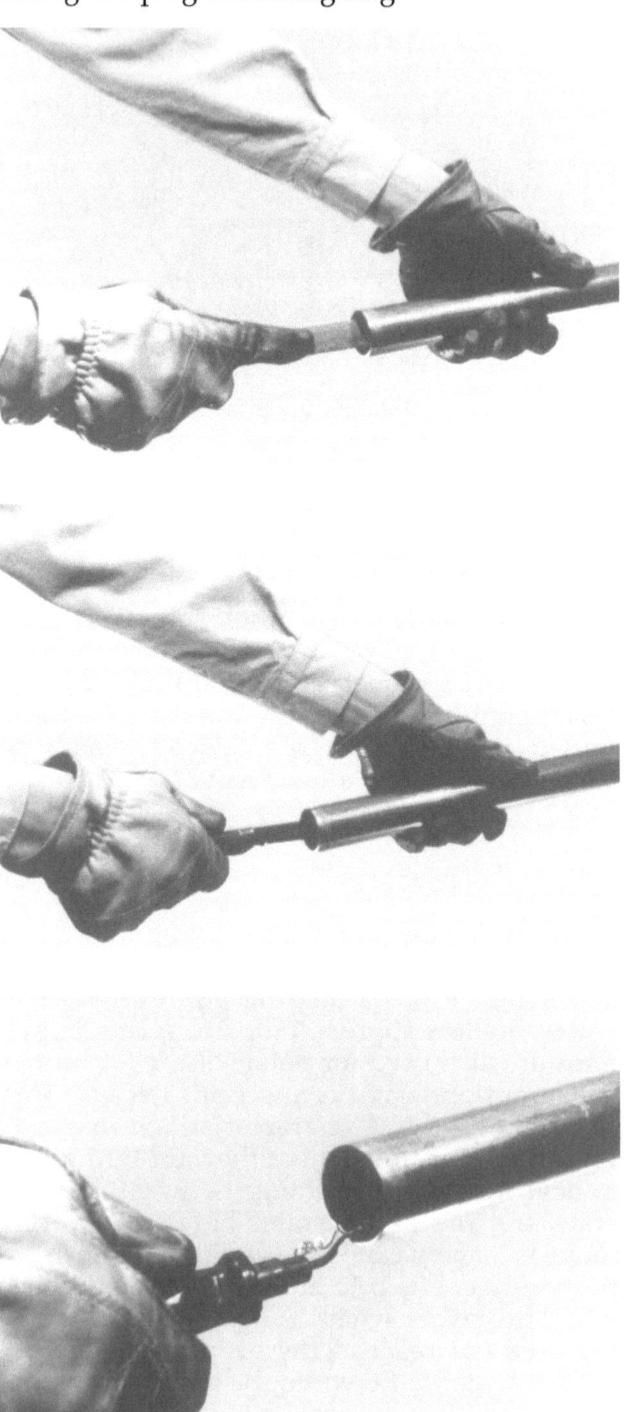

105F12.EPS

Figure 12 Methods of deburring.

> **Did You Know?**
>
> To solder copper fittings, you must heat the fitting until the soldering paste starts to melt. This melting paste looks like beads of sweat on the pipe, which led to the name "sweat fittings."

5.3.3 Cleaning

The removal of all oxides and surface soil from the tube ends and fitting cups is crucial to proper flow of filler metal into the joint. Failure to remove them can interfere with capillary action and may lessen the strength of the joint and cause failure. The capillary space between the tube and fitting is approximately 0.004 inch. Filler metal fills this gap via capillary action. This spacing is critical because it determines whether there is a proper flow of the filler metal into the gap, ensuring a strong joint.

Lightly abrade (clean) the tube ends using sand cloth or nylon abrasive pads (*Figure 13*) for a distance slightly more than the depth of the fitting cup. Clean the fitting cups by using abrasive cloth, abrasive pads, or a properly sized fitting brush.

Copper is a relatively soft metal. If too much material is removed from the tube end or fitting cup, a loose fit may occur, resulting in a poor joint. Chemical cleaning may be used if the tube ends and fittings are thoroughly rinsed after cleaning according to the procedure furnished by the chemical manufacturer. Do not touch the cleaned surface with bare hands or oily gloves. Skin oils, lubricating oils, and grease impair the adherence of the filler metal.

> **WARNING**
>
> Some types of flux produce harmful fumes and/or contain acid. Always make sure the work area is properly ventilated. Wear appropriate personal protective equipment, including safety glasses or goggles. Avoid eye and skin contact with flux. Wash flux off immediately if contact is made.

5.4.0 Preparing Tubing and Fittings for Soldering

Patience and self-discipline are required for good soldering results. Where multiple joints are to be soldered, all the joints on the fitting must be prepared at the same time so that only

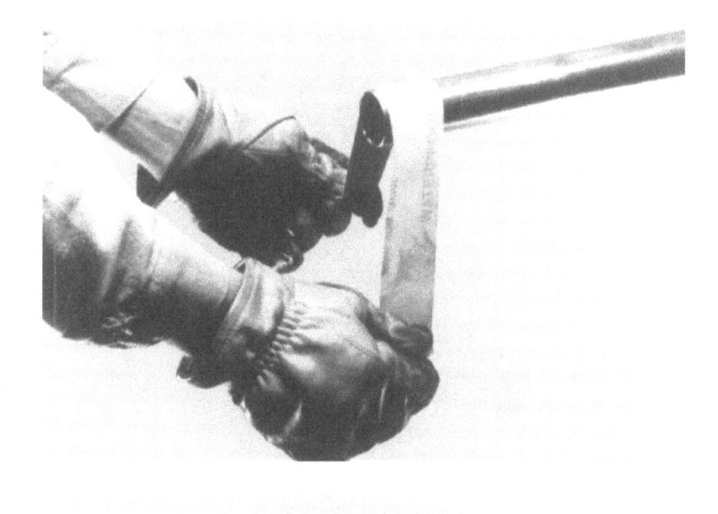

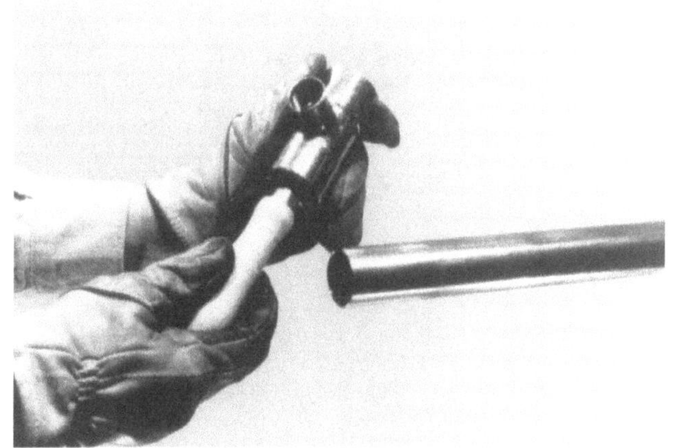

105F13.EPS

Figure 13 Cleaning tools.

one heating is required. Heating a second joint can remelt the first joint and cause leaks. Follow these steps to prepare the tubing and fittings for soldering:

Step 1 Ensure that a fire extinguisher is easily accessible in case the heat source creates a fire.

Step 2 Measure the distance between the faces of the two fittings, using the face-to-face method (*Figure 14*).

Step 3 Determine the cup depth engagement of each of the fittings. The cup depth engagement is the distance that the tubing penetrates the fitting. This measurement can be found by measuring the fitting or by using a manufacturer's makeup chart. *Table 3* shows a manufacturer's makeup chart.

Step 4 Add the cup depth engagement of both fittings to the measurement found in Step 2 in order to find the length of tubing needed.

Step 5 Cut the copper tube to the correct length, making sure that the cut is square to the tube.

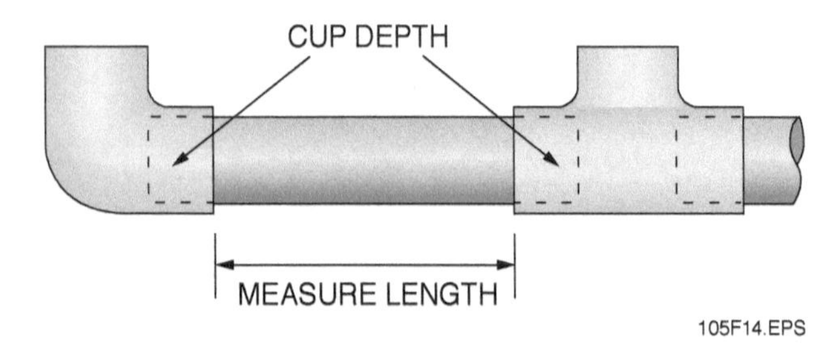

Figure 14 Face-to-face measuring method.

Table 3 Manufacturer's Makeup Chart

Pipe Size	Depth of Cup	Pipe Size	Depth of Cup
1/4	5/16	2	1 11/32
3/8	3/8	2 1/2	1 15/32
1/2	1/2	3	1 21/32
5/8	5/8	3 1/2	1 29/32
3/4	3/4	4	2 5/32
1	29/32	5	2 21/32
1 1/4	31/32	6	3 3/32
1 1/2	1 3/32	–	–

NOTE: All dimensions in inches.

105T03.EPS

Step 6 Ream the inside and outside of both ends of the copper tube using a deburring tool, file, or reamer.

Step 7 Test the fit of the two items to be soldered. The fitting should slide over the tube, but the fit should be tight. Loose fits produce weak joints.

Step 8 Clean the tubing and the fitting using an emery cloth or a special copper-cleaning tool.

CAUTION

When cleaning copper tubing and fittings, be careful to remove all of the oxides on the copper without removing a large amount of metal. Do not touch or brush away filings from the tube or fitting with your fingers because your fingers will contaminate the freshly cleaned metal.

Step 9 Apply flux to the outside of the copper tubing and to the inside of the copper fitting socket immediately after cleaning them (*Figure 15*).

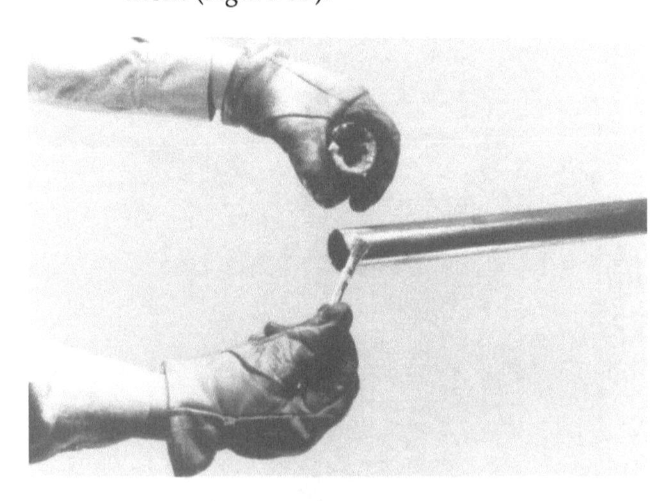

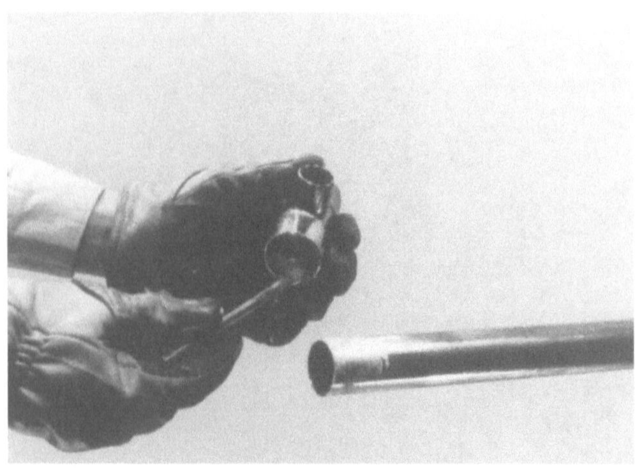

105F15.EPS

Figure 15 Applying flux.

CAUTION

Brazing flux and soldering flux are not the same. Do not allow these fluxes to become mixed or interchanged. Carelessness can ruin quality work.

Step 10 Insert the tube into the fitting socket with a slight twisting motion until the tube touches the inside shoulder of the fitting.

Step 11 Wipe away any excess flux from the joint.

Step 12 Check the tube and fitting for proper alignment and support before soldering.

5.5.0 Heating Joints and Applying Alloy

Electric resistance heating is the preferred method of heating a copper joint. However, this kind of heating equipment is difficult to obtain for field use. The more common method of heating on a construction site is with a torch. Air/fuel torches using propane, acetylene, or other gases are common. Torches using oxygen instead of air generally produce excessive temperatures and are not as convenient to use.

WARNING

Always solder in a well-ventilated area. Fumes from the flux can irritate your eyes, nose, throat, and lungs.

The use of torches and welding equipment are major sources of construction fires. Always follow these guidelines when soldering:

- Ensure that a fire extinguisher is close by when using a torch.
- Shield the structure from the flame and excessive heat.
- Keep the heat and the flame directed only towards the piece to be heated.
- Use wet rags wrapped around the tube block to confine the heat to the joint.

The temperature produced by the heating source and the heat capacity of the source are critical. Torches produce temperatures well in excess of what is needed for soldering. The torch tip needs to be sufficiently large so that the volume of heat generated offsets the heat loss down the tube and into the atmosphere. Move the torch flame around to produce uniform heating and to avoid extreme hot spots that result in premature vaporization or burning of the flux. Hot spots are a problem, especially with the oxyacetylene torch. Oxyacetylene torches are designed for welding and produce much higher temperatures than needed for soldering.

5.5.1 Heating

Obtaining a proper solder joint requires the filler metal to be heated to the liquidus temperature. Melting of the solder alone is not adequate. Check the temperature of the tube as it is being heated using stick or wire solder. Touch the solder to the metal at a point away from the flame to determine when the metal is hot enough to melt the solder. Melting solder with the flame does not help. The bonding (sweat joint) will not occur until the metal itself is hot enough to melt the solder. Begin heating with the flame perpendicular to the tube (*Figure 16*). The copper tube conducts the initial heat into the fitting cup for even distribution of heat in the joint area. The extent of this preheating depends on the size of the joint.

Preheating of the assembly should include the entire circumference of the tube in order to bring the entire assembly up to a suitable preheat condition. However, for joints in the horizontal position, avoid directly preheating the top of the joint to avoid burning the soldering flux. The natural tendency for heat to rise will ensure adequate preheat of the top of the assembly. Experience will indicate the amount of heat and the time needed.

Next, move the flame onto the fitting cup. Sweep the flame alternately between the fitting cup and the tube a distance equal to the depth of the fitting cup. Again, while preheating the circumference of the assembly as described above, with the torch at the base of the fitting cup, touch the solder to the joint. If the solder does not melt, remove it and continue heating.

CAUTION

Do not overheat the joint or direct the flame into the face of the fitting cup. Overheating could burn the flux, which will destroy its effectiveness, and the solder will not enter the joint properly.

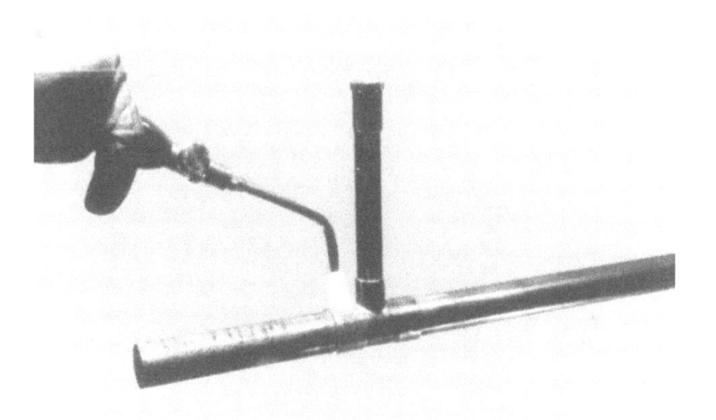

105F16.EPS

Figure 16 Initial heating.

When the solder melts, apply heat to the base of the cup to aid capillary action in drawing the molten solder into the cup toward the heat source.

5.5.2 Applying Solder

Solder joints depend on capillary action, in which free-flowing molten solder is drawn into the narrow clearance between the fitting and the tube. Molten solder metal is drawn into the joint by capillary action regardless of whether the solder flow is upward, downward, or horizontal. Capillary action is most effective when the space between the surfaces to be joined is between 0.002 inch and 0.005 inch. A certain amount of looseness of fit can be tolerated, but too loose a fit can cause difficulties with larger fittings.

The following is an overview of the soldering process. Because soldering requires relatively low heat, heating equipment that mixes acetylene, butane, or propane directly with air is all that is needed. Only one tank of gas and a torch are required for soldering. Follow these steps to solder a joint:

Step 1 Apply heat blocks or heat shields to the tube and between the tube and structure as appropriate.

Step 2 Obtain either an acetylene tank and related equipment or a propane bottle and torch.

Step 3 Set up the equipment according to the manufacturer's instructions.

Step 4 Light the heating equipment according to the type of equipment being used and the manufacturer's instructions.

Step 5 Preheat the tubing first, and then move the flame onto the fitting (*Figure 17*). For joints in the horizontal position, start applying the solder metal slightly off-center at the bottom of the joint (*Figure 18*).

Step 6 Move the flame to the back of the fitting cup.

Step 7 Touch the end of the solder to the area between the fitting and the tube. When the solder begins to melt from the heat of the tube and fitting, the solder will be drawn into the joint by capillary action. The solder can be fed upward or downward into the joint while keeping the torch at the base of the fitting and slightly ahead of the point of application of the solder. If the solder does not melt on contact with the joint, remove the solder and heat the joint more. Do not melt the solder with the flame.

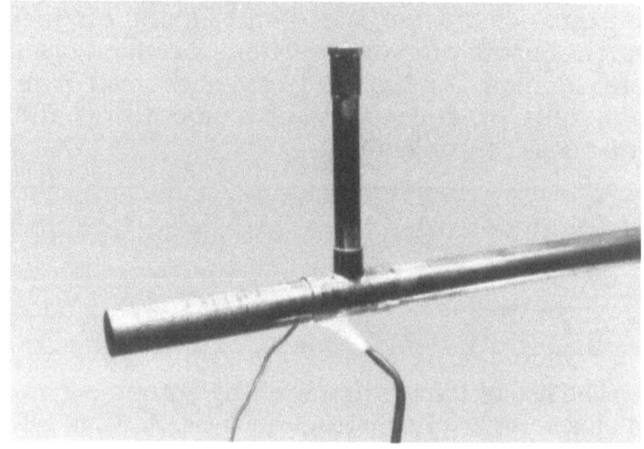

Figure 18 Applying solder to the heated joint.

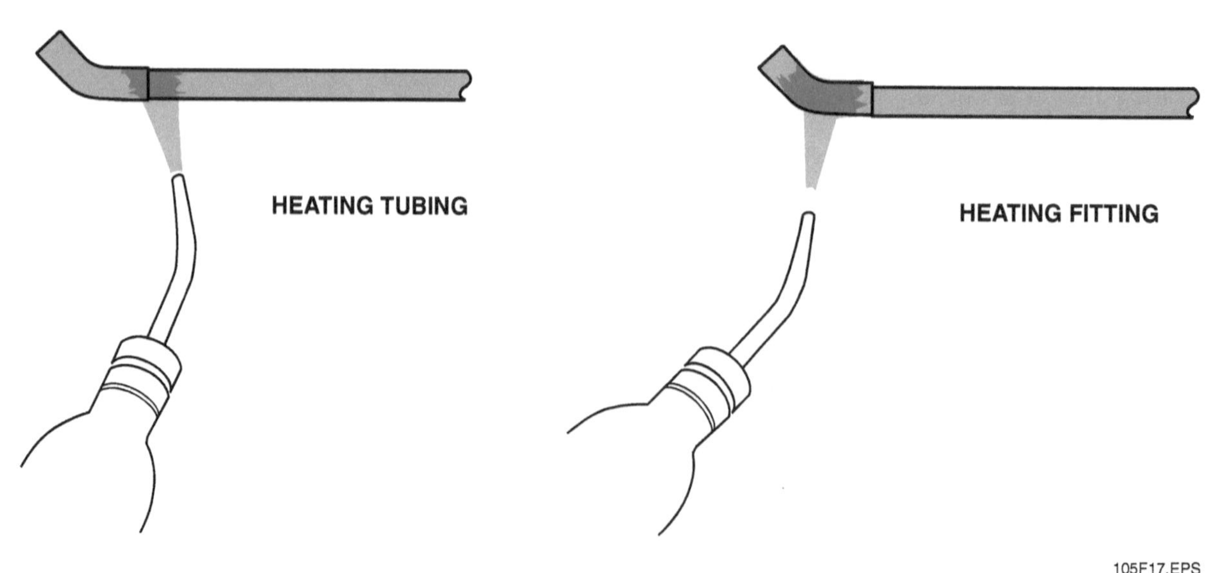

Figure 17 Heating the tubing and fitting.

Step 8 Continue to feed the solder into the joint until a ring of solder appears around the joint, indicating that the joint is filled. The now-solidified solder at the bottom of the joint has created an effective dam that will prevent the solder from running out of the joint as the sides and top of the joint are being filled.

Step 9 Return to the point of beginning, overlapping slightly, and proceed up the uncompleted side to the top, again overlapping slightly. While soldering, small drops may appear behind the point of solder application, indicating that the joint is full to that point and will take no more solder. Throughout this process, you are using all three physical states of the solder: solid, paste, and liquid. For joints in the vertical position, make a similar sequence of overlapping passes, starting wherever is convenient.

Step 10 Allow the joint to cool without moving it.

Step 11 Wipe the joint clean with a cloth after the joint has cooled.

Step 12 Flush the interior of the tube if possible, to remove residual flux.

For joining copper tube to solder cup valves, follow the manufacturer's instructions. The valve should be in a partially open position before applying heat, and the heat should be applied primarily to the tube. Commercially available heat-sink materials can also be used to protect temperature-sensitive components during the joining operation. The amount of solder consumed when adequately filling the capillary space between the tube and either wrought or cast fittings depends on the size of the tube. A 1" tube would use approximately 1.5 pounds of solder for 100 joints. The flux requirement is usually 2 ounces per pound of solder.

5.6.0 Cooling and Cleaning

Allow the completed joint to cool naturally. Shock-cooling with water may stress or crack the joint. When cool, clean off any remaining flux residue with a wet rag (*Figure 19*). Whenever possible, based on end use, completed systems should be flushed to remove excess flux and debris.

6.0.0 Introduction to Brazing

Soldering takes place at temperatures at or below 840°F. Brazing takes place at temperatures above 840°F but below the melting point of the base metal. The melting temperature of copper tube is 1,981°F; brazing alloys used to braze copper tube and fittings typically melt between 1,200°F and 1,500°F. As with soldering, brazing is the joining of two pieces of metal, using a nonferrous, dissimilar metal that has a lower melting temperature than the pieces being joined.

Brazing produces mechanically strong, pressure-resistant joints. The strength of a brazing joint results from the ability of the filler metal to penetrate the base metal. Penetration can only occur if the base metals are properly cleaned, the proper flux and filler metals are selected, and the clearance gap between the outside of the tubing and the inside of the fitting is 0.002 to 0.005 inch.

6.1.0 Code Restrictions

Brazing is the required process for joining copper tubing in a dry system. It is also the approved method for installing an outlet or branch when creating a tee on the run with a T-drill or other related methods of extruding an outlet or a tee branch.

6.2.0 Brazing Metals and Fluxes

The two approved brazing formulations are BCuP-3 and BCuP-4. They contain copper, phosphorus, and silver as their main components. The melting temperatures of these two brazing alloys are as follows:

	Solidus	Liquidus
BCuP-3	1,190°F	1,485°F
BCuP-4	1,190°F	1,335°F

Filler metals used to join copper tubing are of two groups: alloys that contain 28 percent to 56 percent silver (the BAg series) and copper alloys that contain phosphorous (the BCuP series). When joining copper tubing, any of these filler metals can be used; however, *NFPA 13* specifically men-

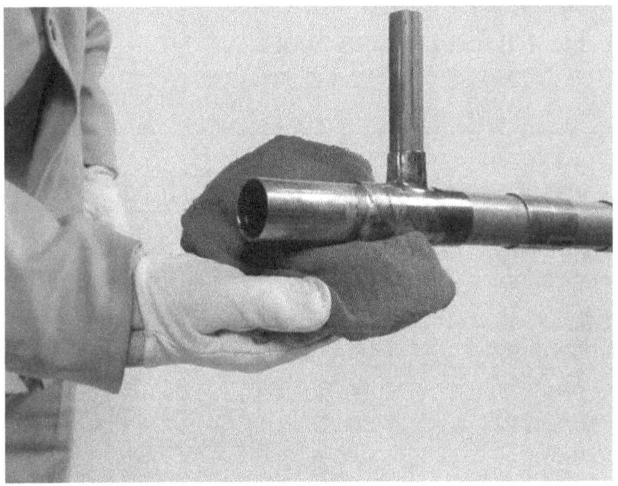

105F19.EPS

Figure 19 Cleaning the solder joint.

tions BCuP-3 and BCuP-4. *Table 4* shows brazing filler materials.

Brazing fluxes can be applied using the same methods as soldering fluxes. Brazing fluxes operate at much higher temperatures than soldering fluxes, so care must be taken never to mix a soldering flux with a brazing flux. For best results, use the flux recommended by the manufacturer of the brazing filler metals.

When copper tubing is joined to wrought copper fittings with copper-phosphorus alloys (BCuP series), flux can be omitted because the copper-phosphorus alloys are self-fluxing on copper. However, fluxes are required for joining all cast fittings.

6.3.0 Fluxing

BCuP filler metal may be successfully used without flux when using copper tube and wrought copper fittings, but proper mechanical cleaning of the tube and fittings is required before starting the brazing operation. The phosphorus in the BCuP alloys provides a self-fluxing capability on standard copper tube. If brass or cast copper alloy is used, brazing flux is required because there is no phosphorus in the cast brass fitting. To braze cast fittings (as opposed to wrought fittings), the entire fitting is sometimes coated with flux to prevent loss of zinc (dezincification).

Brazing fluxes are chemically active at brazing temperatures and are usually borates. Borates are a class of white, water-soluble crystalline salts, the familiar Borax being one of them. Brazing fluxes are available in paste, powder, or liquid form. The usual method of flux application is to brush it on the inside of the fitting cup and the tube end with a brush in the same manner as soldering flux is applied.

Where flux is required, the residues may be removed after the brazing operation by cleaning with hot water and brushing with a stainless steel brush.

6.4.0 Preparing Tubing and Fittings for Brazing

To prepare tubing and fittings for brazing, follow the same procedures for preparing tubing and fittings for soldering. It is critical that proper cleaning techniques be used in order to produce a solid, leak-proof joint. Follow these steps to prepare the tubing and fittings for brazing:

Step 1 Measure the distance between the faces of the two fittings.

Step 2 Determine the cup depth engagement of each of the fittings. The cup depth required for brazing is much shallower than that for soldering because brazing alloy penetration that is three times the thickness of the tube with a well developed fillet is all that is needed to provide a joint that is stronger than the tube (3-T Rule). Short cup brazing fittings are not readily available and most brazed joints will be made with solder-depth fittings. The cup depth being used can be found by measuring the fitting or by using a manufacturer's makeup chart. *Figure 20* shows a manufacturer's brazing fitting makeup chart.

Step 3 Add the cup depth engagement of both fittings to the measurement found in Step 1 in order to find the length of tubing needed.

Step 4 Cut the copper tubing to the correct length, making sure the cut is square to the tube.

Table 4 Brazing Filler Materials

	Percent of Principal Element					
AWS Classification	**Silver**	**Phosphorus**	**Zinc**	**Cadmium**	**Tin**	**Copper**
BCuP-2	–	7–7.50	–	–	–	Balance
BCuP-3	4.75–5.25	5.75–6.25	–	–	–	Balance
BCuP-4	5.75–6.25	7–7.50	–	–	–	Balance
BCuP-5	14.5–15.50	4.75–5.25	–	–	–	Balance
BAg-1	44–46	–	14–18	23–25	–	14–16
BAg-2	34–36	–	19–23	17–19	–	25–27
BAg-5	44–46	–	23–27	–	–	29–31
BAg-7	55–57	–	15–19	–	4.5–5.5	21–23

105T04.EPS

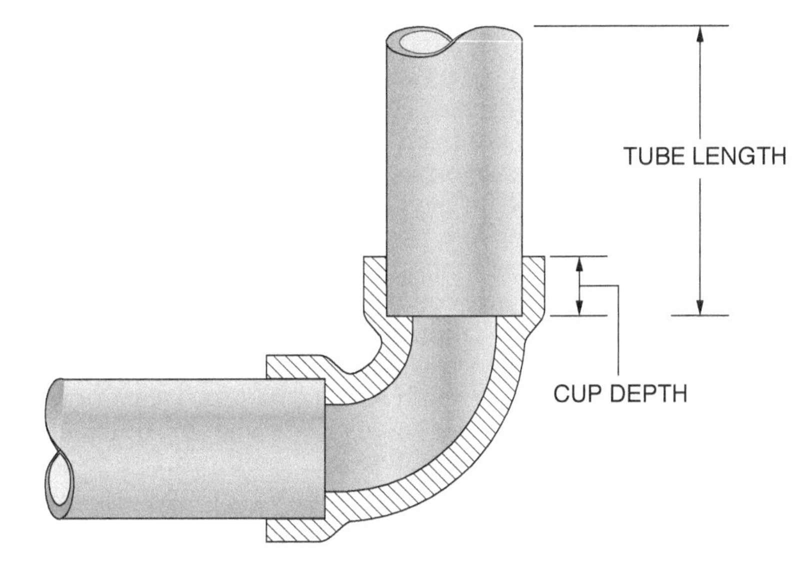

TUBE SIZE (in)	CUP DEPTH (in)
1/4	17/64
3/8	5/16
1/2	3/8
3/4	13/32
1	7/16
1 1/4	1/2
1 1/2	5/8
2	21/32
2 1/2	25/32
3	53/64
3 1/2	7/8
4	29/32
5	1
6	1 7/64
7	1 7/32
8	1 5/16

105F20.EPS

Figure 20 Manufacturer's brazing fitting makeup chart.

Step 5 Deburr the pieces being joined to eliminate any interference to the waterway and the joint itself.

Step 6 Ream the inside and outside of both ends of the copper tubing using a deburring tool, file, or reamer.

Step 7 Clean the tubing and the fitting using an emery cloth or a special copper-cleaning tool.

Step 8 Manually check the fit of the two items to be joined. The fitting should slide over the tube, but the fit should be tight. Loose fits produce weak joints.

CAUTION

Care must be taken when cleaning the tubing and fittings to remove all oxides on the copper without removing a large amount of metal. Do not touch or brush away filings from the tube or fitting with your fingers because your fingers will contaminate the freshly cleaned metal.

Step 9 Apply flux to the copper tubing and to the inside of the copper fitting socket immediately after cleaning them.

Step 10 Insert the tube into the fitting socket with a slight twisting motion until the tube touches the inside shoulder of the fitting.

Step 11 Wipe away any excess flux from the joint.

Step 12 Check the tube and fitting for proper alignment and support before brazing.

6.5.0 Setup of Brazing Equipment

The brazing heating procedure differs from soldering in that different equipment is required to raise the temperature of the metals to be joined above 840°F. Because of the higher temperatures needed, oxygen/acetylene (oxyacetylene) or oxygen/LP (oxyfuel) brazing equipment is used. The flame is produced by burning a fuel gas mixed with pure oxygen. *Figure 21* shows oxyacetylene brazing equipment. Oxyfuel equipment is similar. Only the regulator for the fuel gas is different.

6.5.1 Handling Oxygen and Acetylene Cylinders

Working with oxyacetylene brazing equipment requires special safety precautions. Oxygen and acetylene are compressed and shipped under medium to high pressures in cylinders. Because their use is so common, technicians often get careless about handling them. Oxygen is supplied in cylinders at pressures of about 2,000 psi. Acetylene cylinders are pressurized at about 250 psi. Cylinders must not be moved unless the protective caps are in place.

WARNING

Dropping a cylinder without the cap installed may result in breaking the valve off the cylinder. This allows the pressure inside to escape, causing the cylinder to propel like a rocket. Never transport cylinders with the gauges attached. Never mix cylinder types in storage. Keep cylinder types separate.

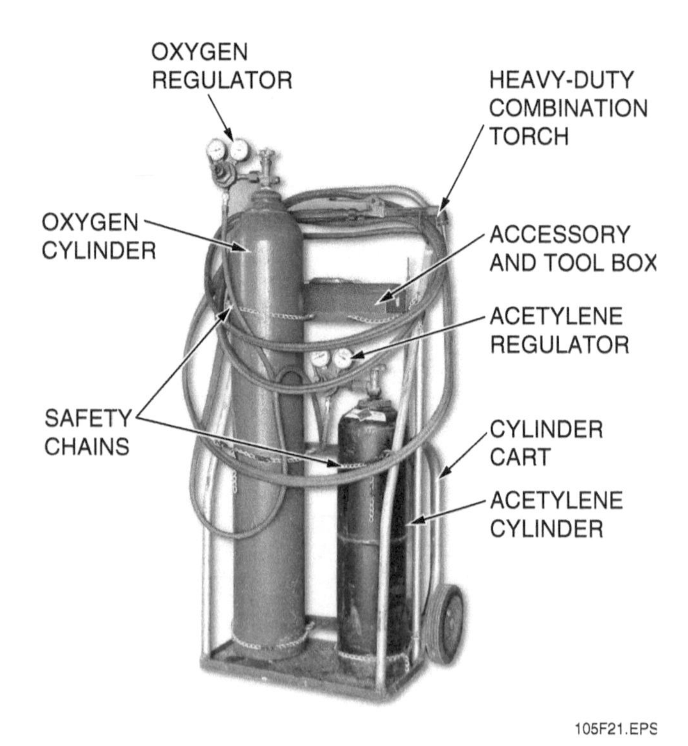

Figure 21 Oxyacetylene or oxygen/LP gas brazing equipment.

During use, transportation, and storage, oxygen and acetylene cylinders must be secured in the upright position with a stout cable or chain to prevent them from falling and injuring people or damaging equipment. When stored at the job site, oxygen and acetylene cylinders must be stored separately with at least 20 feet between them, or with a firewall separating them. Store empty cylinders away from partially full or full cylinders and make sure they are properly marked to clearly show that they are empty.

Oxygen can cause ignition even when no flame or spark is around to set it off, especially when it comes in contact with oil or grease. Never handle oxygen cylinders with oily hands or gloves. Keep grease away from the cylinders and do not use oil or grease on cylinder attachments or valves. Never use an oxygen regulator for any other gas or try to use a regulator with oxygen that has been used for another service.

A pressure-reducing regulator set for not more than 15 psig must be used with acetylene. Acetylene becomes unstable and volatile above 15 psig. The valve wrench should be left in position on open acetylene valves. This enables quick closing in an emergency. It is good practice to open the acetylene valve as little as possible, but never more than 1¼ turns.

6.5.2 Initial Setup of Oxyacetylene Equipment

Follow this procedure to set up oxyacetylene brazing equipment. The setup for oxyfuel equipment is similar.

WARNING

Do not handle acetylene and oxygen cylinders with oily hands or gloves. Keep grease away from the cylinders and do not use oil or grease on cylinder attachments or valves. The mixture of oil and oxygen will cause an explosion.

Make sure that the protective caps are in place on the cylinders before transporting or storing the cylinders.

Step 1 Install and securely fasten the oxygen and acetylene cylinders in a bottle cart or in an upright position.

On Site

Acetylene Hazards and Costs

As indicated previously, working with acetylene is hazardous. When possible, always use oxyfuel gas equipment for brazing. The equipment is safer to use and the gas is considerably cheaper. On smaller tubing and fittings, the heat generated from an air/LP torch and a single fuel gas tank is generally adequate for brazing as well as soldering.

An accumulation of an oxygen and fuel gas mixture can result in a dangerous explosion when ignited. A 2-inch balloon filled with an oxygen and acetylene mixture has the explosive power of an M80 firecracker (one-quarter stick of dynamite). The mixed gases from 100 feet of ¼-inch twin hose will fill a 9-inch balloon. A 9-inch balloon filled with mixed oxygen and acetylene gases has an explosive power that is 90 times more powerful than an M80 firecracker.

WARNING

Do not allow anyone to stand in front of the oxygen cylinder valve when opening because the oxygen is under high pressure (about 2,000 psig) and could cause severe injury when released.

Step 2 Install the oxygen regulator on the oxygen cylinder.

- Remove the cylinder protective cap.
- Open (crack) the oxygen cylinder valve just long enough to allow a small amount of oxygen to pass through the valve, then close it.
- Turn the adjusting screw on the oxygen regulator (*Figure 22*) counterclockwise until it is loose. This will shut off the regulator output and prevent overpressurizing of the hose and torch during hookup.
- Using a suitable wrench, install the oxygen regulator on the cylinder. Oxygens cylinders and regulators have right-hand threads. Tighten the nut snugly, but be careful not to overtighten the nut because this may strip the threads.

WARNING

Acetylene gas is flammable. Do not allow open flames near it.

Step 3 Install the acetylene regulator on the acetylene cylinder.

- Remove the cylinder protective cap.
- Open (crack) the acetylene cylinder valve, using the cylinder key, just long enough to allow a small amount of acetylene to pass through the valve, then close it.
- Turn the adjusting screw on the acetylene regulator counterclockwise until it is loose. This will shut off the regulator output and prevent overpressurizing of the hose and torch during hookup.
- Using a suitable wrench, install the acetylene regulator on the cylinder. Acetylene cylinders and regulators have left-hand threads. Tighten the nut snugly. Be careful not to overtighten the nut because this may strip the threads.

On Site

Portable Oxyacetylene or Oxyfuel Equipment

The equipment shown in *Figure 21* is mounted on a hand truck and is very heavy. This type of equipment is typically used in a shop or on a job site when extensive brazing must be accomplished. For normal installation or service work, portable equipment that can be hand-carried by one person is generally used.

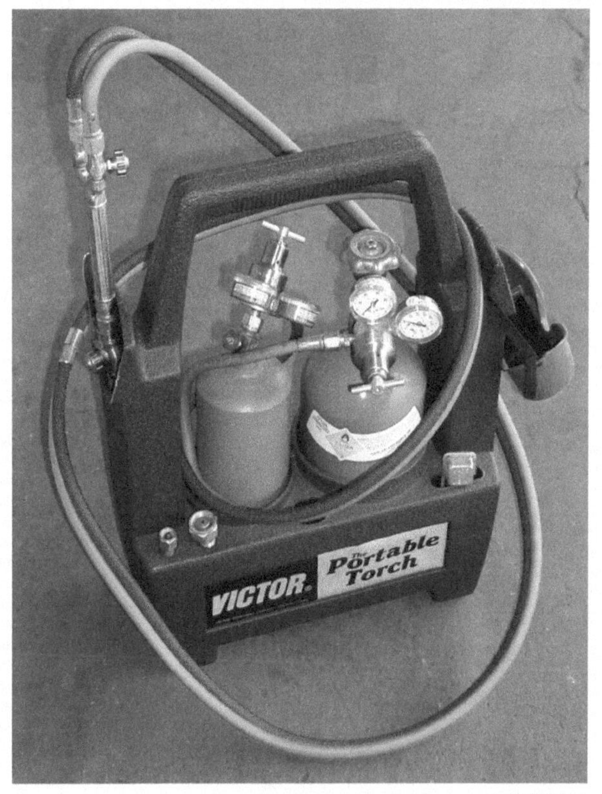

105SA02.EPS

Step 4 Install the hoses and brazing torch.

- Install flashback arrestors on the oxygen and acetylene regulators.
- Connect the green hose to the oxygen gauge and the red hose to the acetylene gauge. Tighten the hoses snugly. Be careful not to overtighten the fittings because this may strip the threads.

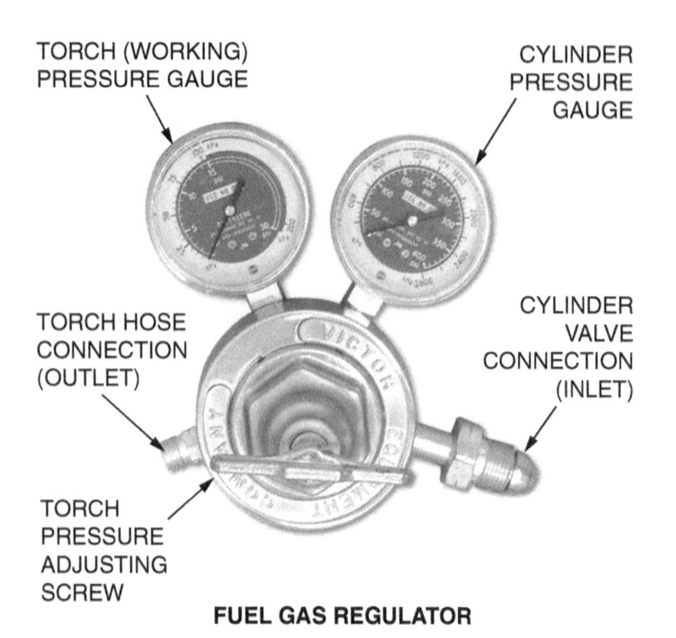

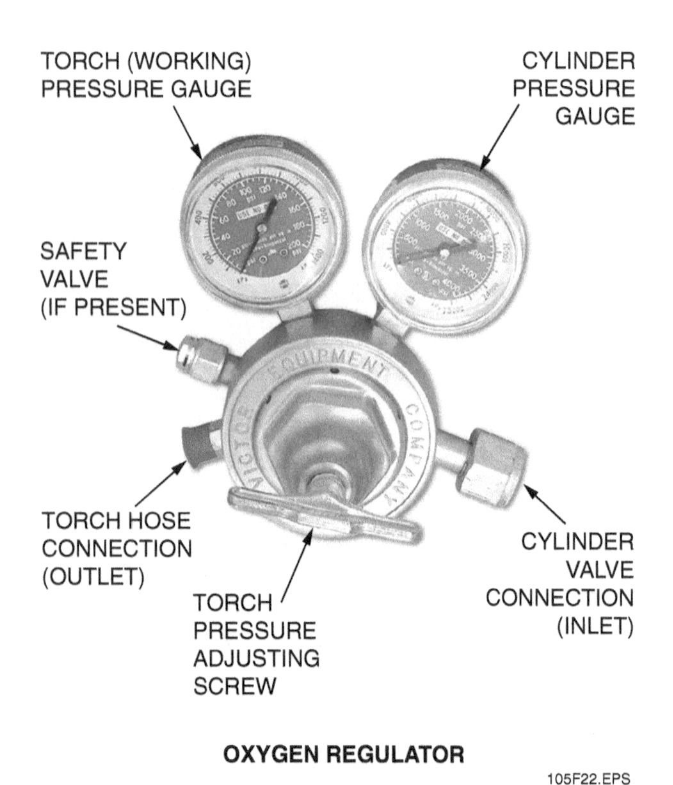

Figure 22 Oxygen and fuel gas regulators.

WARNING

Do not stand in front of the oxygen gauge because the pressure may blow the face of the gauge outward, causing personal injury.

CAUTION

Open the oxygen cylinder valve slowly because a sudden release of pressure could damage the gauges.

Step 5 Purge (clean) the oxygen hose.

- Open the oxygen cylinder valve slowly until a small amount of pressure registers on the oxygen high-pressure gauge; then slowly open the valve completely.

On Site

Protective Valve Caps

The protective valve cap is one of the most important pieces of safety equipment associated with brazing.

105SA01.EPS

- Turn the oxygen regulator adjusting screw clockwise until a small amount of pressure shows on the oxygen working-pressure gauge. Allow a small amount of pressure to build up and purge the oxygen hose, cleaning it.
- Turn the oxygen regulator adjusting screw counterclockwise until it is loose. This will shut off the regulator output.

Step 6 Purge (clean) the acetylene hose.

- Open the acetylene cylinder valve slowly until a small amount of pressure registers on the acetylene high-pressure gauge. Then open it about ½ turn.
- Turn the acetylene regulator adjusting screw clockwise until a small amount of pressure shows on the acetylene working-pressure gauge (*Figure 22*). Allow a small amount of pressure to build up and purge the acetylene hose, cleaning it.
- Turn the acetylene regulator adjusting screw counterclockwise until it is loose. This will shut off the regulator output.

Step 7 Install flashback arrestors on the torch.

Step 8 Install the brazing torch on the ends of the hoses and close the valves on the torch.

WARNING

Never adjust the acetylene regulator higher than 15 psig because acetylene becomes unstable and volatile above this pressure.

Step 9 Check the oxyacetylene equipment for leaks.

- Adjust the acetylene regulator adjusting screw for 10 psig on the working-pressure gauge.
- Adjust the oxygen regulator adjusting screw for 40 psig on the working-pressure gauge.

WARNING

Do not use oil-based soap for leak testing because the mixture of oil and oxygen may cause an explosion.

- Close the oxygen and acetylene cylinder valves and check for leaks. If the working-pressure gauges remain at 10 and 40 psig, there are no leaks in the system. If the readings drop, there is a leak. Use a soap solution or commercial leak detection fluid to check the oxygen or acetylene connections for leaks.
- Open both valves on the torch to release the pressure in the hoses. Watch the working-pressure gauges until they register zero, then close the valves on the torch.
- Turn the oxygen and acetylene regulator valves counterclockwise until they are loose. This will release the spring pressure on the regulator diaphragms and completely close the regulators.

Step 10 Coil the hoses and hang them on the hose holder.

6.5.3 Lighting the Oxyacetylene Torch

After the oxyacetylene brazing equipment has been properly set up, the torch can be lit and the flame adjusted for brazing. There are three types of flames: neutral, carburizing (reducing), and oxidizing (*Figure 23*). The neutral flame burns equal amounts of oxygen and acetylene. The inner cone is bright blue in color, surrounded by a fainter blue outer flame envelope that results when the oxygen in the air combines with the superheated gases from the inner cone. A neutral flame is used for almost all fusion welding or heavy brazing applications.

A carburizing (reducing) flame has a white feather created by excess fuel. The length of the feather depends on the amount of excess fuel in the flame. The outer flame envelope is brighter than that of a neutral flame and is much lighter in color. The excess fuel in the carburizing flame produces large amounts of carbon. The carburizing flame is cooler than the neutral flame and is used for light brazing to prevent melting the base metal.

An oxidizing flame has an excess amount of oxygen. Its inner cone is shorter, with a bright blue edge and a lighter center. The cone is also more pointed than the cone of a neutral flame. The outer flame envelope is very short and often fans out at the ends. The hottest flame, it is sometimes used for brazing cast iron or other metals.

CARBURIZING FLAME

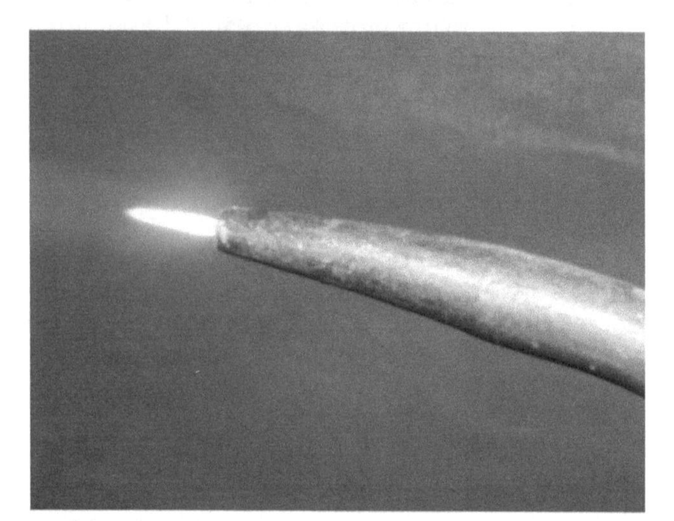

NEUTRAL FLAME

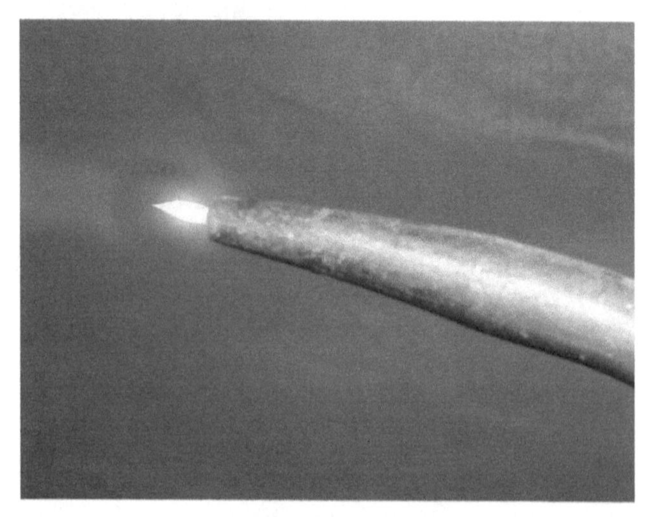

OXIDIZING FLAME

105F23.EPS

Figure 23 Types of flames.

> **WARNING**
>
> Do not use a match or a gas-filled lighter to light a torch. This could result in severe burns and/or could cause the lighter to explode.

Always use a friction lighter (*Figure 24*), also known as a striker or spark-lighter, to ignite the cutting torch. The friction lighter works by rubbing a piece of flint on a steel surface to create sparks.

Use the following procedure to light an oxyacetylene torch:

Step 1 Set up the oxyacetylene torch as discussed previously. Make sure the correct tip is installed before lighting the torch. (Refer to *Table 5* for recommended tip sizes.) Adjust the regulators for the pressure settings recommended by the torch manufacturer.

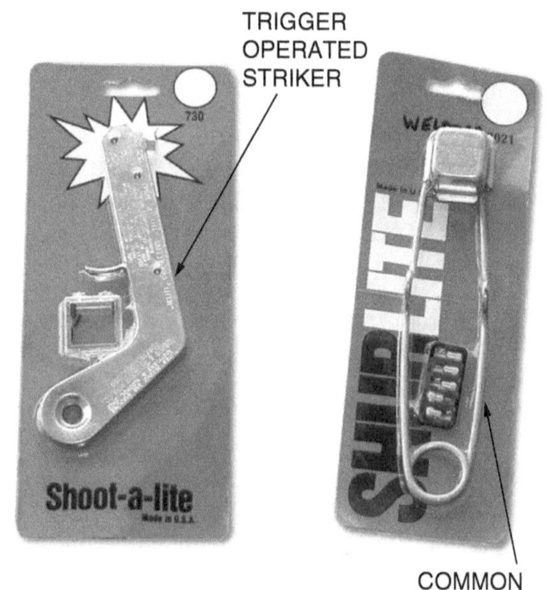

Figure 24 Typical friction lighters.

Table 5 Tip Sizes Used for Common Pipe Sizes

Tip Size (No.)	Rod Size (inches)	Pipe and Fitting Diameter (inches)
4	3/32	1/4–3/8
5	1/8	1/2–3/4
6	3/16	1–1 1/4
7	1/4	1 1/2–2
8	3/16	2–2 1/2
9	3/8	3–3 1/2
10	7/16	4–6

105T05.EPS

NOTE

Pressures are not standardized for oxyacetylene torches. Refer to the manufacturer's instructions for recommended gas pressures for the pipe size being brazed.

Step 2 Adjust the torch oxygen system.

WARNING

Do not stand in front of the oxygen gauge because the pressure may blow the face of the gauge outward, causing personal injury.

CAUTION

Open the oxygen cylinder valve slowly because a sudden release of pressure could damage the gauges.

- Open the oxygen cylinder valve slightly until pressure registers on the oxygen high-pressure gauge; then open the valve fully.
- Turn the oxygen regulator adjusting screw clockwise until pressure shows on the oxygen working-pressure gauge.
- Open the oxygen valve on the torch handle.
- Turn the oxygen regulator adjusting screw clockwise until about 20 to 25 psig registers on the oxygen working-pressure gauge. Always adjust the pressure with the torch valve open. When it is closed, the pressure may register higher.
- Close the oxygen valve on the torch handle.

Step 3 Adjust the torch acetylene system. Be sure to leave the valve wrench on the acetylene cylinder valve so that the valve can be closed quickly in case of an emergency.

- Open the acetylene cylinder valve slightly until pressure registers on the acetylene high-pressure gauge; then open the valve about ½ turn.
- Turn the acetylene regulator adjusting screw clockwise until pressure shows on the acetylene working-pressure gauge.
- Open the acetylene valve on the torch handle.
- Turn the acetylene regulator adjusting screw clockwise until about 5 psig registers on the acetylene working-pressure gauge.
- Close, then open the torch acetylene valve about ½ turn.

On Site

Flashback Arrestors

Flashback arrestors are special one-way valves that prevent a pressurized flame from traveling back up the tip and into the hoses and regulators. A flashback (back-burn) can sometimes happen if either the oxygen or the fuel gas flow rate is inadequate, or if both the oxygen and fuel gas are turned on and then ignited at the welding tip. To be effective, flashback arrestors must be installed with the flow arrow pointing toward the torch handle.

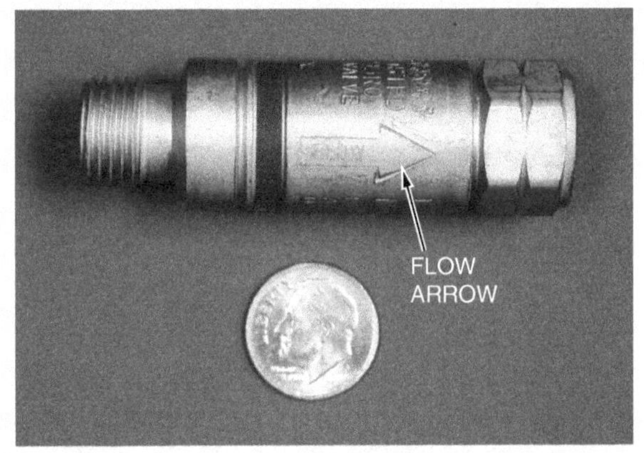

105SA03.EPS

WARNING

When lighting the torch, be sure to:

- Wear gloves and goggles.
- Hold the striker near the end of the torch tip. Do not cover the tip with the striker. Always use a striker to light the torch. Never use matches, cigarettes, or an open flame because this could result in severe burns or cause the lighter to explode. Also, make sure the torch is not pointed toward people or toward any flammable material.
- Light the fuel gas first, then open the oxygen valve on the torch.
- Shut off the gas immediately any time a flame appears from a leak in a hose.

Step 4 Light the oxyacetylene torch.

- Hold the striker in one hand and the torch in the other hand. Strike a spark in front of the escaping acetylene gas.
- Open the acetylene valve on the torch until the flame jumps away from the tip about $1/16$ inch. Then close the valve until the flame just returns to the tip. This sets the proper fuel gas flow for the size tip being used.
- Open the oxygen valve on the torch slowly to add to the burning acetylene.

> **NOTE**
>
> Observe the luminous cone at the tip of the nozzle and the long, greenish envelope around the flame, which is excess acetylene that represents a carburizing flame. As you continue to add oxygen, the envelope of acetylene should disappear. The inner cone will appear soft and luminous, and the torch will make a soft, even, blowing sound. This indicates a neutral flame, which is the ideal flame for brazing. If too much oxygen is added, the flame will become more pointed and white in color, and the torch will make a sharp snapping or whistling sound. For brazing thin materials, a cooler, carburizing flame can be used to help prevent accidental melting of the base metal.

Step 5 Shut off the torch when brazing is finished.

- First, shut off the acetylene valve on the torch. This will cause the torch to pop and prevent carbon buildup at the tip.
- Shut off the oxygen valve on the torch.
- Shut off both the oxygen and acetylene cylinder valves completely.
- Open both valves on the torch to release the pressure in the hoses. Watch the working-pressure gauges until they register zero, then close the valves on the torch.
- Turn the oxygen and acetylene regulator valves counterclockwise until loose. This will release the spring pressure on the diaphragms in the regulators.

Step 6 Coil the hoses and hang them on the hose holder.

6.6.0 Heating Techniques

Applying the proper amount of heat and reaching the required temperature with brazing is more difficult than with soldering. Brazing temperatures can exceed 1,200°F to 1,500°F. Apply heat to the parts to be joined, preferably with an oxyfuel torch with a neutral flame. Air-fuel is sometimes used on smaller sizes. Heat the tube first, beginning about one inch from the edge of the fitting, sweeping the flame around the tube in short strokes at right angles to the axis of the tube. It is very important that the flame be kept in motion and not remain on any one point long enough to damage the tube. The result will be a smooth joint with no globs of unmelted brazing material present. *Figure 25* shows the tube and fitting being preheated and brazed.

The higher brazing temperatures require more precautions against fire than are required at soldering temperatures. Be certain that a proper fire extinguisher is at hand. Always provide a shield to intercept drops of brazing material and deflect flame away from structural elements and heat blocks to limit the temperature of the tube where it penetrates walls, ceilings, and joists. These protective devices must be standard equipment on any copper installation. Sheet metal deflects flame, but will transfer heat readily. A piece of sheet rock (gypsum drywall) insulates, but does not have good flame resistance. Sheet rock with protective sheet metal between the flame and the sheet rock is an easily fabricated protective method. Wet cloths or wipers wrapped around tubing can be used to block conductive heat travel.

The use of a torch in close spaces can reduce the oxygen in the air below a safe level for breathing. Fumes and products of combustion produced in the brazing process can foul the air being breathed. Brazing must be done only in an area with adequate ventilation.

> **On Site**
>
> ## Cup-Type Striker
>
> When using a cup-type striker to ignite a welding torch, hold the cup of the striker slightly below and to the side of the tip, parallel with the fuel gas stream from the tip. This prevents the ignited gas from deflecting back toward you from the cup and reduces the amount of carbon soot in the cup. The flint in a striker can be replaced.
>
> 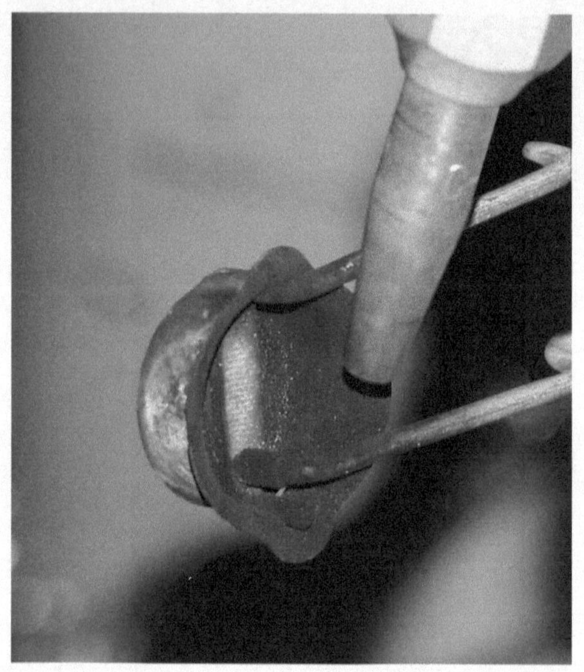
>
> 105SA04.EPS

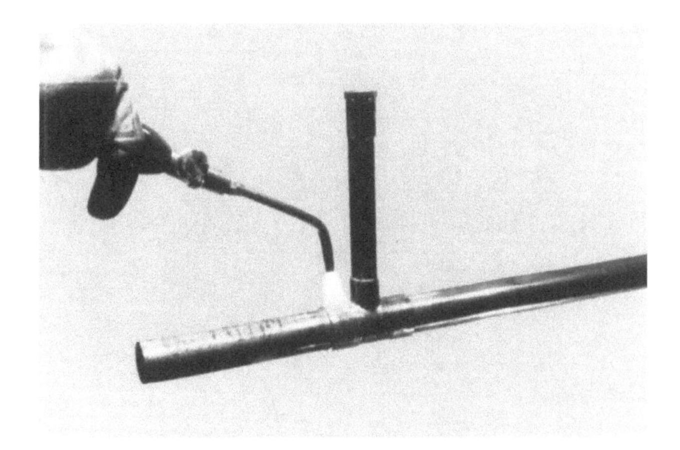

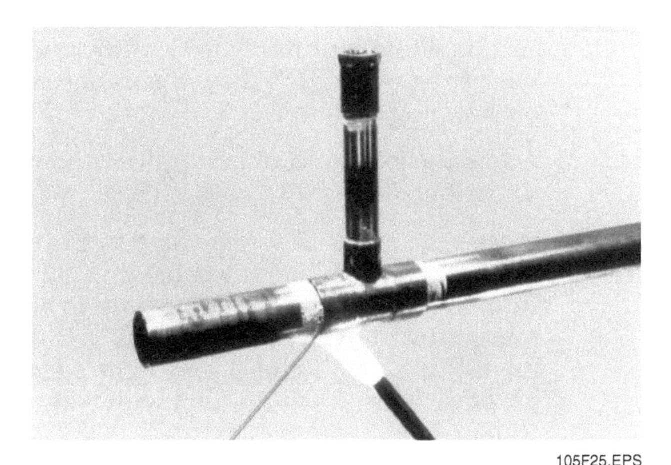

105F25.EPS

Figure 25 Preheating and brazing of tube and fitting.

6.7.0 Brazing Joints

Except for the temperature involved, brazing is very much like soldering. Capillary action is required to produce a satisfactory joint. Follow these steps to braze a joint:

Step 1 Ensure that a fire extinguisher is easily accessible in case the heat source creates a fire.

Step 2 Set up the oxyacetylene brazing equipment.

On Site

Adjusting the Fuel Gas Flame

When properly adjusted, the gas flame should return to the tip.

105SA05.EPS

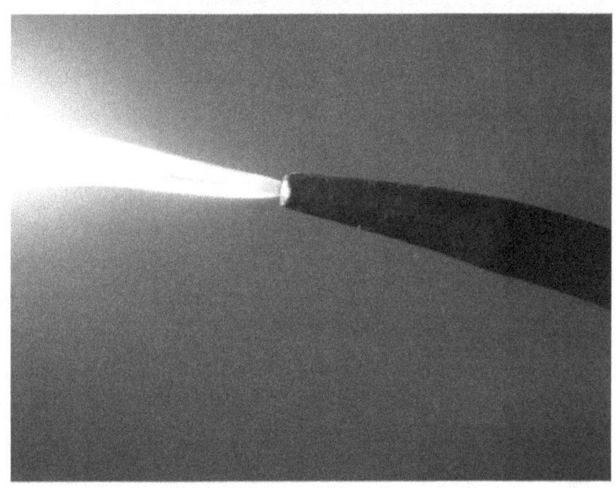

105SA06.EPS

Step 3 Ensure that the tubing to be brazed is properly supported.

Step 4 Ensure that the tubing and fitting are properly aligned and supported. Restrain the two parts so they remain in the proper place while brazing.

Step 5 Apply heat blocks/shields to the tube and between the tube and structure as appropriate.

Step 6 Apply heat to the tubing about one inch from the joint and, if brazing flux is used, observe the flux. The flux will bubble and turn white and then melt into a clear liquid.

Step 7 Switch the flame to the fitting at the base of the cup. Heat uniformly, sweeping the flame alternately from the fitting to the tube until the flux stops bubbling. Avoid excessive heating of cast fittings, due to the possibility of cracking. When the flux appears liquid and transparent, start sweeping the flame back and forth along the axis of the joint to maintain heat on the parts to be joined, especially toward the base of the cup of the fitting. The flame must be kept moving to avoid melting the tube or fitting.

Step 8 Continue to move the heat back and forth over the tubing and the fitting.

> **NOTE**
>
> Allow the fitting to receive more heat than the tubing by pausing at the fitting while continuing to move the flame back and forth. Concentrate the heat to the back of the fitting cup. For one-inch tube and larger, it may be difficult to bring the whole joint up to temperature at one time. It will often be desirable to use an oxyfuel, multiple-orifice heating tip (rosebud) to maintain a more uniform temperature over large areas. A mild preheating of the entire fitting is recommended for larger sizes, and the use of a second torch to retain a uniform preheating of the entire fitting assembly may be necessary in the largest diameters.

Step 9 Touch the filler metal rod to the joint to test whether the filler metal will melt. Keep the flame away from the filler metal itself. The temperature of the tube and fitting at the joint should be high enough to melt the filler metal. Do not use the flame to melt the filler metal.

> **NOTE**
>
> If the filler metal melts upon contact with the joint, the brazing temperature has been met and you should proceed to Step 10. If the filler metal does not melt upon contact, continue to heat and test the joint until the filler metal melts.

Step 10 Hold the filler metal rod to the joint (*Figure 26*) and allow the filler metal to enter into the joint while holding the torch slightly ahead of the filler metal and directing most of the heat to the shoulder of the fitting.

Step 11 Apply brazing material around the entire joint and continue applying heat until wicking takes place. Wicking is the capillary action that draws the brazing metal into the joint between the tube and the fitting.

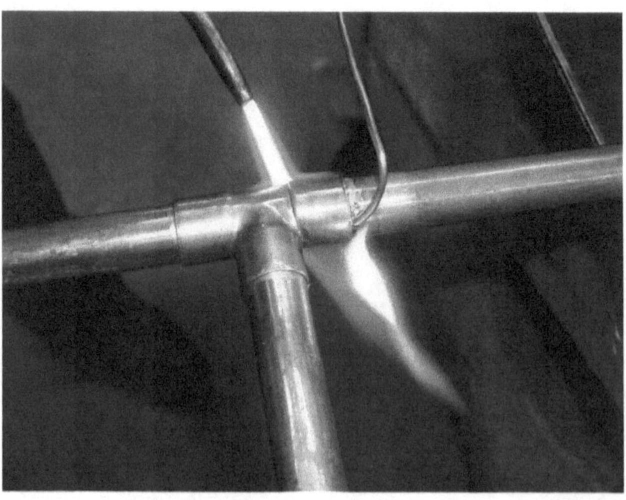

105F26.EPS

Figure 26 Heating a fitting with a torch.

Step 12 Continue to fill the joint with the filler metal until the filler metal inside the joint is at least three times the thickness of the tubing. This will not fill the joint, but a joint with this amount of penetration will be stronger than the tubing. If the tubing is two inches or more in diameter, two torches can be used to evenly distribute the heat. For larger joints, small sections of the joint can be heated and brazed. Be sure to overlap the previously brazed section while continuing to work around the fitting. *Figure 27* shows working in overlapping sectors.

Step 13 Allow the joint to cool naturally without disturbing it.

Step 14 Wash the joint with warm water to clean excess, dried, or hardened brazing flux from the joint. If the flux is too hard to be removed with water, chip the excess flux off using a small chisel and light peen hammer. Then wash the joint with warm water.

Step 15 Test all completed assemblies for joint integrity. Follow the testing procedure prescribed by applicable codes governing the intended service.

6.8.0 Capillary Action

Capillary action carries the brazing material into the joint between the tube and the fitting to provide the required bond. The standard gap is 0.001 inch to 0.005 inch. Wicking action is the key to a successful joint. This is not a process of piling brazing metal on the top of the joint, but one of causing it to penetrate between the two base metals to produce the molecular bond in a manner

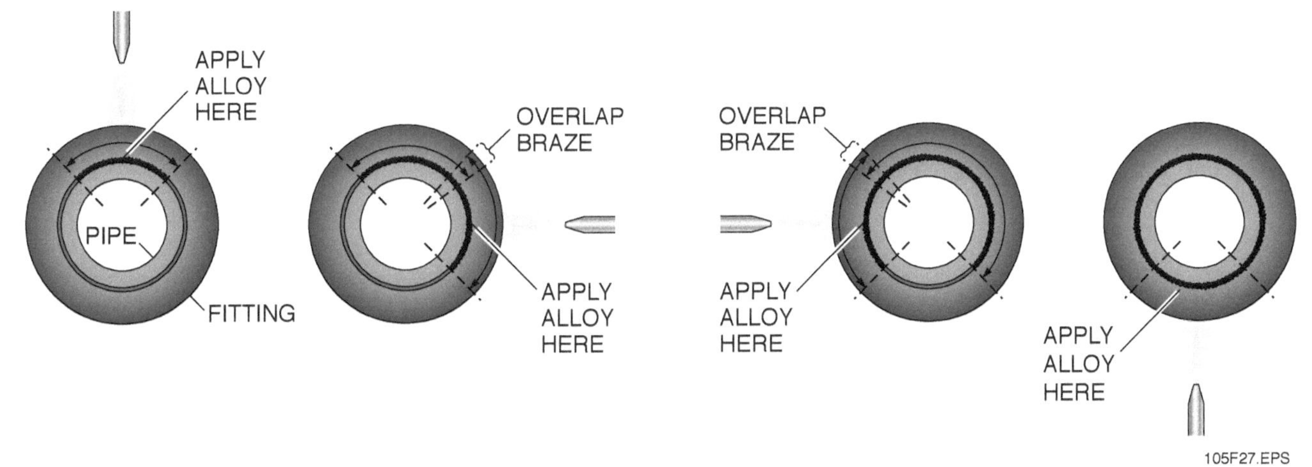

Figure 27 Working in overlapping sectors.

similar to soldering. Adequate and uniform heat must be applied to bring the base metals to the liquidus temperature of the brazing material.

If the filler metal fails to flow or has a tendency to ball up, it indicates oxidation on the metal surfaces or insufficient heat on the parts to be joined. If the tube or fitting start to oxidize during heating, there is too little flux. If the filler metal does not enter the joint and tends to flow over the outside of either member of the joint, it indicates that one member is overheated or the other is underheated.

7.0.0 Support Spacing for Copper Tube

Support spacing for copper tube is different than for other tubing materials because of the differences in the tube itself. The following sections discuss the spacing limitations for copper tube and the protection of copper tubing from hangers.

7.1.0 Spacing Limitations for Copper Tube

Strength requirements are based on maximum spacing between hangers for copper tube. *NFPA 13* recognizes that the mechanical strength of copper tube is less than that of steel pipe. This means hangers must be spaced closer together on copper tube. *Table 6* shows the maximum allowable hanger spacing for copper tube.

When certified by a registered professional engineer, closer spacing may be used to overcome the strength limitations of some hangers.

Two other differences need to be recognized between copper tube and steel pipe support requirements. For maximum operating pressures over 100 psi, the maximum unsupported distance between an end sprinkler (pendent position or drop nipple) and the last hanger on the branch line is limited to 12 inches or less for steel pipe, or 6 inches or less for copper tube.

In an armover situation with maximum pressures less than 100 psi, the maximum unsupported armover to a sprinkler, sprinkler drop, or sprig-up must be 24 inches or less for steel pipe, or 12 inches or less for copper tube.

With maximum pressure exceeding 100 psi, maximum unsupported armover for steel pipe is 12 inches and 6 inches for copper tube.

If these configurations present a problem, the branch line can always be extended to pick up the next structural member.

7.2.0 Protecting Copper Tubing from Hangers

Any time two dissimilar metals such as copper and steel come in contact with each other and moisture is present, a galvanic cell may be formed. This cell is similar to a battery in that it can generate electric current. The electric current that passes between the two materials can produce a galvanic action that may cause corrosion.

Table 6 Hanger Spacing for Copper Tube (Table 9.2.2.1(a))

Copper Tube Size	Maximum Hanger Spacing
¾"–1"	8'
1¼"–1½"	10'
2"–3"	12'
3½"–8"	15'

For SI Units: 1" = 25.4mm; 1' = 0.3048m

105T06.EPS

Although the magnitude of the current is small, over an extended period of time the damage can be significant.

A way to prevent this action is to use galvanized (zinc-coated) hanger material. The galvanic action produced between galvanized material and copper is minimal.

Another way to prevent this problem is to insulate the ferrous hanger so there is no electrical contact between the tube and the ferrous hanger material. Coating the contact area with plastic or using insulating tape are acceptable ways to provide protection.

> **NOTE**
> *NFPA 13* requires that hanger material attaching directly to pipe be Listed by a recognized testing laboratory.

8.0.0 Grooved and Compression Couplings for Copper Tube

Grooved couplings and fittings specifically Listed for copper tube may be used to join copper tubing. These are Listed by UL and FM Global to 175 psi operating pressure for joining 2" through 6" copper tubing. Grooving the copper tube requires a special copper roll groover. It follows the conventional methods of roll grooving of steel pipe. *Figure 28* shows a roll grooved coupling.

The dimensions pointed out by the letters in *Figure 28* must comply with engineering specifications. The A dimension is the distance from the pipe end to the groove and provides the gasket seating area. This area must be free from indentations, projections, or roll marks to provide a leak-proof sealing seat for the gasket. The B dimension is the groove width. It controls expansion and angular deflection based on its distance from the tube end and its width in relation to the width of the coupling housing key. The C dimension is the proper diameter of the base of the groove. This dimension must be within the given diameter tolerance and concentric with the outside diameter (OD) of the pipe. The D dimension is the

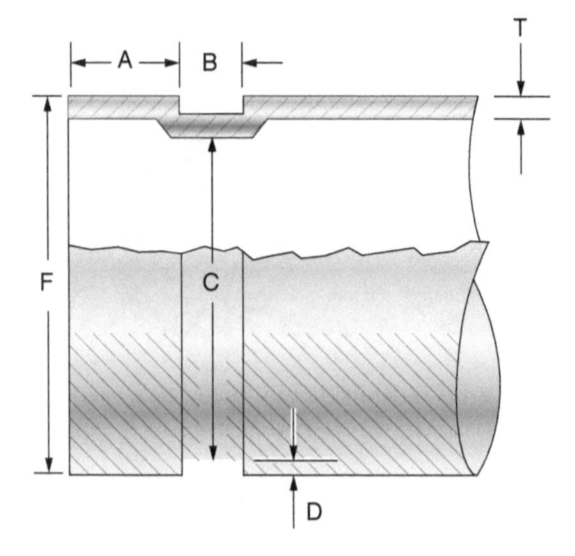

Figure 28 Roll grooved.

nominal depth of the groove and is used as a trial reference only. This dimension must be changed if necessary to keep the C dimension within the stated tolerances. The F dimension is used with the standard roll only and gives the maximum allowable pipe end flare. The T dimension is the lightest grade or minimum thickness of pipe suitable for roll or cut grooving.

Press fittings specifically listed for copper tube may be used to join copper tubing. These are listed to 175 psi operating pressure for joining ¾" through 4" copper tubing.

Examine the copper tubing and fittings for defects, sand holes, or cracks. Do not use tubing or fittings with defects. Cut copper tubing with a wheeled tubing cutter or approved copper tubing cutting tool. Cut the tubing square to permit proper joining with the fittings. Fully insert the tubing into the fitting and mark the tubing at the shoulder of the fitting. Check the fitting alignment against the mark on the tubing to ensure that the tubing is fully engaged (inserted) in the fitting. Press the joints using the tool(s) approved by the manufacturer.

Summary

Soldering is the joining of two pieces of metal with a dissimilar metal. The melting point for soldering is above 840°F. The solder permitted in wet pipe fire sprinkler systems is the 95/5 (tin-antimony) alloy. A proper solder joint requires a tight fit between mating pieces, as well as clean surfaces, proper fluxing, and heating of the pieces being joined to the liquidus temperature of the solder. Soldering has occurred when the required amount of solder metal wicks into the joint via capillary action. To ensure the desired seal against liquids, the assembly must remain still until the solder has cooled. Good craftsmanship requires the removal of excess flux as the final act in soldering.

Brazing has many common characteristics with soldering. The joining temperatures and the fluxing operation are higher. The need for fire and personal safety precautions increase at these higher temperatures. Successful joints depend on proper deburring, a tight fit between the two parts, and raising the temperature to the proper level to secure the wicking or capillary action that carries the brazing alloy into the joint.

Review Questions

1. Which of these types of copper tube is permitted for use in fire sprinkler systems by *NFPA 13, 13R,* and *13D*?
   a. DWV, ACR, and K
   b. K, L, and ACR
   c. K, L, and M
   d. ACR, M, and L

2. Galvanic action, which can lead to corrosion of copper tube, is caused by _____.
   a. overheating the pipe
   b. leaving the pipe out in the rain
   c. joining dissimilar metals
   d. using the wrong flux

3. Tube cutters can be used to cut pipe up to _____.
   a. 1" in diameter
   b. 2" in diameter
   c. 3" in diameter
   d. 4" in diameter

4. In the soldering process established by *ASTM B828,* which of these groupings is in the correct sequence?
   a. Reaming, cleaning, fluxing
   b. Heating, cleaning, fluxing
   c. Heating, reaming, cutting
   d. Heating, reaming, fluxing

5. Cleaning to remove oxides and surface soil from tube ends is done because oxides and surface soil can interfere with capillary action.
   a. True
   b. False

6. The filler metal used for brazing usually melts at a temperature of _____.
   a. 375°F to 500°F
   b. 750°F to 850°F
   c. 850°F to 1,200°F
   d. 1,200°F to 1,500°F

7. When using oxyacetylene brazing equipment, the acetylene regulator should never be adjusted for a pressure greater than _____.
   a. 15 psig
   b. 40 psig
   c. 100 psig
   d. 250 psig

8. Almost all brazing applications use a(n) _____.
   a. oxidizing flame
   b. carburizing flame
   c. feather flame
   d. neutral flame

9. During the brazing process, the flame is applied directly to the filler metal in order to melt it.
   a. True
   b. False

10. Grooved couplings and fittings are approved for use on copper tubing from 2" to _____.
    a. 4" in diameter
    b. 5" in diameter
    c. 6" in diameter
    d. 7" in diameter

Trade Terms Quiz

Fill in the blank with the correct trade term that you learned from your study of this module.

1. A(n) _________ is a soft solder joint, or a joint soldered below 840°F.
2. _________ refers to water that is fit for human consumption.
3. A descriptive term for solder that is flowing readily and that has been melted by heating is _________.
4. A(n) _________ is a nonconductor of electricity.
5. Any substance made up of two or more metals is called a(n) _________.
6. _________ is an alloy that is melted and used to join metallic surfaces.
7. The process by which the oxygen in the air combines with metal to produce tarnish and rust is _________.
8. _________ refers to direct-current electricity, especially when produced chemically.
9. _________ is a method of fusing metals with a nonferrous filler metal using heat above 840°F, but below the melting point of the base metals being joined.
10. _________ is a method of joining metals with a nonferrous filler metal using heat below 840°F and below the melting point of the base metals being joined.
11. A mixture of oxygen and acetylene that is used to produce an extremely hot flame for cutting, welding, and brazing metal is _________.
12. _________ is a chemical substance that aids the flow of solder and removes and prevents the formation of oxides on the pieces to be joined by soldering.
13. Boric acid salts used as a flux are called _________.
14. Copper tube fittings that are soldered or brazed as opposed to threaded are called _________.
15. The release of compressed gas to the atmosphere through some part or parts for the purpose of removing contaminants is called a(n) _________.
16. A slight surface depression is called a(n) _________.
17. To _________ is to draw solder into a joint.
18. _________ is the movement of a liquid along the surface of a solid in a kind of spreading action.
19. A descriptive term for solder in the solid or stable phase is _________.

Trade Terms

Alloy
Borates
Brazing
Capillary action
Dielectric
Dimple
Flux
Galvanic
Liquidus
Oxidation
Oxyacetylene
Potable
Purge
Solder
Soldering
Solidus
Sweat joint
Wick
Wrought fittings

Cornerstone of Craftsmanship

Shaun Rushlo

Cox Fire Protection, Inc.
Foreman

When Shane Rushlo first applied for a job at Cox Fire Protection, the company owner wasn't sure what to expect from him. But Shane worked hard and soon won the gold medal at the Associated Builders and Contractors (ABC) Sprinkler Fitter Championships. Five years later he is a foreman, and his boss is very proud of his success.

How did you choose a career in the sprinkler fitting field?
At the time I just needed a job. I was dating a girl, and there was an opening at her father's company. I applied and they put me to work. I have been here ever since.

What types of training have you been through?
I started with the ABC apprenticeship program. I have also taken the OSHA 10-hour safety course, and forklift, scissor lift, and off-road vehicle training courses. There were several courses offered by manufacturers that showed me how to work with different types of pipe and tools. I think you should take a class before using a new tool or specialty products. It is important to learn the proper use of the tool and how to operate it safely.

What kinds of work have you done in your career?
I have installed many different types of sprinkler systems, both residential and commercial. I also installed fire pumps and specialty systems. Now I am a foreman and have additional responsibilities.

What do you like about your job?
One of my favorite things about being a foreman is training new people. I like to teach people to do things the right way.

I also like doing specialty systems, like double interlock systems. I like the challenge of learning how to install new types of systems.

I really enjoyed the trip to Las Vegas for the competition.

What factors have contributed most to your success?
Be a leader and not a follower. Nobody wants to be a helper forever. Be open to new ideas. Listen to your elders, people with more experience. Listen to their way, compare it to your way, and come up with the best way.

What advice would you give to those new to the sprinkler fitter field?
Listen to other people. Don't be stubborn and think you know everything. Being stubborn definitely does not work in this business. Learn something new every day. If you aren't learning something new every day, you are doing something wrong.

Don't get frustrated. Many people think that the program is a little slow in the first year and quit. ABC apprentice training is a good program. Stick with it, because it picks up in the second and third year. And then you are off and running.

Learn to do new things like specialty systems at a young age. That can give you an advantage, understanding new things that the older people don't know.

Trade Terms Introduced in This Module

Alloy: Any substance made up of two or more metals.

Borates: Boric acid salts used as a flux.

Brazing: A method of fusing metals with a nonferrous filler metal using heat above 840°F, but below the melting point of the base metals being joined.

Capillary action: The movement of a liquid along the surface of a solid in a kind of spreading action.

Dielectric: A nonconductor of electricity.

Dimple: A slight surface depression, or the act of creating one.

Flux: A chemical substance that aids the flow of solder and removes and prevents the formation of oxides on the pieces to be joined by soldering.

Galvanic: Refers to direct-current electricity, especially when produced chemically.

Liquidus: The state of solder when it has melted by heating and is flowing readily.

Oxidation: The process by which the oxygen in the air combines with metal to produce tarnish and rust

Oxyacetylene: A mixture of oxygen and acetylene that is used to produce an extremely hot flame for cutting, welding, or brazing metal.

Potable: Refers to water that is fit for human consumption.

Purge: The release of compressed gas to the atmosphere through some part or parts, such as a hose or pipeline, for the purpose of removing contaminants.

Solder: An alloy that is melted and used to join metallic surfaces.

Soldering: A method of joining metals with a nonferrous filler metal using heat below 840°F and below the melting point of the base metals being joined.

Solidus: The solid or stable phase of solder.

Sweat joint: A soft solder joint, or a joint soldered below 840°F.

Wick: To draw solder into a joint.

Wrought fittings: Copper tube fittings that are soldered or brazed as opposed to threaded.

Additional Resources

This module presents thorough resources for task training. The following resource material is suggested for further study.

Cast Copper Solder-Joint Pressure Fittings, ASME B16.18, Current Edition. New York, NY: American Society of Mechanical Engineers.

How to Successfully Install Copper Fire Sprinkler Systems, 1989. Video. American Fire Sprinkler Association.

NFPA 13, Standard for the Installation of Sprinkler Systems, Latest Edition. Quincy, MA: National Fire Protection Association.

Specification for Seamless Copper Water Tube, ASTM B88, Latest Edition. West Conshohocken, PA: American Society for Testing and Materials International.

Specification for Seamless Copper Pipe, ASTM B42, Latest Edition. West Conshohocken, PA: American Society for Testing and Materials International.

Standard for the Installation of Sprinkler Systems in One- and Two-Family Dwellings and Manufactured Homes, NFPA 13D, Latest Edition. Quincy, MA: National Fire Protection Association.

Standard for the Installation of Sprinkler Systems in Residential Occupancies Up to and Including Four Stories in Height, NFPA 13R, Latest Edition. Quincy, MA: National Fire Protection Association.

Victaulic Field Assembly and Installation Instruction Pocket Handbook I-100. Easton, PA: Victaulic Company of America.

Wrought Copper and Copper Alloy Solder-Joint Pressure Fittings, ASME B16.22, Latest Edition. New York, NY: American Society of Mechanical Engineers.

Figure Credits

Copper Development Association, Module opener, 105F12, 105F13, 105F15, 105F16, 105F18, 105F19, 105F25

National Fire Protection Association, 105T01, 105T06

Reprinted with permission from *NFPA 13-2013, Installation of Sprinkler Systems*, Copyright © 2012, National Fire Protection Association, Quincy, MA 02169. This reprinted material is not the complete and official position of the National Fire Protection Association on the referenced subject which is represented only by the standard in its entirety.

Capitol Manufacturing Company, 105F02

T-DRILL Industries Inc., 105F03, 105F04

Ridge Tool Co./Ridgid®, 105F05, 105F07 (chop saw), 105F09, 105F11 (tube cutter, fitting brush)

DeWalt Industrial Tool Co., 105F07 (portable band saw)

BernzOmatic, 105F11 (soldering torch)

Oatey Co., 105F11 (wire solder, cleaning materials, flux brush and solder paste)

Topaz Publications, Inc., 105F21–105F24, 105F26, 105SA01-105SA06

MODULE 18105-13 — ANSWERS TO REVIEW QUESTIONS

Answer		Section
1.	c	2.0.0
2.	c	2.3.0
3.	d	3.0.0
4.	a	5.3.0
5.	a	5.3.3
6.	d	6.0.0
7.	a	6.5.1
8.	d	6.5.3
9.	b	6.7.0
10.	c	8.0.0

MODULE 18105-13 — ANSWERS TO TRADE TERMS QUIZ

1. Sweat joint
2. Potable
3. Liquidus
4. Dielectric
5. Alloy
6. Solder
7. Oxidation
8. Galvanic
9. Brazing
10. Soldering
11. Oxyacetylene
12. Flux
13. Borates
14. Wrought fittings
15. Purge
16. Dimple
17. Wick
18. Capillary action
19. Solidus

NCCER CURRICULA — USER UPDATE

NCCER makes every effort to keep its textbooks up-to-date and free of technical errors. We appreciate your help in this process. If you find an error, a typographical mistake, or an inaccuracy in NCCER's curricula, please fill out this form (or a photocopy), or complete the online form at **www.nccer.org/olf**. Be sure to include the exact module ID number, page number, a detailed description, and your recommended correction. Your input will be brought to the attention of the Authoring Team. Thank you for your assistance.

Instructors – If you have an idea for improving this textbook, or have found that additional materials were necessary to teach this module effectively, please let us know so that we may present your suggestions to the Authoring Team.

NCCER Product Development and Revision

13614 Progress Blvd., Alachua, FL 32615

Email: curriculum@nccer.org
Online: www.nccer.org/olf

❏ Trainee Guide ❏ AIG ❏ Exam ❏ PowerPoints Other ______________________________

Craft / Level: Copyright Date:

Module ID Number / Title:

Section Number(s):

Description:

Recommended Correction:

Your Name:

Address:

Email: Phone:

Underground Pipe

18106-13

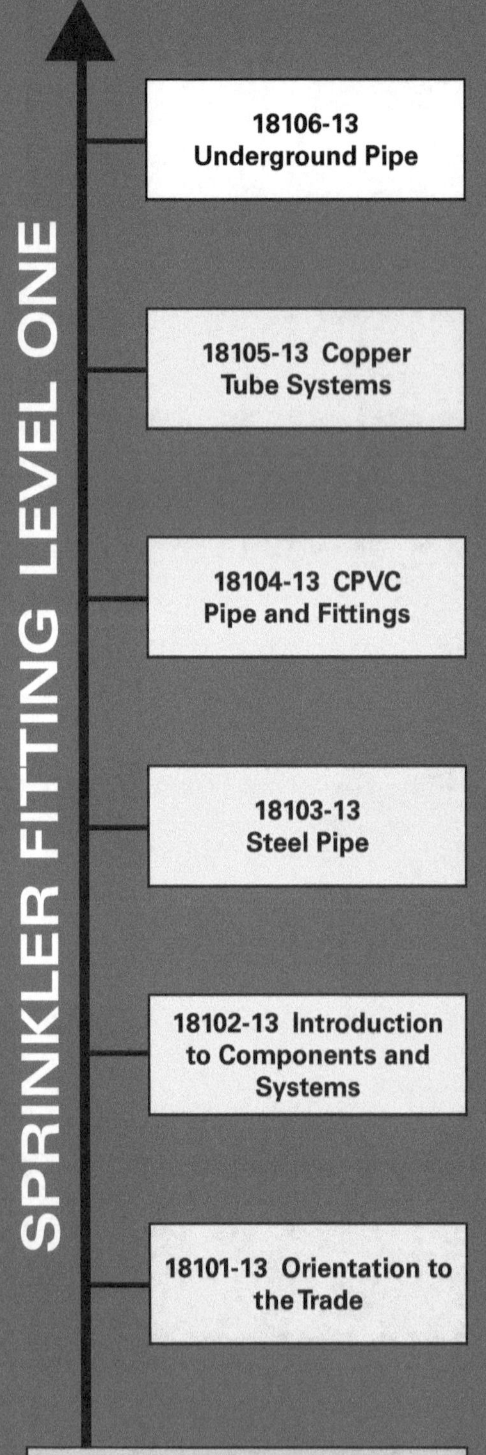

This course map shows all of the modules in *Sprinkler Fitting Level One*. The suggested training order begins at the bottom and proceeds up. Skill levels increase as you advance on the course map. The local Training Program Sponsor may adjust the training order.

Objectives

When you have completed this module, you will be able to do the following:

1. Identify types and properties of soil.
2. Identify trenching safety requirements.
3. Explain sloping requirements for different types of soil.
4. Explain how to dig trenches.
5. Describe excavation support (shoring) systems.
6. Describe types of bedding material.
7. Identify and describe types of underground pipe.
8. Describe thrust blocks and restraints.
9. Identify and describe hydrants, yard valves, hydrant houses, and associated equipment.
10. Explain testing, inspection, and chlorinating of underground pipe.
11. Fill out an underground test certificate.

Trade Terms

Angle of repose
Atmospheric vacuum breaker
Back siphonage
Backflow
Bedding
Bell hole
Benching
Carbonation
Competent person
Compression strength
Cross braces
Cross-connection
Disturbed soil
Ditch breakers
Double-check valve assembly (DCV)
Dual-check valve backflow preventer (DC)
Excavation
Hydration
Intermediate atmospheric vent vacuum breaker
Manufactured air gap
Pressure-type vacuum breaker
Proctor
Protective system
Reduced-pressure zone principle backflow preventer (RPZ)
Shield
Shore
Shoring
Silt
Skeleton
Sloping
Spline
Subsidence
Support system
Test cock
Tight sheeting
Trench
Trench box
Uprights
Wales

Prerequisites

Before you begin this module, it is recommended that you successfully complete *Core Curriculum; Sprinkler Fitting Level One*, Modules 18101-13 through 18105-13.

NOTE

In some locations, sprinkler fitter trainees will be required to perform metric conversions. A metric conversion chart is provided in *Appendix A* at the back of this module.

Contents

Topics to be presented in this module include:

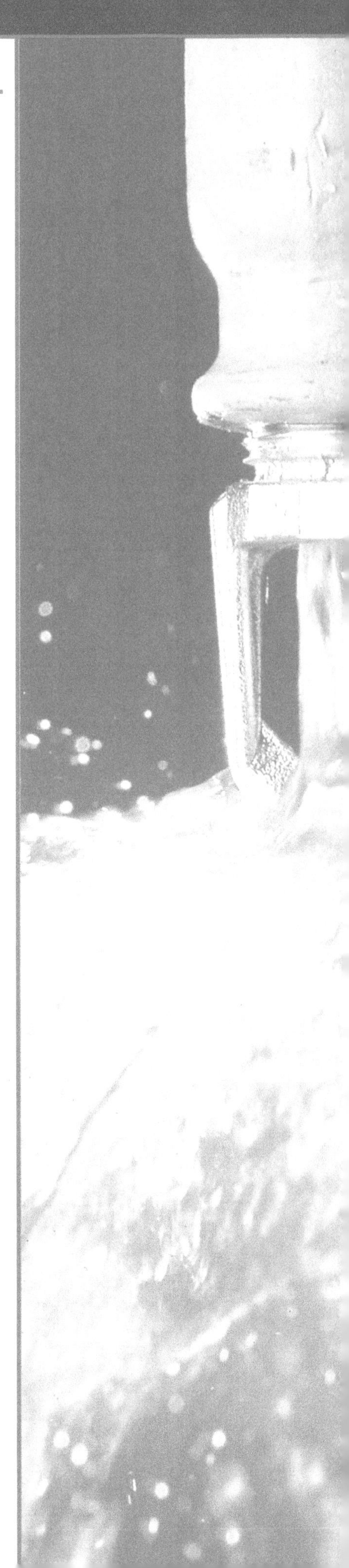

Contents (continued)

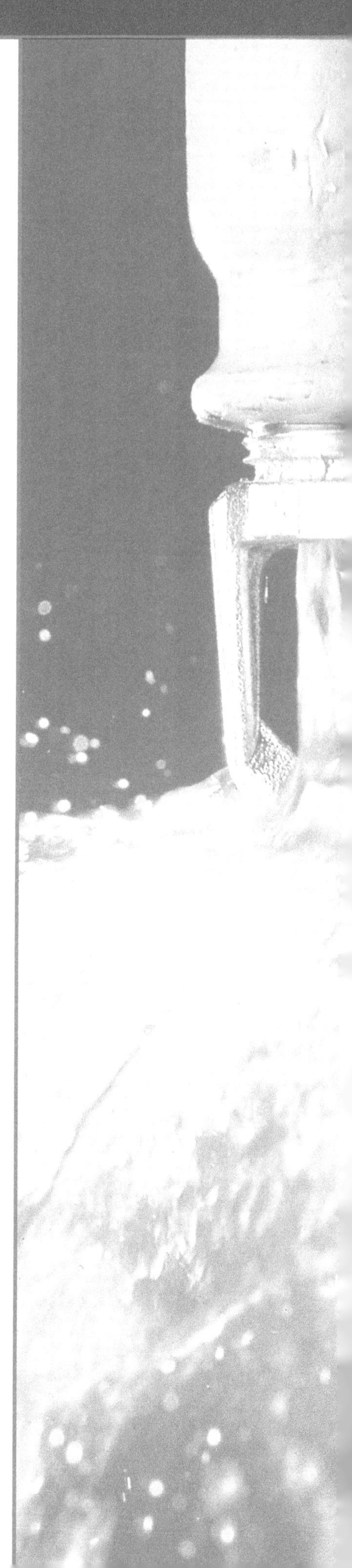

Figures and Tables

1.0.0 Introduction

A primary concern of a sprinkler fitter working in a trench is safety. Always be aware of your surroundings, and watch and listen for possible signs of trouble that might indicate trench failure. Following the procedures and methods described in this module will go a long way toward creating a safe and professional working environment.

The key to safe pipe laying and safe working conditions in general is to understand the materials you work with, including soil properties, trench shoring systems, and piping systems.

This module provides information on how to work with different types of soil so that you can lay pipe safely and competently. You will learn about the various types of soils and how the soil type determines how trenches are dug and supported. You will also learn about trenching hazards, trench failures, and how to make trenches safe for you and your co-workers.

After the trench is dug, pipe must be laid on the proper bedding. This module describes bedding materials and explains proper bedding techniques. You will learn how to use thrust blocks and restraints to control pipe movement in the trench. Types of underground pipe are discussed along with in-building risers and double check-valve assemblies. You will also learn about hydrants, yard valves, hydrant houses, and related components.

The sprinkler fitter must understand the proper way to install underground pipe. Installation includes laying, inspecting, testing, flushing, and chlorinating. Keep in mind that the AHJ responsible for inspecting bedding, backfilling, and compaction may not be the fire department. It could be the water department or public works.

This module will be referencing *NFPA 24, Standard for the Installation of Private Fire Service Mains and Their Appurtenances,* throughout. Although *NFPA 24* is the standard for private fire service mains for systems installed according to the *NFPA 13* standard, if you are installing only sprinkler systems *NFPA 13* is sufficient because it includes all of the relevant tables from *NFPA 24* in addition to standards applicable to underground installations. For anything not covered in *NFPA 13,* refer to *NFPA 24.* Note that *NFPA 24* does not apply to:

- Mains that are under the control of a water utility
- Privately owned mains that provide fire protection and/or domestic water operated as water utilities
- Underground water mains that serve sprinkler systems designed and installed in accordance with *NFPA 13R* and under 4 inches in size, and *NFPA 13D* systems

2.0.0 Trenching Hazards

WARNING

When a trench is being dug, it is possible to encounter buried hazardous materials or old sewer or septic systems. If such an incident occurs, notify the proper authorities because the materials can make you sick. Some of these materials are highly toxic and can be handled only by trained specialists.

Safety is crucial during any excavation job. During an excavation for underground piping, earth is removed from the ground, creating a trench (*Figure 1*). A trench is a narrow excavation cut into the ground. The depth of a trench is greater than the width, and the width does not exceed 15' at the bottom. The soil that is removed from the trench is called spoil. When earth is removed from the trench, extreme pressures may be generated on the trench walls. If the walls are not properly secured by shoring, sloping, or shielding, they will collapse.

The collapse of unsupported trench walls can instantly crush and bury workers. This type of collapse happens because not enough material is available to support the walls of an excavation. One cubic yard of earth weighs approximately 3,000 pounds; that's the weight of a small car. That's more than enough weight to seriously injure or kill a worker. In fact, each year in the United States, more than 100 people are killed and many more are seriously injured in cave-in accidents.

Working in and around excavations is one of the most hazardous jobs you will ever do. Safety precautions must be exercised at all times to prevent injury to yourself and others. Some of the hazards you may encounter during an excavation include the following:

106F01.EPS

Figure 1 Excavation site.

On Site

Underground Utilities

Before beginning to dig, determine if there are any utilities or other structures buried on the site. Utility companies go to great lengths to inform the public and contractors where their lines are to prevent damage to them. The survey crew will set stakes to mark underground utilities. These markers are color-coded using the standard underground color code shown here. When you see these markers, act accordingly. If you do not know what to do, ask your supervisor.

Color	Meaning
Red	Electric power lines, cables, conduit and lighting cables
Yellow	Gas, oil, steam, petroleum, or gaseous material
Orange	Communication, alarm or signal lines, cables or conduits
Blue	Potable water
Green	Sewers and drain lines
Purple	Reclaimed water, irrigation and slurry lines
White	Proposed excavation
Pink	Temporary survey markings

106SA01.EPS

- Cave-ins due to trench failure
- Workers falling into the trench when walking too close to the trench edge
- Trench flooding from broken water or sewer mains
- Electrical shock from coming in contact with electrical cable in the trench, or coming in contact with overhead lines
- Toxic liquid or gas leaks from nearby facilities or pipes
- Excessive vibration from auto traffic, if the excavation site is near a highway
- Falling dirt or rocks from an excavator bucket

2.1.0 Soil Hazards

The type of soil in and around a trench is also a factor that contributes to the collapse of trench walls. *Table 1* shows a sample list of 84 recorded trench failures broken down by the type of soil in which they occurred. Soil type is a major factor to consider in trenching operations. Only a company-assigned competent person has enough training, education, and experience on the job to determine if the soil in and around a trench is safe and stable. However, it is still your responsibility to know the basics about soil and its associated hazards.

WARNING

Never enter a trench unless you have approval from your company-assigned competent person.

2.1.1 Properties of Soil

Soil is comprised of soil particles, air, and water in varying quantities. Soil particles, or grains, are made up of chunks, pieces, fragments, and tiny bits of rock that are released by the weathering of parent rocks. Weathering is a natural process of erosion that can be physical or chemical. Physical processes include freezing and thawing, gravity, and erosion by rivers and rainfall. Chemical processes include oxidation, hydration, and carbonation in which minerals are chemically broken down by the elements and removed via ground and rain water. Properties of a given soil include grain size, soil gradation, and grain shape.

Table 1 Failures by Soil Type

Type of Soil	Number of Failures
Clay and/or mud	32
Sand	21
Wet dirt (probably silty clay)	10
Sand, gravel, and clay	8
Rock	7
Gravel	4
Sand and gravel	2

106T01.EPS

2.1.2 Types of Soils

The soil that is found on most construction sites is a mixture of many mineral grains coming from several kinds of rocks. Average soils are usually a mixture of two or three materials, such as sand and silt, or silt and clay. The type of mixture determines the soil characteristics. For example, sand with small amounts of silt and clay may compact well and provide a very good excavation soil.

2.1.3 Soil Behavior

Each of the various soil types, depending on the condition of the soil at the time of the excavation, will behave differently (*Table 2*). Sandy soil tends to collapse straight down. Wet clays and loams tend to slab off the side of the trench. These two conditions are shown in *Figure 2*.

Firm, dry clays and loams tend to crack. Wet sand and gravel tend to slide. These conditions are shown in *Figure 3*. You should be aware of the type of soil you are working in and know how it behaves. When you are working in or near a trench, stay alert to changes in the trench. Watch for developing cracks, moisture, or small movements in the trench material. Alert your supervisor and co-workers to any changes you notice in the trench walls. Changes in trench walls may be an early indication of a more severe condition.

2.1.4 Making a Soil Ribbon

Table 2 lists some common properties to look for when identifying soil type. One way to do this is to see if the soil will ribbon. Grab a handful of soil. Does it feel gritty or smooth? Squeeze the soil in your hand. Does it form a cast (clump)? Lightly wet the soil and try it again. Try to make a soil ribbon (*Figure 4*) by lightly wetting the soil until it feels like moist putty. Then, try to squeeze it between your thumb and forefinger upwardly into a thin flat ribbon. If a ribbon forms, the soil contains clay. The longer the ribbon, the more clay it contains.

3.0.0 Guidelines for Working In and Near a Trench

When working in or around any excavation or trench, you are responsible for personal safety. You are also responsible for the safety of others in the work trench. The following guidelines must be enforced to ensure everyone's safety.

- Never enter an excavation until the excavation has been inspected by the OSHA-approved competent person.
- Never enter an excavation without the approval of the OSHA-approved competent person on site.
- Inspect the excavation daily for changes in the excavation environment, such as rain, frost, or severe vibration from nearby heavy equipment.
- Wear protective clothing and equipment, such as hard hats, safety glasses, work boots, and gloves. Use respirator equipment if necessary.
- If you are exposed to vehicle traffic, wear traffic warning vests that are marked with, or made of, reflective or highly visible material.
- If water starts to accumulate in the trench, get out of the trench immediately.
- Do not walk under loads being handled by power shovels, derricks, or hoists.
- Stay clear of any vehicle that is being loaded.
- Be alert. Watch and listen for possible dangers.
- Do not work above or below a co-worker on a sloped or benched excavation wall.
- Barricade access to excavations to protect pedestrians and vehicles (*Figure 5*).

Case History

Worker Injured Attempting Rescue

Employees were laying sewer pipe in a 15'-deep trench. The sides of the trench, 4' wide at the bottom and 15' wide at the top, were not shored or protected to prevent a cave-in. Soil in the lower portion of the trench was mostly sand and gravel, and the upper portion was clay and loam. The trench was not protected from vibration caused by heavy vehicle traffic on the road nearby. To leave the trench, employees had to exit by climbing over the backfill. As they attempted to leave the trench, there was a small cave-in covering one employee to his ankles. When the second employee went to his co-worker's aid, another cave-in occurred, covering the first worker to his waist. The first employee died on the scene, due to a rupture of the right ventricle of his heart. The second employee suffered a hip injury.

The Bottom Line: Always make sure you have a safe route out of a trench.

Source: The Occupational Safety and Health Administration (OSHA)

- Check with your supervisor to see if workers entering the excavation need excavation entry permits.
- Ladders and ramps used as exits must be located every 25' in any trench that is over 4' deep.
- Make sure someone is on top to watch the walls when you enter a trench.
- Make sure you are never alone in a trench. Two people can cover each other's blind spots.
- Keep tools, equipment, and the excavated dirt at least 2' from the edge of the excavation.

Table 2 Field Method for Identification of Soil Texture

Soil Texture	Visual Detection of Particle Size and General Appearance of Soil	Squeezed in Hand	Soil Ribboned Between Thumb and Finger
Sand	Soil has a granular appearance in which the individual grain sizes can be detected. It is free-flowing when in a dry condition.	*When air dry*: Will not form a cast and will fall apart when pressure is released. *When moist*: Forms a cast which will crumble when lightly touched.	Cannot be ribboned.
Sandy Loam	Essentially a granular soil with sufficient silt and clay to make it somewhat coherent. Sand characteristics dominate.	*When air dry*: Forms a cast which readily falls apart when lightly touched. *When moist*: Forms a cast which will bear careful handling without breaking.	Cannot be ribboned.
Loam	A uniform mixture of sand, silt, and clay. Grading of sand fraction quite uniform from coarse to fine. It is mellow, has a somewhat gritty feel, yet is fairly smooth and slightly plastic.	*When air dry:* Forms a cast which will bear careful handling without breaking. *When moist*: Forms a cast which can be handled freely without breaking.	Cannot be ribboned.
Silt Loam	Contains a moderate amount of the finer grades of sand and only a small amount of clay. Over half of the particles are silt. When dry it may appear quite cloddy which can be readily broken and pulverized to a powder	*When air dry*: Forms a cast which can be freely handled. Pulverized, it has a soft, flourlike feel. *When moist*: Forms a cast which can be freely handled. When wet soil runs together and puddles.	It will not ribbon, but has a broken appearance, feels smooth, and may be slightly plastic.
Silt	Contains over 80% silt particles, with very little fine sand and and clay. When dry, it may be cloddy, and readily pulverizes to powder with a soft flourlike feel.	*When air dry*: Forms a cast which can be handled without breaking. When moist: Forms a cast which can be freely handled. When wet, it readily puddles.	It has a tendency to ribbon with a broken appearance, feels smooth.
Clay Loam	Fine textured soil breaks into very hard lumps when dry. Contains more clay than silt loam. Resembles clay in a dry condition; identification is made on the physical behavior of moist soil.	*When air dry*: Forms a cast which can be handled freely without breaking. *When moist*: Forms a cast which can be freely handled without breaking. It can be worked into a dense mass.	Forms a thin ribbon which readily breaks, barely sustaining its own weight.
Clay	Fine textured soil breaks into very hard lumps when dry. Difficult to pulverize into a soft flourlike powder when dry Identification is based on cohesive properties of the moist soil.	*When air dry*: Forms a cast which can be freely handled withoutbreaking. *When moist*: Forms a cast which can be freely handled without breaking.	Forms long, thin, flexible ribbons. Can be worked into a dense, compact mass. Considerable plasticity.
Organic Soils	Identification based on the high organic content. Muck consists of thoroughly decomposed organic material with considerable amount of mineral soil finely divided with some fibrous remains. When considerable fibrous material is present, it may be classified as peat. The plant remains or sometimes the woody structure can easily be recognized. Soil color ranges from brown to black. They occur in lowlands, swamps, or swales. They have high shrinkage upon drying.		

106T02.EPS

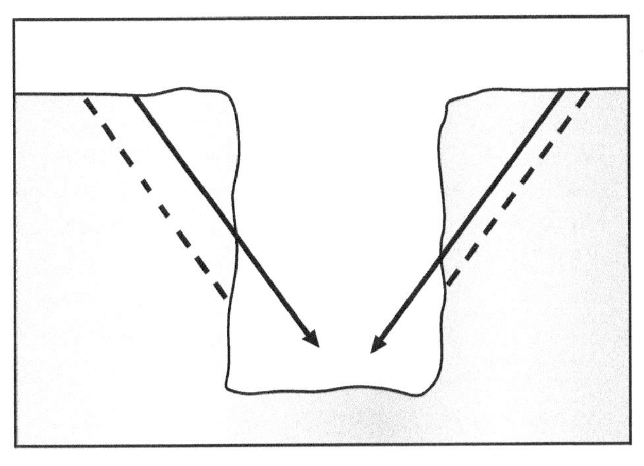

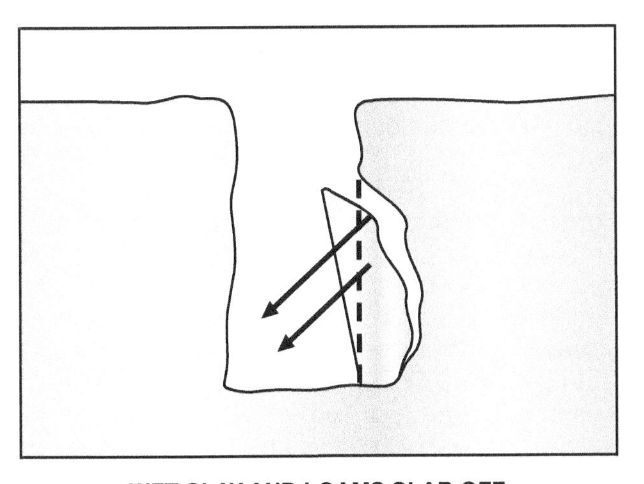

106F02.EPS

Figure 2 Behavior of sandy soil and wet clay and loams.

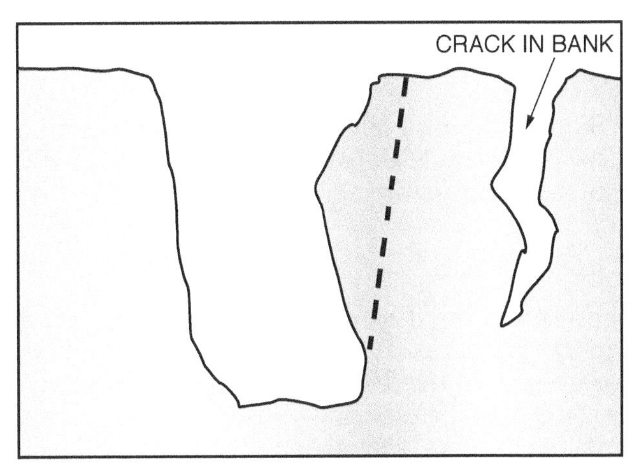

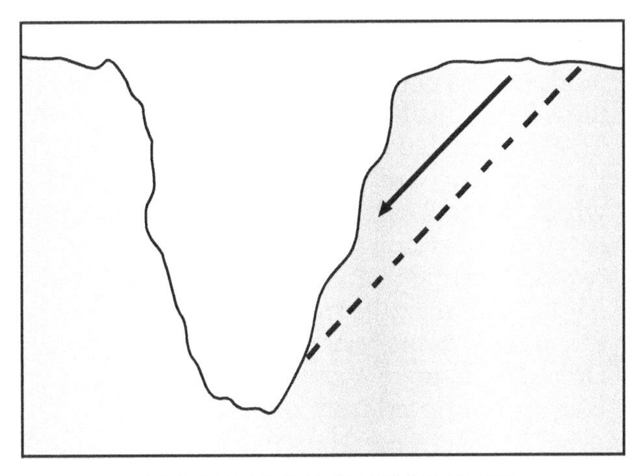

106F03.EPS

Figure 3 Behavior of dry clay and loams, and wet sand and gravel.

106F04.EPS

Figure 4 Making a soil ribbon.

106F05.EPS

Figure 5 Barricaded trench.

- Make sure shoring, trench boxes (*Figure 6*), benching, or sloping are used for excavations and trenches over 5' deep.
- Stop work immediately if there is any potential for a cave-in. Make sure any problems are corrected before starting work again.

3.1.0 Ladders

There must be at least one method of entering and exiting all excavations over 4' deep. Ladders are generally used for this purpose (*Figure 7*). Ladders must be placed within 25' of each worker.

When ladders are used, there are a number of requirements that must be met.

- Ladder side rails must extend a minimum of 3' above the landing.
- Ladders must have nonconductive side rails if work will be performed near equipment or systems using electricity.
- Two or more ladders must be used where 25 or more workers are working in an excavation in which ladders serve as the primary means of entry and exit, or where ladders are used for two-way traffic in and out of the trench.
- All ladders must be inspected before each use for signs of damage or defects.
- Damaged ladders should be labeled DO NOT USE and removed from service until repaired.
- Use ladders only on stable or level surfaces.
- Secure ladders when they are used in any location where they can be displaced by excavation activities or traffic.
- While on a ladder, do not carry any object or load that could cause you to lose your balance.
- Exercise caution whenever using a trench ladder.

4.0.0 Indications of an Unstable Trench

A number of stresses and weaknesses can occur in an open trench or excavation. For example, increases or decreases in moisture content can affect the stability of a trench or excavation. The following sections discuss some of the more frequently identified causes of trench failure. These conditions are illustrated in *Figure 8*.

Tension cracks usually form at a depth of one-quarter to one-half of the way down from the top of a trench. Sliding or slipping may occur as a result of tension cracks. In addition to sliding, tension cracks can cause toppling. Toppling occurs when the trench's vertical face shears along the tension crack line and topples into the excavation. An unsupported excavation can create an unbalanced stress in the soil, which in turn causes subsidence at the surface, and bulging of the vertical face of the trench. If uncorrected, this

106F06.EPS

Figure 6 Trench box.

106F07.EPS

Figure 7 Ladder in a trench.

condition can cause wall failure and trap workers in the trench, or greatly stress the protective system. Bottom heaving is caused by downward pressure created by the weight of adjoining soil. This pressure causes a bulge in the bottom of the cut. Heaving and squeezing can occur even when shoring and shielding are properly installed.

CAUTION

Protective systems are designed for even loads of earth. Heaving and squeezing can place uneven loads on the shielding system and may stress particular parts of the protective system.

Another indication of an unstable trench is boiling. Boiling occurs when water flows upward into the bottom of the cut. A high water table is one of the causes of boiling. Boiling can happen quickly and can occur even when shoring or trench boxes are used. If boiling starts, stop what you are doing and leave the trench immediately.

5.0.0 Digging Trenches

A string line is a standard method used to create a truly straight trench. Underground installations are based on straight runs of pipe and specific angles where bends are necessary. It is important

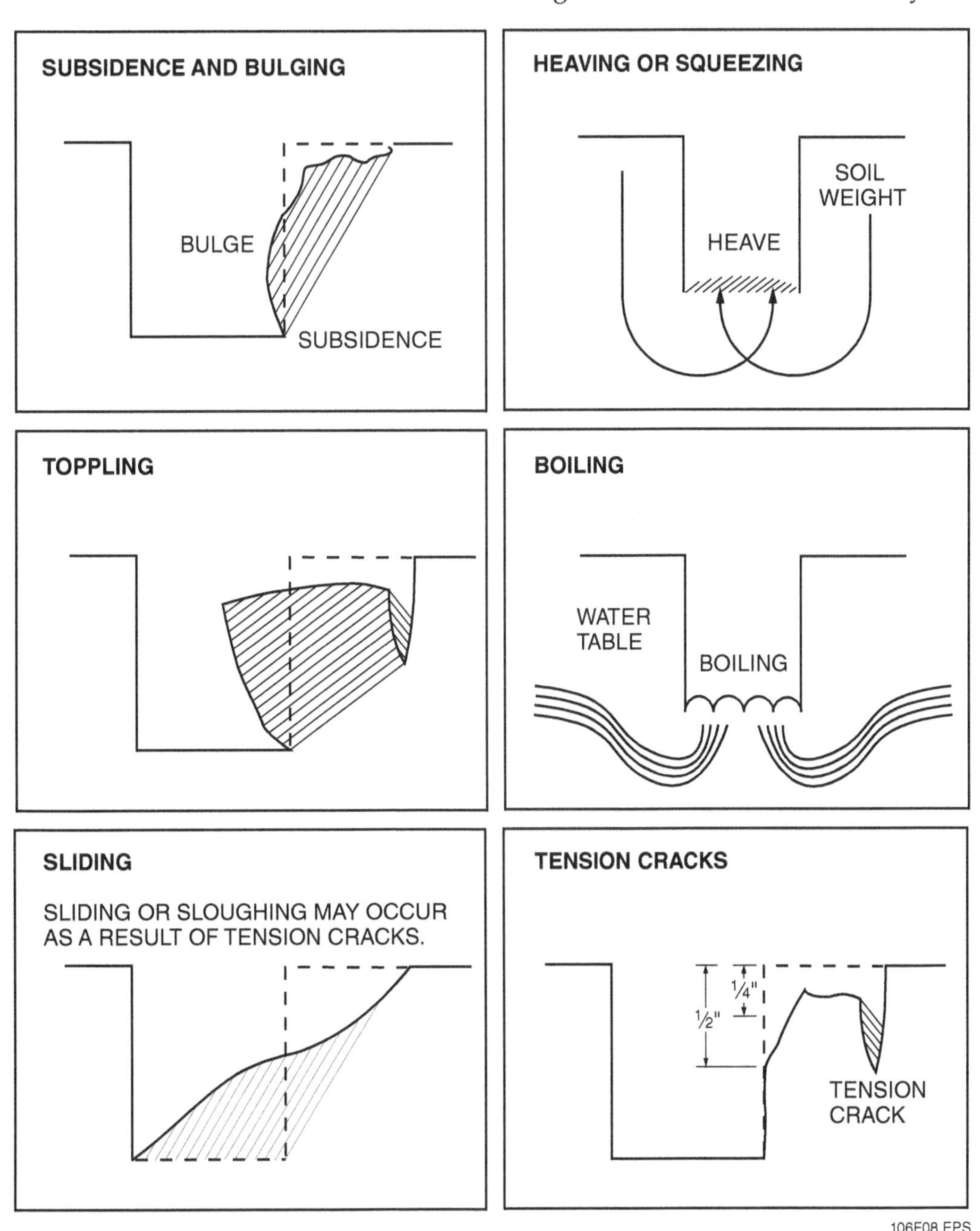

Figure 8 Indications of an unstable trench.

Case History

Trenching Equipment

An operator working on a trenching job tried to position a backhoe to backfill an electrical trench on the top outer edge. As the operator steered into softer, excavated soil, the equipment tilted onto its side and slid downward, stopping only when it engaged the chain link fence 10' below the trench. The operator was wearing a seatbelt and no injuries resulted from this incident.

The Bottom Line: If the employee had not been wearing his seatbelt and the fence had not been in place, this incident could have easily resulted in a very serious injury.

Source: The Occupational Safety and Health Administration (OSHA)

that the trench be straight, not only for initial installation, but for future excavations when repairs are necessary. A grade line may also be required to set the trench depth.

Some joints between underground pipe and fittings are mechanical push-on joints. These joints are principally held together by thrust blocks. Thrust blocks are masses of concrete poured between the fittings and undisturbed soil. Thrust blocks have the ability to keep the pipe and fittings together by resisting the force of thrust. For this reason, as little soil as possible should be disturbed.

Any change in the direction of the pipe also creates the need for a fitting and thrust blocks. Unless the forces are applied along the pipe, joints can come apart due to ground shift, and a blow-out results. For this reason, any change in direction of the pipe must be made with standard bends and with thrust blocking requirements in mind.

5.1.0 Trenching Equipment

There are sizes and types of trenching equipment appropriate for any trenching job. Two types of basic trench digging equipment are the backhoe and wheel trencher. Equipment operators must be properly trained and licensed (if required) to operate trenching equipment.

5.1.1 Backhoe

The backhoe is the first choice for trenching equipment in the sprinkler industry because it is both cost-effective and portable. It is usually positioned to straddle the ditch line while digging. When the trench is wide or the walls of the trench need to be sloped, the backhoe should be used. Several backhoes are sometimes used to expedite the trenching operation. They work very well on short lengths of trench and where obstacles such as existing piping and buried power lines must be jumped. The backhoe is versatile and can be used to trench, supply bedding material, set pipe, and backfill. As with any piece of machinery used, all safety devices must be in working order.

5.1.2 Wheel Trencher

Wheel trenchers are manufactured in a variety of sizes and models. The wheel trencher functions best when excavating depths 6' and under, and when there are no utility lines or other obstacles. While a trencher is generally more expensive to move to the small job site, it works faster than a backhoe or a dragline. The wheel trencher leaves very little dirt along the top of the trench but does leave loose dirt behind the wheel. This has to be removed by hand or with a backhoe.

Case History

Overhead Power Line Severed by Dump Truck

A subcontractor delivered a load of backfill to a construction area. After dumping the load, the driver failed to lower the dump bed before moving forward. The dump bed contacted and cut a 2,400V overhead power line. A spotter was in place while the truck was dumping, but turned away as the truck started to drive away. As the power line was cut, it fell clear of the truck, and no one was in the area where the end of the severed line fell. Luckily, there were no injuries.

The Bottom Line: Always be aware of overhead and underground power lines when excavating.

Source: The Occupational Safety and Health Administration (OSHA)

The surface around the trench must be level. Otherwise, the wheel trencher cuts slanted sides, causing the trench to be out of plumb. Trenches out of plumb are dangerous to the crew laying the pipe. In addition, if the trencher is leaning, it puts unnecessary stress on the machinery, contributing to breakdowns and excessive operating costs.

5.2.0 Trench Failure

The most common hazard during an excavation is trench failure or cave-in. Using common sense and following all applicable safety precautions will make the trench a much safer place to work.

> **NOTE**
>
> Excavations 5' or deeper require manufactured and engineered trench shielding and shoring devices. If shielding or shoring devices are not used, trench walls may be sloped to reduce the risk of cave-ins. Each of these protective systems should be used in accordance with OSHA regulations.

To understand the seriousness of trench failure, consider what can happen when there is a shift in the earth that surrounds an unsupported trench. Workers could be buried when:

- One or both edges of the trench cave in
- One or both walls slide in
- One or both walls shear away and collapse

Failure of unsupported trench walls is not the only cause of worker burial. Tons of dirt can be dumped on the workers if the spoil pipe or excavated earth slides into the trench. Such slides occur when the pile is placed too close to the edge of the trench, or when the ground beneath the pile gives way. There must be a minimum of 2' between the trench wall and the spoil pile. This area must also be kept free of any tools and materials.

The following conditions will likely lead to a trench cave-in. If you notice any of these conditions, immediately inform your supervisor. These conditions are listed in order of seriousness:

- Disturbed soil from previously excavated ground
- Trench intersections where large corners of earth can break away
- A narrow right-of-way causing heavy equipment to be too close to the edge of the trench
- Vibrations from construction equipment, nearby traffic, or trains

Did You Know?

Three main classes of equipment are used in trenching: backhoes, excavators, and trenching machines. The trenching machine shown here is specifically designed for trench excavation.

106SA03.EPS

On Site

Excavations

Whether you're digging an excavation for a foundation or for some other purpose, federal law requires that you first contact the designated authority at least 72 hours before the start of the excavation. This is necessary in order to check for the existence and/or location of underground utility lines or cables. Failure to notify the designated authority as required prior to the excavation can result in your employer being liable for any damage caused to utility lines or cables, or for any personal injury to workers and others that can occur as a result of damaging any such lines or cables. In some localities, there may be an additional requirement that you obtain a digging permit before doing any excavation.

106SA02.EPS

- Increased subsurface water that causes soil to become saturated, and therefore unstable
- Drying of exposed trench walls, causing the natural moisture-binding soil particles to be lost
- Inclined layers of soil dipping into the trench, causing layers of different types of soil to slide one upon the other, causing the trench walls to collapse

6.0.0 Making the Trench Safe

There are several ways to make the trench a safer place to work (*Figure 9*). Trench shoring, shielding, and sloping are different methods used to protect workers and equipment. It is important that you recognize the differences between them.

- *Shoring* – Shoring a trench supports the walls of the excavation and prevents their movement and collapse. Shoring does more than provide a safe environment for workers in a trench. Because it restrains the movement of trench walls, shoring also stops the shifting of adjacent soil formations containing buried utilities, or the shifting of adjacent soil on which sidewalks, streets, building foundations, or other structures are built.
- *Trench shields* – Trench shields, also called trench boxes, are placed in unshored excavations to protect personnel from excavation wall collapse. They provide no support to trench walls or surrounding soil, but for specific depths and soil conditions they will withstand the side weight of a collapsing trench wall.
- *Sloping* – Sloping an excavation means cutting the walls of the excavation back at an angle to the trench floor. This angle must be cut at least to the angle of repose for the type of soil being used. The angle of repose is the greatest angle above the horizontal plane at which a material will rest without sliding.

6.1.0 Shoring Systems

Perhaps the most critical aspect of excavation is the shoring. This is what holds up the soil along the sides of the trench to prevent cave-ins. Sometimes the soil will remain in place if the walls are sloped, but some soils will not stand on even a moderate slope. The type of shoring required is determined by three things: trench depth, trench width, and type of soil.

Shoring is not required in solid rock, hard shale, or hard slag. The only other excavations exempt from shoring are trenches that will not be entered, trenches less than 4' deep, or trenches in which the walls are sloped to within 4' of the bottom of the trench and the slope is less than 45 degrees from the horizontal. That is, the trench does not exceed 1' of vertical rise to each foot of horizontal run.

Trenches 4' deep or more must have a ladder or some other means of escape or entry. The ladder has to be located so that workers in the trench must travel no more than 25' (lateral travel) to the trench entrance or exit.

Shoring must always extend a minimum of 1' above the trench edge. This is to prevent soil from falling back into the trench and possibly injuring anyone working in the trench. If the trench is deeper than 25', or wider than 12', the shoring must be designed by a professional engineer.

Shoring systems are metal, hydraulic, mechanical, or timber structures that provide a

106F09.EPS

Figure 9 Shoring methods.

Did You Know?

Federal and state safety regulations have established standards for protecting those who work in excavations. *OSHA 29 CFR 1926* defines the trench protective devices that are acceptable. OSHA mandates how the protective devices must be used. For excavations 5' deep or deeper, OSHA regulations require appropriate manufactured and engineered trench shielding and shoring devices, or trench walls sloped to angles that reduce the risk of cave-ins (other than in a stable location).

framework to support excavation walls. Shoring uses uprights, wales, and cross braces to support walls. *Figure 10* shows a shoring system in place.

6.1.1 Hydraulic Shores and Spreaders

Hydraulic shores, shown in *Figure 11*, can be installed and removed quickly. Each shore consists of two vertical rails connected by a hydraulic cylinder. The shores are placed in the trench and the hydraulic cylinders are pumped up to push the vertical rails against the wall. Note that this system uses the hydraulic cylinders as cross bracing. This arrangement is commonly referred to as a skeleton shoring system. Hydraulic spreaders, shown in *Figure 12*, can be added to a skeleton shoring system to provide additional cross bracing.

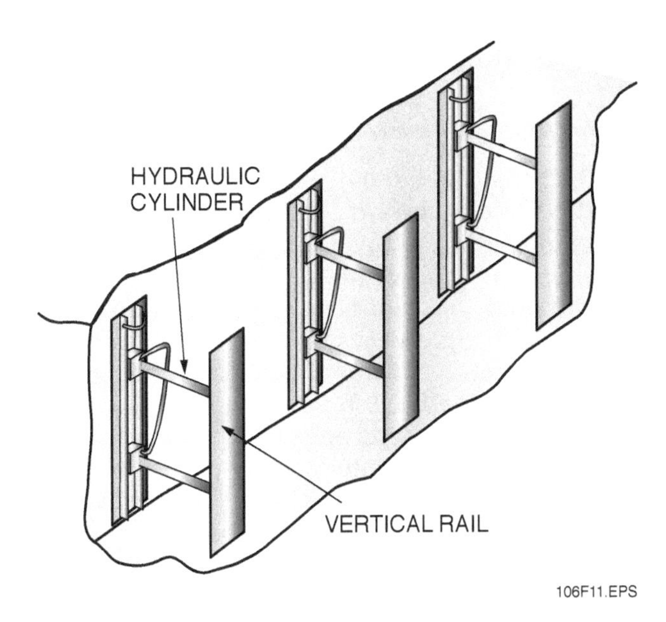

Figure 11 Hydraulic skeleton shoring system.

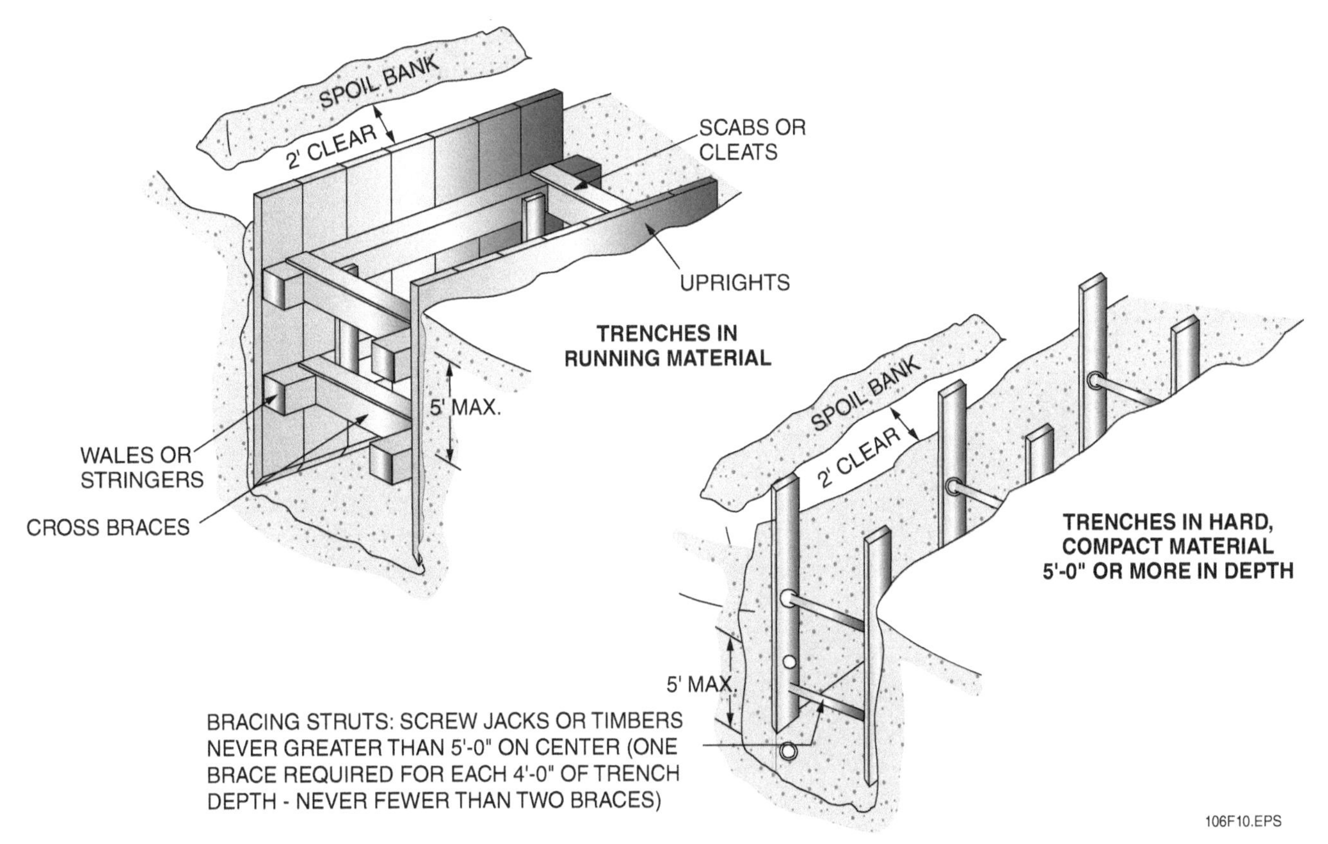

Figure 10 Shoring system in place.

6.1.2 Vertical Sheeting

When excavating near existing structures, or performing long-term excavations, vertical sheeting may be used and supported with hydraulic wales, as shown in *Figure 13*. This method is referred to as tight-sheet shoring. Other types of spreaders, including screw jacks, trench jacks, and hydraulic spreaders, are also available.

6.1.3 Interlocking Steel Sheeting

Interlocking steel sheeting (*Figure 14*) may be specified under certain conditions such as deep excavations, and excavations near buildings or building foundations. Interlocking steel sheeting is commonly used on Department of Transportation (DOT) right-of-ways. It will prevent damage to sub-base pavement caused by vibration from vehicle traffic. Interlocking steel sheeting is required when working in waterways. Steel sheeting consists of interlocking panels of steel reinforced with cross members. It is similar in design to tight sheeting. Steel sheeting is engineered for a particular application. It must be installed precisely in accordance with the engineer's specifications. Steel sheeting is commonly installed by driving it into the ground using a vibrating hydraulic hammer. It can also be driven with a drop hammer or backhoe bucket.

> **NOTE**
> The systems just described identify the common components used in shoring systems. You may encounter job-built systems, aluminum or wooden sheeting, or other approved methods.

6.1.4 Shoring Safety Rules

To avoid accidents and injury when shoring an excavation, special safety rules must be followed.

- Never enter an excavation before the shoring is in place.
- Do not install the shoring while you are inside the trench. All shoring must be installed from the top of the trench.
- The cross braces must be level across the trench.
- The cross braces should exert equal pressure on each side of the trench.
- The vertical uprights must be flat against the excavation wall.
- All materials used for shoring must be thoroughly inspected before use, and must be in good condition.
- Shoring is removed by starting at the bottom of the excavation and going up.
- Vertical supports are pulled out of the trench from above.
- Every excavation must be backfilled immediately after the support system is removed.

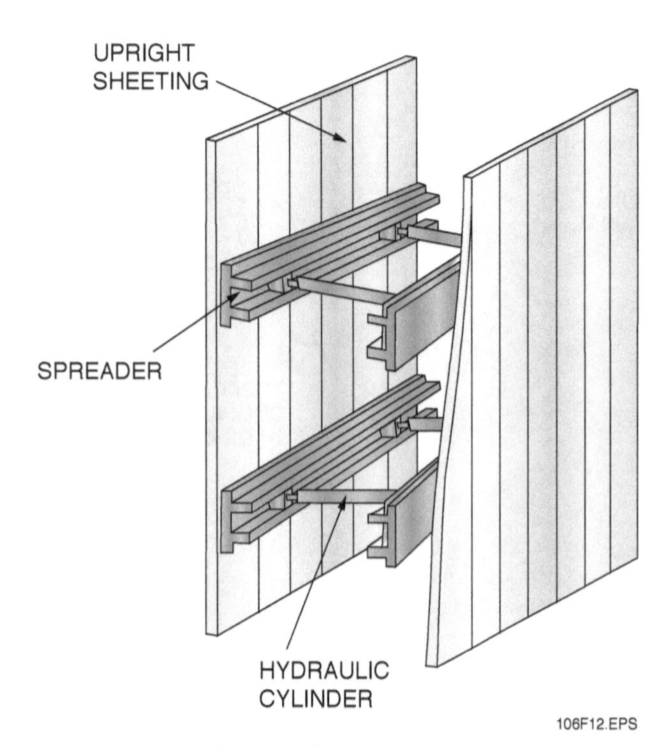

Figure 12 Hydraulic spreaders.

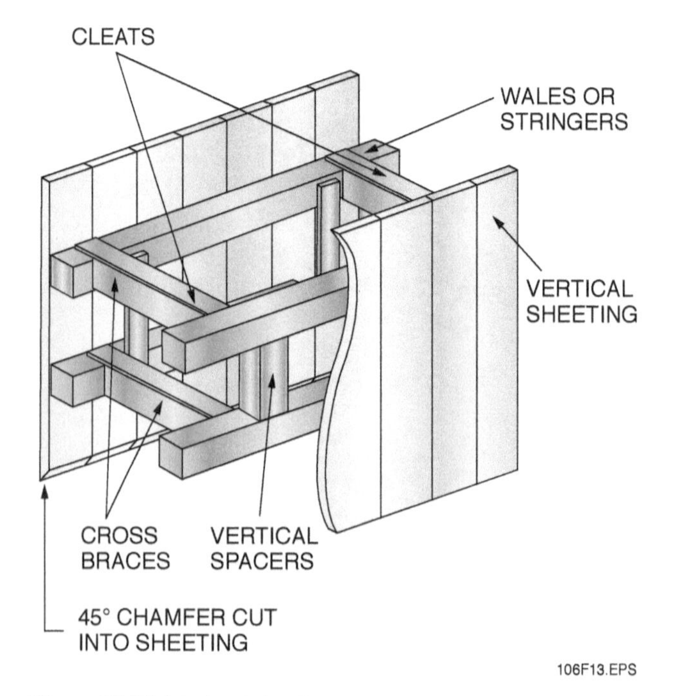

Figure 13 Tight-sheet shoring.

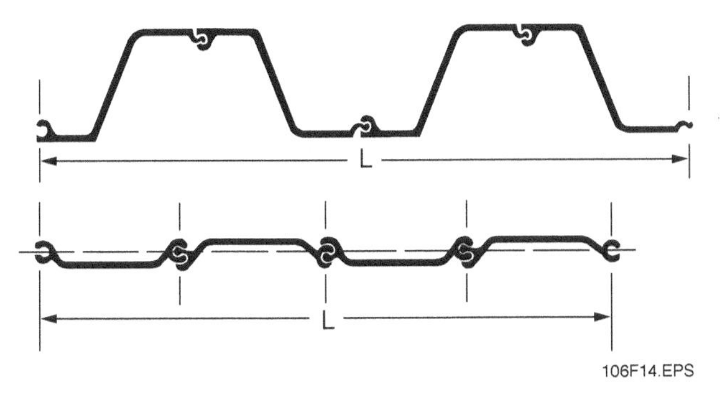

Figure 14 Interlocking steel sheeting.

6.2.0 Shielding Systems

A shielding system is a structure that is able to withstand the forces imposed on it by a cave-in, and protect employees within the structure. Shields can be permanent structures, or can be portable and moved along as work progresses. The shielding system is also known as a trench box or trench shield. A trench box is shown in *Figure 15*. If the trench will not stand long enough to excavate for the shield, the shield can be placed high and pushed down as material is excavated.

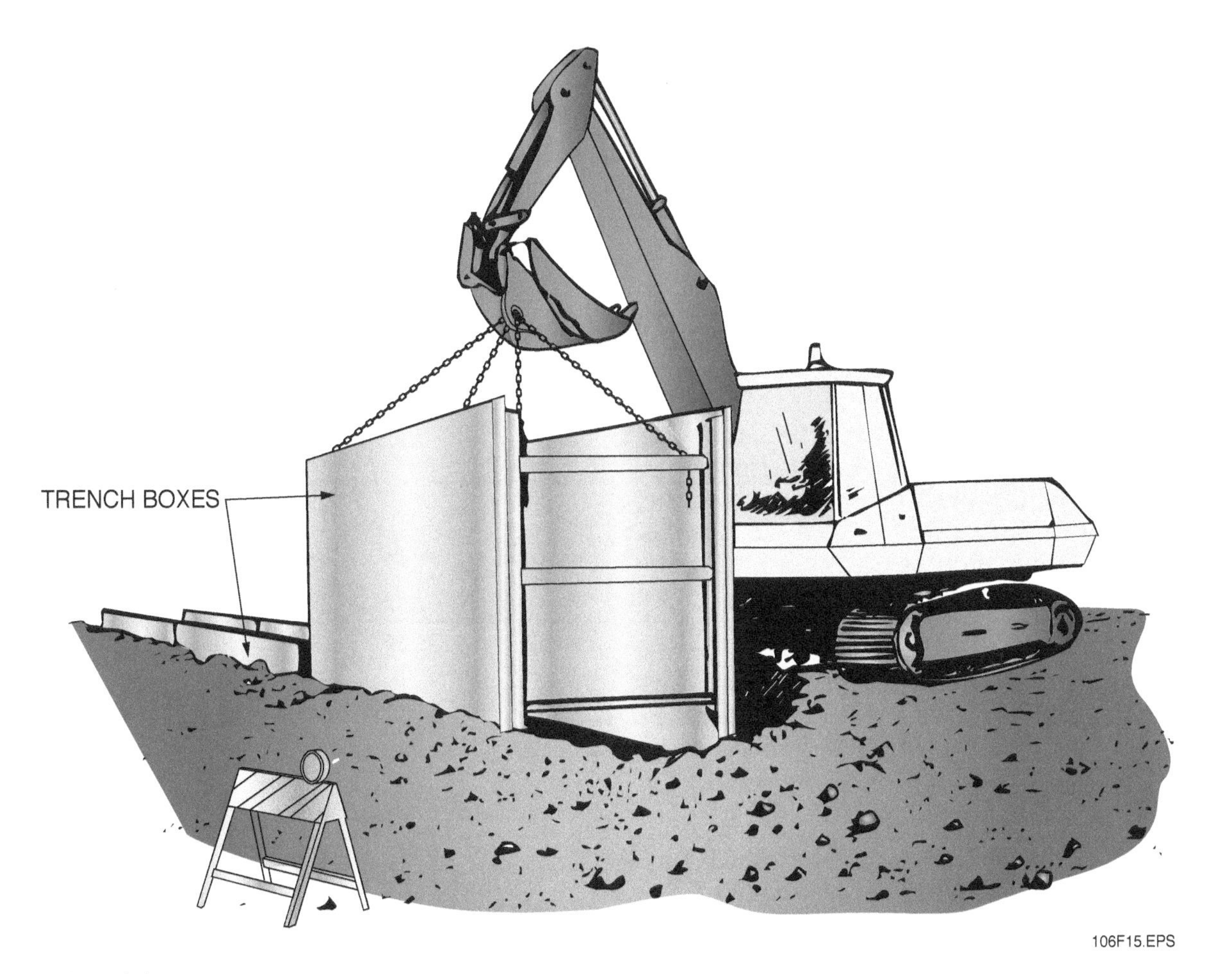

Figure 15 Trench boxes.

Case History

No Way Out

Two employees were laying pipe in a 12'-deep trench when one of the employees saw the bottom face of the trench move. He jumped out of the way along the length of the trench. The other employee was fatally injured as the wall caved in.

The Bottom Line: Always ensure that the walls of a trench are sloped, and an emergency exit is provided.
Source: The Occupational Safety and Health Administration (OSHA)

Case History

Cave-In Caused by Water

Four employees were laying a lateral sewer line at a building site. The foreman and a laborer were laying an 8"-diameter, 20'-long plastic sewer pipe in the bottom of a trench that was 36" wide, 9' deep, and approximately 50' long. The trench was neither sloped nor shored, and there was water entering it along a rock seam near the bottom. The west side of the trench caved in near the bottom, burying one employee to his chest and completely covering the other. The rescue operation took two hours for the first man and four hours for the second. Both men died.

The Bottom Line: Working in an excavation can be fatal. Always ensure your trench is safe.

Source: The Occupational Safety and Health Administration (OSHA)

The excavated area between the outside of the trench box and the face of the trench should be as small as possible. The space between the trench box and the excavation side must be backfilled to prevent side-to-side movement of the trench box. Remember that job site and soil conditions, as well as trench depths and widths, determine what type of trench protection system will be used. A single project can include several depth or width requirements, and varying soil conditions. This may require that several different protective systems be used for the same site. A registered engineer must certify shields for the existing soil conditions. The certification must be available at the job site during assembly of the trench box. After assembly, certifications must be made available upon request.

6.2.1 Trench-Box Safety

If used correctly, trench boxes protect workers from the dangers of a cave-in. All safety guidelines for excavations also apply to trench boxes. Follow these safety guidelines when using a trench box:

- Be sure that the vertical walls of a trench box extend at least 18" above the lowest point where the excavation walls begin to slope. *Figure 16* shows proper trench box placement in each of three soil types. The types of soil are discussed in later sections.
- Never enter the trench box during installation or removal.
- Never enter an excavation behind a trench box after the trench box has been moved.
- Backfill the excavation as soon as the trench box has been removed.
- An exit from the trench box and the excavation must be located within 25' of each worker.

6.3.0 Sloping Systems

A sloping system is a method of protecting workers from cave-ins. Sloping is accomplished by inclining the sides of an excavation. The angle of incline varies according to such factors as the soil type, environmental conditions of exposure, and nearby loads. There are three general classifications of soil types, and one type of rock. For

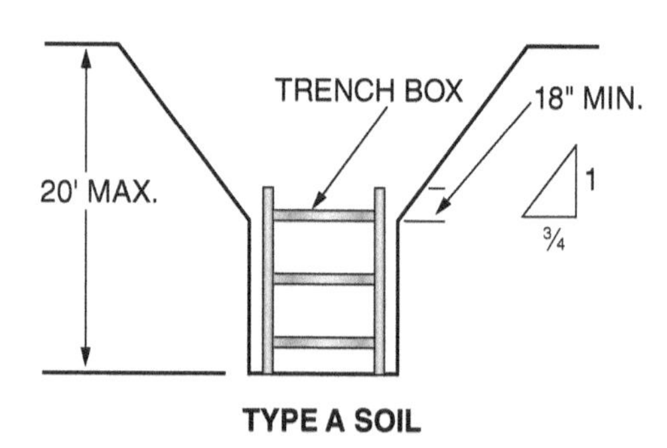

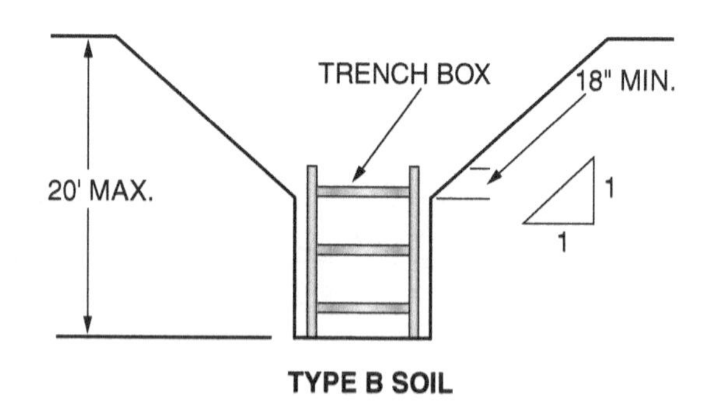

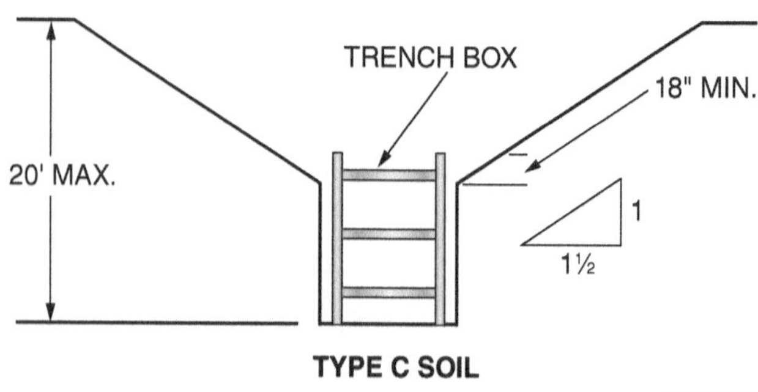

Figure 16 Proper trench box placement.

each classification of soil type, OSHA defines maximum angles for the slope of the walls, as shown in *Table 3*. The designation and selection of the proper sloping system is far more complex than described in this section. Factors such as the depth of the trench, the amount of time the trench is to remain open, and other factors will affect the maximum allowable slope.

Step-back, also known as benching, is another method of sloping the excavation walls without the use of a support system. Step-back uses a series of steps that must rise on the approximate angle of repose for the type of soil being used. The safety rules that apply to step-back excavations also apply to sloped excavations.

6.4.0 Sloping Requirements for Different Types of Soils

Many excavations are started with a vertical cut. Although some soils will stand to considerable depths when cut vertically, most will not. When vertical slopes fall to a more stable angle, large amounts of material usually fall into the excavation. *Figure 17* shows methods of failure in excavation walls.

The slope of the excavation walls must be angled so that material will not fall into the excavation. The slope is figured by angle from the

Table 3 Maximum Allowable Slopes

Soil or Rock Type	Maximum Allowable Slopes for Excavations Less Than 20' Deep
Stable rock	Vertical (90 degrees)
Type A	$\frac{3}{4}$:1 (53 degrees)
Type B	1:1 (45 degrees)
Type C	$1\frac{1}{2}$:1 (34 degrees)

106T03.EPS

TENSION CRACK

FORMATION OF TENSION CRACKS

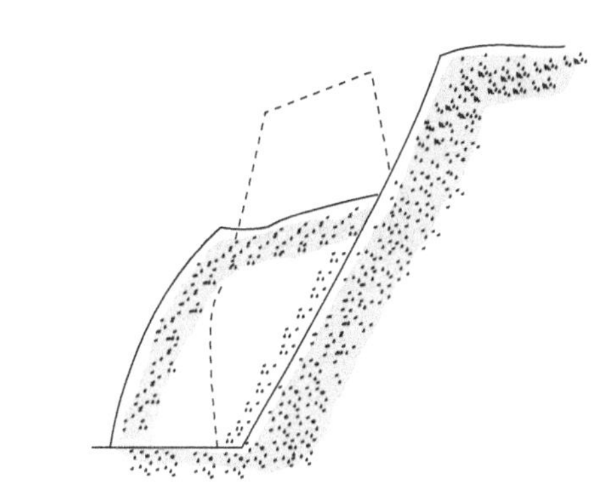

SLIDING OF SOIL INTO EXCAVATION

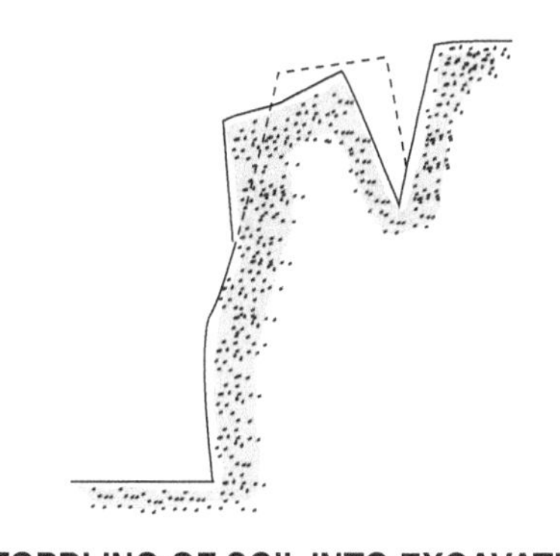

TOPPLING OF SOIL INTO EXCAVATION

106F17.EPS

Figure 17 Methods of failure in excavation walls.

Case History

Trencher Rollover

While operating a trencher, an operator was changing the equipment position by backing down a side slope leading to the edge of an empty ditch. The trenching bar was in an elevated position. As the operator shifted gears, the equipment became top heavy, tipped over sideways, rolled one-half turn into the ditch, and came to rest on its top. Silt fences were being installed and during the installation, it was necessary to cut a shallow trench. The Activity Hazard Analysis (AHA) was reviewed with the operator; however, the AHA did not include precautions for operation on sloped surfaces. The accident occurred when the operator was returning to the beginning of the cut. Stability of the equipment was verified and the operator was removed uninjured.

The Bottom Line: Accidents can happen at any time. Use extreme caution when operating equipment near the edge of an excavation.

Source: The Occupational Safety and Health Administration (OSHA)

horizontal plane and by horizontal run to vertical rise. For example, a 45-degree slope would be a 1:1 slope, meaning that for each foot measured horizontally, the slope rises 1' vertically. The allowed slope of the excavation walls varies according to the type and characteristics of the soil being used. Four basic classifications of soil have been identified. The classification is based on the stability of the soil and the maximum allowable slopes for excavations in each of the four types of soils. These four classifications, in decreasing order of stability, are stable rock, Type A, Type B, and Type C. *Figure 18* shows maximum allowable slopes for each soil type.

6.4.1 Stable Rock

Stable rock refers to natural solid mineral matter that can be excavated with vertical sides and remain intact while exposed. Stable rock includes solid rock, shale, or cemented sands and gravels. *Figure 19* shows an excavation cut in stable rock.

6.4.2 Type A Soil

Type A soil refers to solid soil with a compression strength of at least 1.5 tons per square foot. Cohesive soils are soils that do not crumble, are plastic when moist, and are hard to break up when dry. Examples of cohesive soils are clay, silty clay, sandy clay, and clay loam. Cemented soils, such as caliche and hardpan, are also considered Type A. No soil can be considered Type A if any of the following conditions exist:

- The soil is fissured. Fissured means a soil material that has a tendency to break along definite planes of fracture with little resistance, or a material that has cracks, such as tension cracks, in an exposed surface.
- The soil can be affected by vibration from heavy traffic, pile driving, or other similar effects.
- The soil has been previously disturbed.

All sloped excavations that are 20' deep or less must have a maximum allowable slope of ¾ horizontal to 1 vertical. *Figure 20* shows a simple slope excavation in Type A soil.

An exception to this rule occurs when the excavation is 12' deep or less and will remain open 24 hours or less. These excavations in Type A soil can have a maximum allowable slope of ½ horizontal to 1 vertical. *Figure 21* shows a simple slope, short-term excavation in Type A soil.

6.4.3 Type B Soil

Type B soil refers to cohesive soils with a compression strength of greater than 0.5 ton per square foot, but less than 1.5 tons per square foot. It also refers to granular soils, including angular gravel, which are similar to crushed rock, silt, sandy loam, unstable rock, and any unstable or fissured Type A soils. Type B soils also include previously disturbed soils, except those that would fall into the Type C classification. Excavations made in Type B soils have a maximum allowable slope of 1 horizontal to 1 vertical. *Figure 22* shows simple slope excavations in Type B soil.

6.4.4 Type C Soil

Type C soil is the most unstable soil type. Type C soil refers to cohesive soil with a compression strength of 0.5 ton per square foot or less. Gravel, loamy soil, sand, any submerged soil or soil from which water is freely seeping, and unstable sub-

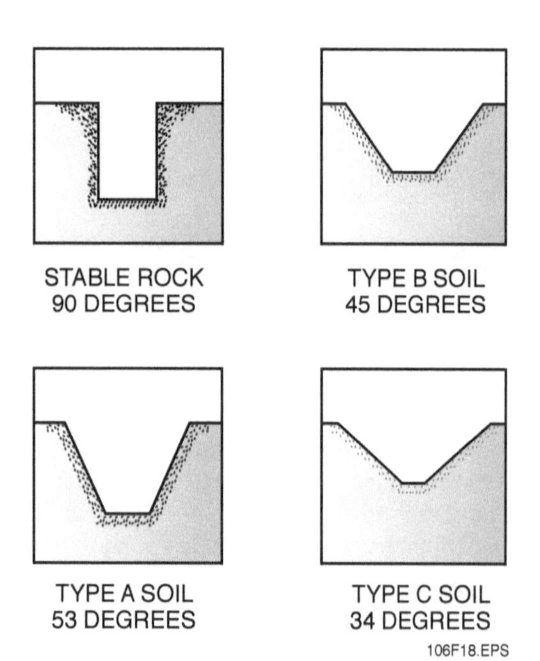

Figure 18 Maximum allowable slopes for soil types.

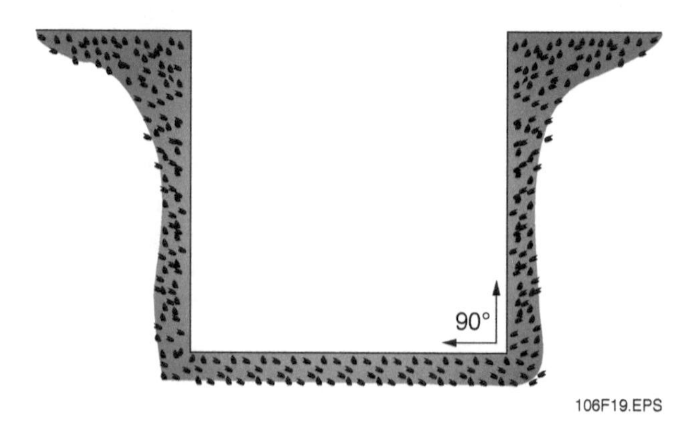

Figure 19 Excavation cut in stable rock.

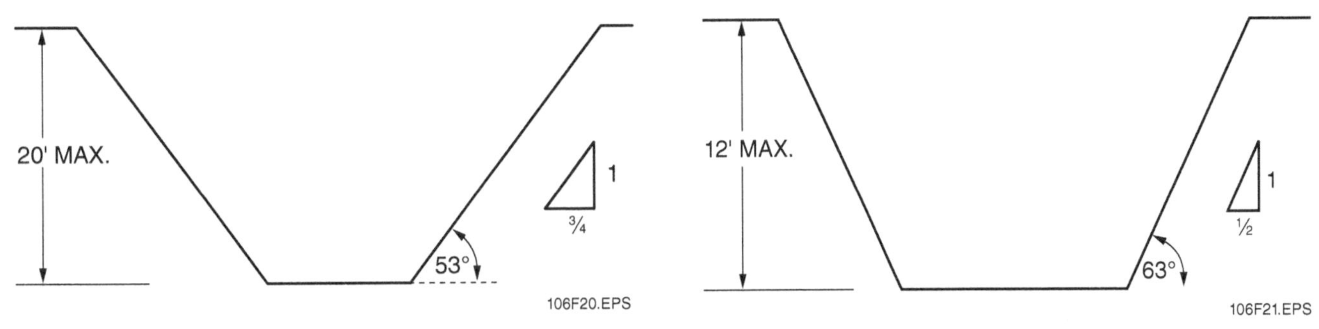

Figure 20 Simple slope excavation in Type A soil.

Figure 21 Simple slope, short-term excavation in Type A soil.

THIS BENCH ALLOWED IN COHESIVE SOIL ONLY.

20' MAX.

4' MAX.

1

1

SIMPLE BENCH

20' MAX.

45°

1

1

THIS BENCH ALLOWED IN COHESIVE SOIL ONLY.

20' MAX.

4' MAX.

4' MAX.

1

1

MULTIPLE BENCH

106F22.EPS

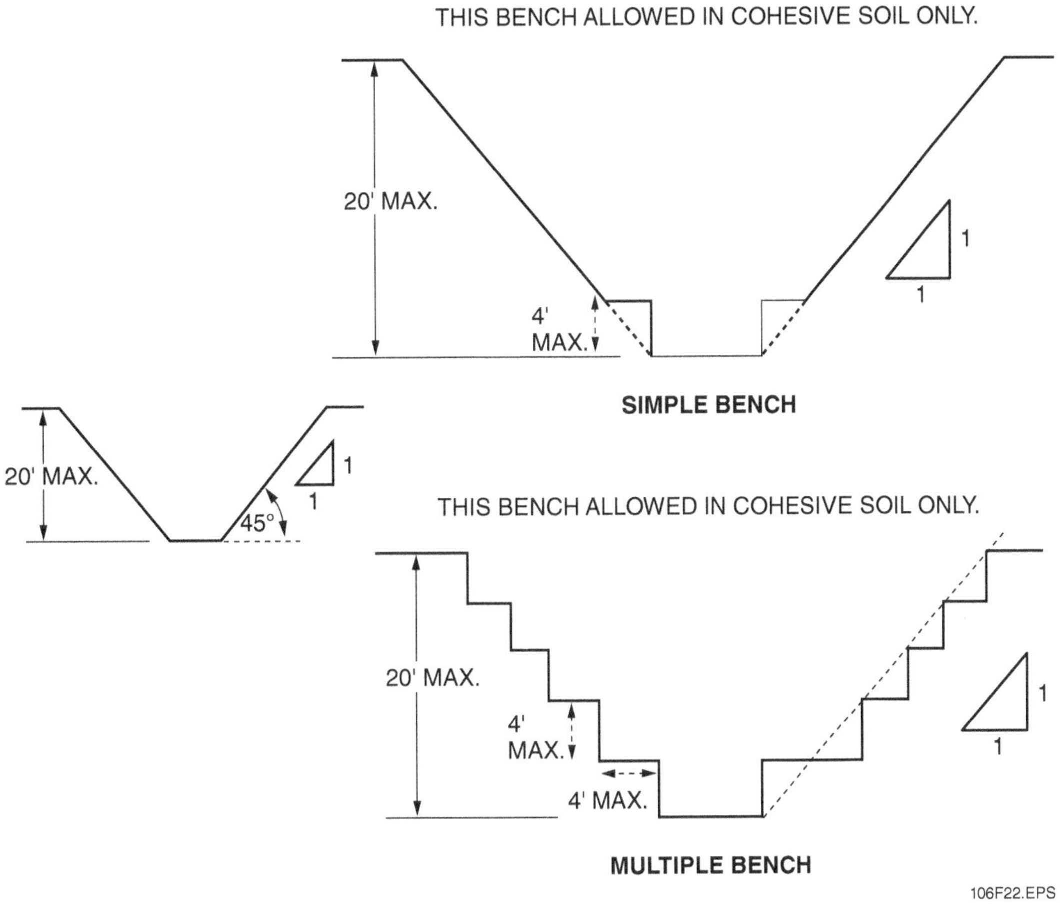

Figure 22 Simple slope excavations in Type B soil.

merged rock are considered Type C soils. Excavations made in Type C soils have a maximum allowable slope of 1½ horizontal to 1 vertical. *Figure 23* shows a simple slope excavation in Type C soil.

6.5.0 Combined Shoring Systems

A cross between trench boxes and steel sheeting is a slide-rail system. These systems are designed to be used in shallow pits, tunnel pits, and trenches, virtually anywhere that a trench box or

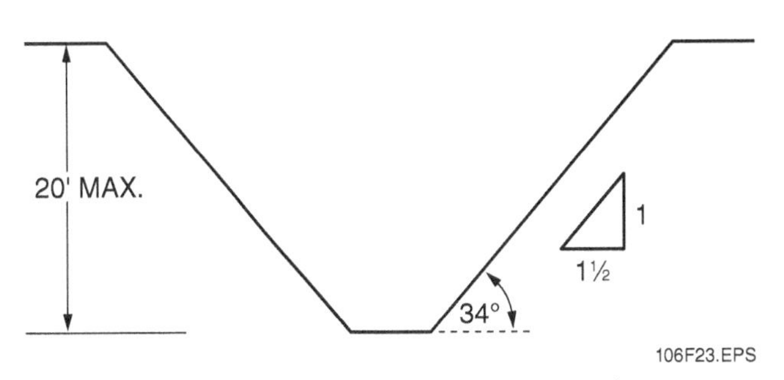

Figure 23 Simple slope excavation in Type C soil.

sheeting systems can be used. The slide-rail system shown in *Figure 24* consists of the following components:

- *Slide rails* – The slide rails are the horizontal components of the system. They have a cavity to accept the lining plates and are fitted with cross braces. Slide rails are available in single-, double-, and triple-rail configurations. The triple-rail system is used at the greatest depth.
- *Lining plates* – Lining plates are used to provide the trench sidewall support. Lining plates are installed in the slide rails and are pushed into the ground as the excavation proceeds.
- *Cross braces* – Cross braces connect between the slide rails to provide lateral support to the system, usually at the end of the slide rail.

7.0.0 BEDDING AND BACKFILLING

Job specifications call for certain conditions to be met regarding how the pipe is bedded and backfilled. Bedding material refers to the material directly under and surrounding the pipe. If laying procedures are not specified, do not proceed until they are known. The bedding material requirements depend on the type of pipe being laid in the trench. These requirements vary with job specifications and locations.

The purpose of proper embedment is to protect the pipe by cradling it in compacted soil to prevent movement, settling, floating, or loads being placed on the coupling or joint. Embedment and backfilling require more expertise than just throwing the dirt back in the trench. The initial backfill, or embedment, must be done carefully.

In certain geographic areas, and especially where the municipal utility piping can be directly affected, the municipality determines the material to be used for the embedment, or initial backfill.

Regardless of the type of embedment used, it must conform to standards set by *ANSI A21.1, ASTM D-2774, ASCE Manual 37,* or *WPCF Manual 9.*

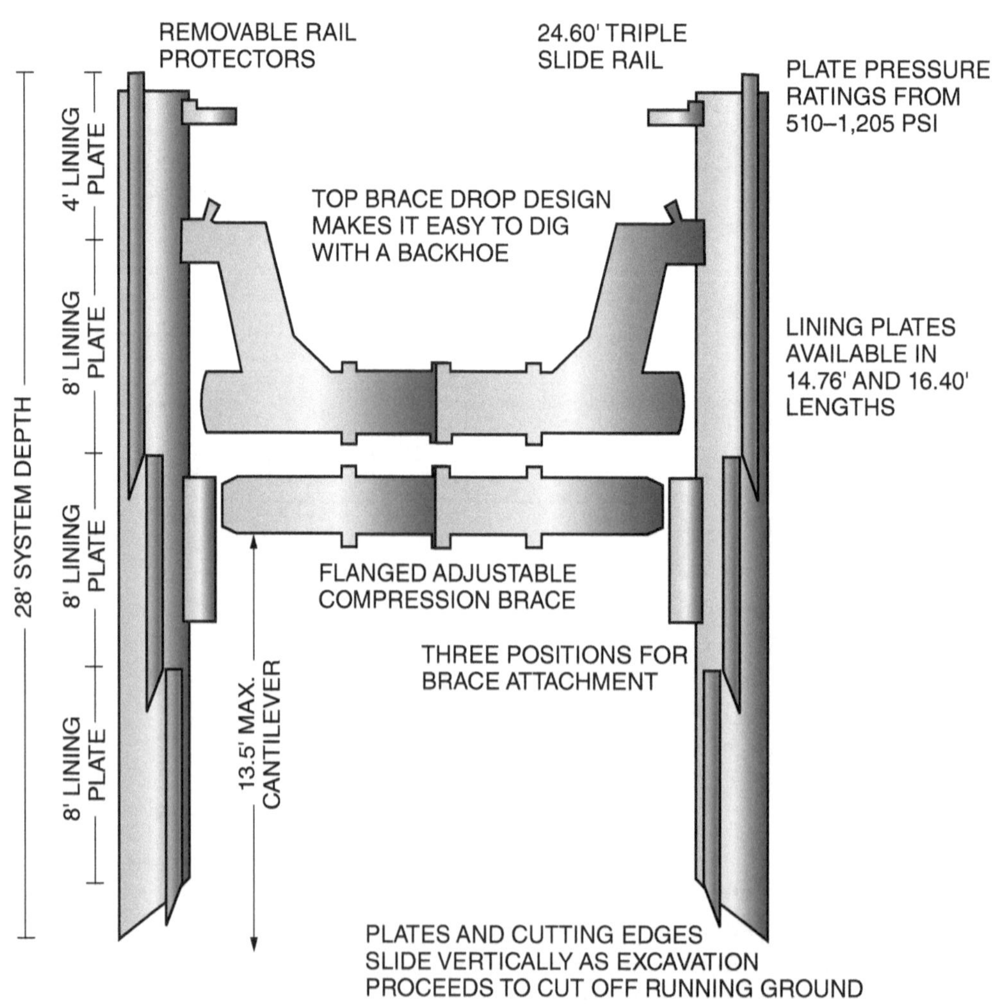

Figure 24 Slide-rail system.

7.1.0 Placing Bedding

In preparing the ditch for replacement of the pipe, pipeline specifications should be followed. A sound pipe installation begins with a stable foundation. This foundation may include a layer of stabilization material to support the bedding, a layer of bedding material to cushion and support the pipe, and initial backfill to blanket the pipe. These three elements can be critical to the loadbearing ability of the pipeline.

7.1.1 Stabilization

Some soil conditions may require a stabilization layer to support the pipe bedding. Unless the pipe is moved to a different trench, the need for such a layer should be evident in the existing trench. If the need is evident, the stabilization layer under the exposed pipe may need to be repaired after excavation.

If the pipeline is laid in a new trench, the bottom should be inspected to determine the need for stabilization. A worker can walk in the trench, shifting weight from one foot to the other and determine if the soil underneath moves, shifts, or gives. If it does not feel solid, stabilization may be needed.

Crushed stone is often used, in sizes ranging from less than 1" to large rock (rip-rap). The material is poured into the trench and spread to a uniform depth.

7.1.2 Bedding

Bedding should be sufficient to provide uniform, firm support for the pipe and to maintain pipe grade. The type of bedding support is dictated by specifications. The loadbearing capacity of the pipeline is determined from the strength of the pipe itself and the type of bedding.

Bedding extends from the top of the stabilization layer, if present, to the bottom of the pipe. Some materials used include native soil, sand, stone, or concrete. Specifications commonly require 4" to 6" of bedding material. The material is poured into the trench and leveled to grade along the full length of the trench. Lasers are often used to check grade.

7.1.3 Initial Backfill

After laying the pipe on the bedding, the initial backfill is added. This is material placed from the bedding or the bottom of the pipe to some specified level above the pipe. The purpose of this initial backfill is to protect the pipe. You must use only clean fill material free of rocks, clumps, foreign objects, or organic materials. Initial backfill may include any of the following:

- Haunching, tamping, or filling to the spring line
- Tamping or filling to the top of the pipe
- Tamping or filling to some point above the pipe

Haunching is the process of placing and consolidating material in areas that may remain void or loose when the rest of the backfill is added. Because bedding only supports a narrow strip along the bottom of the pipe, sand or dirt is haunched in along the bottom sides, or haunches, of the pipe. This material is compacted with a hand tamp or other such tool. *Figure 25* shows some different pipe laying conditions.

Type 1 is a flat-bottomed trench, which means that the soil on which the pipe rests is undisturbed. The soil that the pipe rests on is not backfill. This method is not recommended when laying pipe larger than 30" in diameter. In Type 2, the pipe rests on undisturbed, flat-bottomed ground and is surrounded to its center line with backfill that is lightly tamped. Type 3 shows the pipe laying on at least 4" of loose soil that is leveled to a given grade. Backfill is then placed to the top of the pipe and lightly tamped. In Type 4, the pipe is placed on at least 4" of sand, gravel, or crushed stone, depending on the specifications. Backfill is placed to the top of the pipe and compacted to about 80 percent of the density of the undisturbed soil. In Type 5, the pipe is placed on a bed of at least 4" of compacted granular material. The material is compacted to the center line of the pipe and then backfilled to the top of the pipe. Backfill should be granular material or soil from the original excavation that is free of rocks, frozen earth, or other foreign matter. The backfill is compacted to 90 percent of the undisturbed soil.

7.2.0 Backfilling

A single pipeline frequently requires the use of a variety of backfilling procedures to perform the work effectively and economically while complying with specifications regarding pipe and coating protection. Excavators or dozers are commonly used to move dirt from the spoil back to the ditch.

Adequate fill material must be provided underneath the pipe as well as over it. The spoil bank is the main source for backfill material, but loss through excavation may require additional material. The backfill material around the pipe must not damage the pipe or coating. Additional material should meet the same specifications as original construction.

The backfill is laid around the pipe in such a manner that it supports the pipe evenly, reducing stresses that may later cause pipe damage. The bedding, side fill, and cover should provide a uniform, even blanket around the pipe.

When filling is completed, the surface should be restored as closely as possible to its original condition, including the replacement of topsoil.

7.2.1 Preparing Fill

Any rocks and foreign objects left in the ditch or in the backfill material are potential hazards to the pipe and its coating. Foreign material such as cans, tools, clamps, or scrap metal, if left in the ditch, can weaken corrosion protection, or damage the pipe in that area.

Fill dirt, particularly that used for bedding and close to the pipe, may be screened and sifted to remove rocks, large chunks, and other debris. The spoil from a new ditch made through rock cannot be used as fill because of the amount of rock debris. New fill dirt must be brought in to pad and fill most of the ditch. Sometimes rocky soil may be used in the upper part of the ditch, as long as the rocks are small, well away from the pipe, and

LAYING CONDITION	DESCRIPTION
 TYPE 1	FLAT-BOTTOM TRENCH. LOOSE BACKFILL.
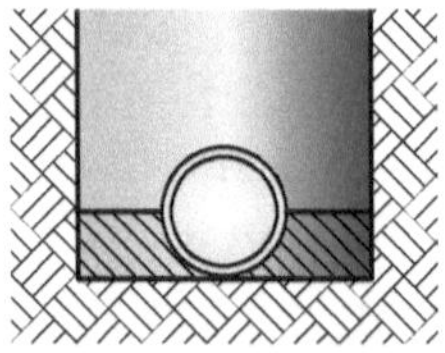 TYPE 2	FLAT-BOTTOM TRENCH. BACKFILL LIGHTLY. CONSOLIDATED TO CENTER LINE OF PIPE.
 TYPE 3	PIPE BEDDED IN 4" MINIMUM LOOSE SOIL OR BEDDING MATERIAL. BACKFILL LIGHTLY CONSOLIDATED TO TOP OF PIPE.
 TYPE 4	PIPE BEDDED IN SAND, GRAVEL, OR CRUSHED STONE TO DEPTH OF ⅛ PIPE DIAMETER, 4" MINIMUM. BACKFILL COMPACTED TO TOP OF PIPE.
 TYPE 5	PIPE BEDDED TO ITS CENTER LINE IN COMPACTED GRANULAR MATERIAL, 4" MINIMUM UNDER PIPE. COMPACTED GRANULAR OR SELECT MATERIAL TO TOP OF PIPE.

NOTE: These options vary with job specifications and locations.

106F25.EPS

Figure 25 Different pipe laying conditions.

only used up to about 1' from the surface. The final layer must be clean fill dirt or topsoil. If excavated soil has been contaminated with product from a leak before or during repair, it should not be used as fill.

Spoil from ditches dug in frozen ground can be as dangerous as rocky soil. The frozen bits of earth are irregular and hard as rock. They may be used if they can be broken up sufficiently to not pose a hazard to the pipe. However, frozen soil is difficult to distribute evenly and, when it thaws, may result in areas of unsupported pipe or uneven fill. In some areas, soil that was unfrozen when excavated can freeze at the surface, particularly overnight. This fill would cause the same problems.

7.2.2 Removing and Positioning Skid Piles and Sandbags

Before backfilling, skids must be removed. This is only done while other support is being provided under the pipe. In rocky rough soils, padding of earth, sandbags, or other approved supports may be laid under the pipe. Synthetic materials are also often used to provide a padded surface on the bottom of the ditch.

7.2.3 Steps for Backfilling

Backfilling a trench following pipeline maintenance is a delicate operation with the goal of protecting the pipeline from damage. Use the following steps as a checklist to properly backfill a trench:

Step 1 Identify soil types and their suitability to support pipe. Some soil types compact better than others, making them more suitable to support the pipeline.

Step 2 Perform a visual inspection of backfill material to identify foreign objects that could damage the pipeline system. For example, use a shovel to spot-check for large objects or foreign material that could be mixed in. If you see any foreign material or objects, make sure they are removed.

Step 3 Determine when backfill material is unsuitable for backfill around the pipeline. Remember that soft material must be used near the coating. The backfill may need to be replaced if there are too many rocks, roots, or other foreign material.

Step 4 Use suitable backfill material within a radius of 6" of the pipe, or per company policy.

Step 5 Fill equally along both sides of the pipe until the pipe is covered. This will help prevent pipe movement and provide additional support for the pipe.

Step 6 Crown the backfill according to company procedures. This is usually performed to compensate for settlement of backfill.

> **NOTE**
> Provide extra cover at road crossings, ditches, and streams. These areas face an increased risk from third-party damage, and extra cover allows some extra protection. In special circumstances like these, you should reference procedural manuals, such as your company manual and *49 CFR Part 195.*

Step 7 Compact the soil as appropriate for proper pipe support, size of support, and spacing of support. Use a tamping tool, backhoe, or other appropriate tool. Be sure that you know when the soil is properly compacted.

> **NOTE**
> Perform extra compaction at road crossings or other locations where heavy loads may cross the pipeline. This provides extra protection for pipe. Less settlement also means that there's less chance of third-party damage.

7.3.0 Performing Soil Compaction

The replaced soil frequently needs to be compacted to avoid a depression caused by later settlement and because loose, uncompacted earth is subject to erosion by rainwater. By consolidating the soil particles in the fill, compaction also strengthens the pipeline cover and increases the protection of the blanket of fill around the pipe.

Compaction can be accomplished by driving a heavy crawler-type tractor over the backfilled ditch. Extra cover is often used at road crossings, ditches, and streams. Extra compaction is allowed at road crossings and other locations where heavy loads are expected to cross the pipeline.

Backfill is normally crowned after compaction to allow for some settlement. Ditch breakers may be required on the slope of a ditch, diversion level, or surface to prevent backfill from washing away.

7.4.0 Laying Pipe in Trenches

Properly unloading the pipe is the first step to laying pipe in the trench. This is done using ropes or machinery, depending on the kind of pipe being laid.

PVC is easily laid by hand in almost all pipe sizes. Iron pipe must be laid with the aid of machinery because of its weight. The best machinery for this is a boom or crane truck. The pipe can be loaded on the truck and the crane used to place the pipe in the trench.

The pipe should lie level in the trench. When installing bell and spigot pipe, the ground must be dug out beneath the bell to prevent the bell from supporting the pipe. If the pipe is joined by means of couplings, the soil must be dug out underneath the couplings to keep the weight of the pipe off the couplings.

At least one manufacturer shows the use of mounds between couplings as an acceptable method of laying pipe. However, digging out provides a stable embedment and may be required by local ordinances. *Figure 26* shows a proper laying method.

7.5.0 Standard Laying Conditions

Figure 27 identifies standard laying conditions for underground pipe, as defined in *Standard ANSI/AWWA-D150/A21.50.* Laying conditions may be included in the contract specifications where private property is concerned. Most municipalities have laying conditions defined for various kinds of underground water service piping where public property is involved.

The defined laying condition is influenced by depth of burial, experience with local soil as far as stability is concerned, and whether or not the piping run is under paved areas. Specific compaction requirements may also be included.

7.6.0 Maximum Depth of Cover

Laying condition requirements depend on job specifications and directly affect the maximum depth of cover. Other factors include the size and type of pipe, wall thickness, and type of soil. *Figure 27* shows the effect of the laying condition on the maximum depth of cover for 12" and 24" ductile iron pipe with cement lining.

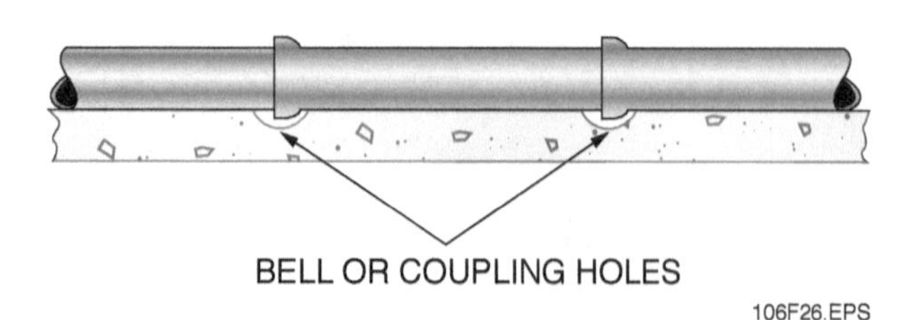

Figure 26 Proper laying method.

Buried piping must be located below the frost line to prevent freezing during the winter months. The piping must also be located deep enough to be protected from other surface loads that could cause mechanical damage. Piping located under driveways, roads, and railroad tracks

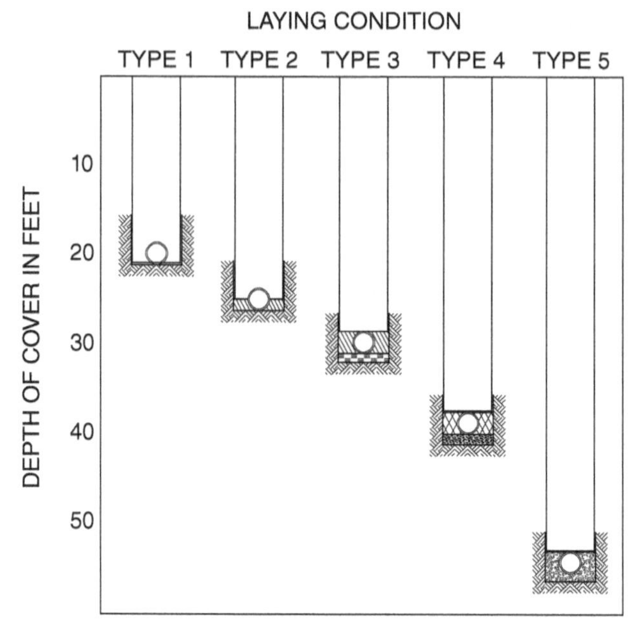

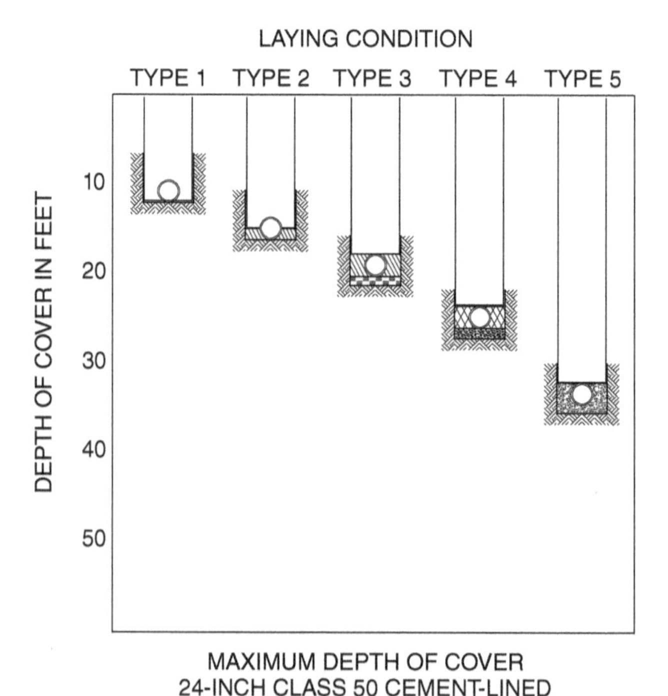

Figure 27 Effect of laying condition on maximum depth of cover.

must be buried deeper to prevent undue loads on the sprinkler piping. See *NFPA 24* for additional information.

> **NOTE**
> In some jurisdictions, authorities may require inspection before the trench is backfilled.

7.7.0 Compaction

After the water line has passed inspection, the final backfilling and compaction can begin. Some municipalities enforce compaction measurements on the backfill.

Generally, there is a specification on achieving 90 to 95 percent of the maximum attainable density in the soil. A typical municipality might require a minimum of 90 percent compaction under all circumstances, with a minimum compaction of 95 percent under streets. This means 90 to 95 percent of the volume is soil or sand and 5 to 10 percent is air. Obviously, coarse material will be more difficult than fine material to compact to these high levels.

The compaction process begins with the first lift or layer placed and tamped in the trench. Each lift should be 8" to 12" deep. Most trenches are tamped using vibratory compacting equipment. They are also called whackers or tampers. *Figure 28* shows two types of compactors.

The contractor requests a soil test from an ASTM-approved testing laboratory. The laboratory tests the soil for its moisture content, and issues a proctor curve. A proctor curve is a moisture density curve. This determines what, if anything, needs to be added to the soil in order to meet the compaction criteria that may be required on the job.

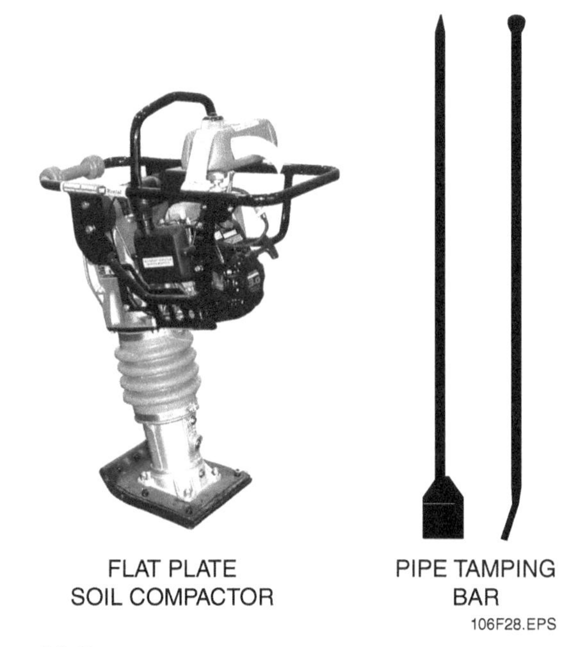

Figure 28 Compactors.

If the soil from the first test passed inspection, subsequent lifts can then be added. Tests have to be made for each lift.

The two most common methods for making this test are the sand cone method and the nuclear test method. The sand cone method takes at least a 24-hour period for each layer of soil. After a layer of soil is compacted, a sample of soil is taken for every 75' to 100' of trench. The sample is then taken to a testing laboratory. The test itself requires 16 hours, as mandated by ASTM. If that layer passes, another layer can be put down, with a soil sample taken to the lab and tested for 16 hours. Since this has to be done layer by layer, an average underground trench could take weeks to be filled.

Nuclear soil tests are not as lengthy as tests using the sand cone method. The inspector is on the job site as each layer goes down. As with the sand cone method, the soil is tested layer by layer every 75' to 100'; however, the soil can be tested on the spot, saving numerous trips back and forth to the testing lab. Nuclear testers work like Geiger counters. A probe containing nuclear particles is placed approximately 6" into the soil, then activated. Impulses are then sent into the soil, and the impulses reflected back to the testing equipment are recorded. The greater the soil compaction, the fewer the impulses reflected back.

The timing of these tests is important. For example, if the sand cone method is being used on a sunny day, the soil may test at 95 percent. But, suppose overnight thunderstorms occur and the soil becomes saturated. This can change the reading to the point that the soil will not pass the compaction requirement and has to dry out before further lifts can be added.

The type of soil used for the backfill also plays a key role in the compaction process. If the soil is sandy, more moisture will be required to meet the percentage specified for the compaction. Clay soil will compact more solidly with less moisture. If the soil is mixed, and the test sample is taken where there are sand pockets or chips of hardpan, the test can fail. That is why it is important that the backfill material be thoroughly mixed, and any hard chips crushed. It is not advisable to mix sand and clay soils for backfill because it is difficult to combine the two without getting too much water in the clay.

8.0.0 Underground Piping Installations

The minimum requirements for pipe and fittings used for underground pipe in fire protection are specified in *NFPA 24*. More detailed specifications may be imposed by the project engineer in his specifications, or by the municipal authority in their general specifications.

The choice of pipe is also affected by how the pipe is used as a private line or as part of the municipal utility system. Other considerations are ease of installation, the cost of the pipe, and local codes. Some of these become construction requirements. Others are options taken by the company performing the installation.

8.1.0 Cast-Iron Pipe

Cast-iron pipe, also known as soil pipe, has been used for water supply lines for over 500 years. Some studies claim that as much as 96 percent of the original cast-iron water mains installed are still in service today.

The process of casting iron pipe has gone through many changes. A major change occurred in 1914. An engineer named De Lavaud developed the centrifugally-spun process. In the De Lavaud process, the pipe is cast centrifugally in rotating, water-cooled molds. This produces a uniform product with superior strength at a cost less than previous methods. Today, cast-iron water pipe is no longer manufactured. For fire sprinkler work it has been replaced with ductile iron.

8.2.0 Ductile (Nodular Cast) Iron Pipe

The manufacture of ductile iron (nodular cast iron) began in 1960. In toughness, ductile iron is intermediate between cast iron and steel. In shock resistance, it is comparable to ordinary grades of mild carbon steel. It maintains good pressure tightness under high stress. It can be welded and brazed and is corrosion resistant.

Did You Know?

Man has understood how to use iron longer than he has understood how to use any other metal. Prehistoric people used iron from meteorites to make tools, weapons, and other objects. Refined iron tools and weapons marked the Iron Age, which began about 1200 B.C.E. in the Near East. Until the 1800s, iron was the main industrial metal. With the advent of the Industrial Revolution in the mid-1800s, manufacturers discovered an economical way of making steel (a mixture of iron and carbon). Steel soon replaced iron as the basic material for making tools and equipment for industry.

8.3.0 Polyvinyl Chloride (PVC) Pipe

PVC is becoming increasingly popular for use in water mains and firelines. Its light weight makes it easy to load, transport and handle. The ease of installation of PVC pipe makes it cost-effective. PVC is not affected by soil or water chemical conditions.

PVC pipe is constructed with a smooth interior that stays smooth over years of service with no loss in carrying capacity. Because it is non-metallic, it does not lose strength due to external galvanic corrosion, soil corrosion, or potable water corrosion. It is available in sizes that are compatible with cast and ductile iron fittings. This allows connections to be made without special adapters or complicated procedures. Refer to the manufacturer's instructions for guidelines.

8.3.1 CPVC

Chlorinated polyvinyl chloride (CPVC) pipe is listed for underground use. Refer to the manufacturer's approved installation instructions for special backfill deflection tolerances and cure times. CPVC does not require thrust blocks.

8.4.0 Restrained Joint PVC

Restrained joint PVC is the first non-metallic restrained joint designed for use with municipal water and fire protection systems. Restrained joint PVC eliminates rodding and additional bolts to restrain the joint. It is intended to eliminate costly concrete thrust blocks in a properly designed water system. PVC pipe must meet the performance requirements of *AWWA C900, Standard for Polyvinyl Chloride (PVC) Pressure Pipe, 4" through 12" for Water Distribution* and will be furnished in cast-iron equivalent outside diameters with restrained rubber gasket joints.

Restrained joint PVC has a working pressure rating of 150 psi and is assembled to ductile iron pipe fittings with restrained mechanical joint gland adapters. This type of PVC pipe and couplings provide a restrained joint by using precision machined grooves on the pipe and in the coupling. When aligned, these grooves allow a **spline** to be inserted, resulting in a fully restrained joint that locks the pipe and coupling together. A flexible O-ring gasket in the coupling provides a hydraulic pressure seal.

Did You Know?

Although plastic materials became popular in the 1950s and 1960s, plastics date back more than a century to the introduction of celluloid in 1870. The 1930s and 1940s saw the introduction of Lucite™, Plexiglas™, and nylon. PVC, the first plastic used for plumbing purposes, was under development in the 1930s.

8.5.0 Glass Filament Reinforced Epoxy Pipe

One type of pipe using epoxy and continuous filament glass fiber reinforcement was once manufactured for underground use. Permastran® was a trade name for this pipe. Its tendency to delaminate with impact made it difficult to handle. The tendency to deform over time in the ditch as a result of careless installation procedures or packing of the fill from the top was another disadvantage. This material is no longer in the marketplace. It can be recognized from its thin wall construction and greenish translucent appearance. In repair and modification of underground pipe, replacement of this kind of pipe with more reliable types should be considered.

8.6.0 Pipe Identification

The working pressure of a pipe is determined by the size and class of the pipe. For example, a 12", Class 50 ductile iron pipe has a working pressure of 250 psi.

The class rating also indicates the wall thickness of the pipe. *Table 4* relates class to wall thickness for ductile iron pipe.

8.7.0 Types of Approved Piping

Various agencies are responsible for the classification and approval of the various types of pipe used in underground installations. The agencies most commonly referred to in this text are American Water Works Association (AWWA), American National Standards Institute (ANSI), and American Society for Testing and Materials International (ASTM). Each of these agencies sets minimum standards for the production of materials used in all industries. These standards are set with the safety, health, and welfare of the consumer as well as the economic needs of the public in mind. *Table 5* shows the standards applicable to the production and use of various types of underground pipe.

NFPA 13 and *NFPA 24* require pipe for fire service mains to be Listed for fire protection service and to comply with AWWA standards, where applicable. *NFPA 13* and *NFPA 24* further require that the type and class of pipe for a particular installation must be determined through consideration of its fire resistance, the maximum working pressure, the laying conditions under which

Table 4A Class and Wall Thickness

Ductile Iron Pipe[1]

Push-on Joint (Pressure Classes)

Pipe Size[2]	Pressure Class	Thickness	Outside Diameter
12	350	0.28	13.20
16	250	0.30	17.40
	300	0.32	17.40
	350	0.34	17.40

Mechanical Joint (Special Classes)

Nominal ID	ANSI Thickness Class	Pipe Outside Diameter
12	50	13.20
	51	13.20
	52	13.20
	53	13.20
	54	13.20
	55	13.20
	56	13.20
16	50	17.40
	51	17.40
	52	17.40
	53	17.40
	54	17.40
	55	17.40
	56	17.40

Push-on Joint (Special Classes)

Nominal ID	ANSI Thickness Class	Pipe Outside Diameter
12	50	13.20
	51	13.20
	52	13.20
	53	13.20
	54	13.20
	55	13.20
	56	13.20
16	50	17.40
	51	17.40
	52	17.40
	53	17.40
	54	17.40
	55	17.40
	56	17.40

106T04A.EPS

Table 4B Class and Wall Thickness

PVC C900 Municipal Water Distribution Pipe[1]				
DR 25 – CLASS 100				
Nominal Size	Pipe Outside Diameter	Minimum Wall Thickness	Inside Diameter	Bell Outside Diameter
4	4.800	0.192	4.416	6.10
6	6.900	0.276	6.348	8.25
8	9.050	0.362	8.326	11.25
10	11.100	0.444	10.212	13.25
12	13.200	0.528	12.144	16.00
DR 18 – CLASS 150				
4	4.800	0.267	4.266	6.20
6	6.900	0.383	6.134	8.25
8	9.050	0.503	8.044	11.25
10	11.100	0.617	9.866	13.25
12	13.200	0.733	11.734	16.25
DR 14 – CLASS 200				
4	4.800	0.343	4.114	6.36
6	6.900	0.493	5.914	8.50
8	9.050	0.646	7.758	11.50
10	11.100	0.793	9.514	13.75
12	13.200	0.943	11.314	16.50

NOTES:

1. For other pipe sizes, and for pipe not listed for fire service, refer to the manufacturer's data sheets. See manufacturer's data sheets for other types of pipe such as asbestos cement, concrete pressure pipe, or CPVC. Also, you may encounter pipe that is no longer manufactured (cast iron). If so, refer to the manufacturer's data sheets for information.
2. All dimensions in inches.

106T04B.EPS

it is to be installed, soil conditions, corrosion, and susceptibility of pipe to external loads. External loads include earth loads, buildings, and traffic or vehicle loads. Pipe used in private fire service must be designed to withstand a working pressure of not less than 150 psi.

NFPA 13 and *NFPA 24* also allow cement lining for cast iron and ductile iron pipe. This lining greatly reduces the tuberculation process on the inside of the pipe. Tuberculation is the buildup of hardness products and rust on the inside of the pipe. In unlined pipes, this material collects over the entire inside diameter of the pipe. The resulting reduction in inside diameter causes drastic pressure and flow losses that can make the pipe system useless.

Cement lining is cast when the pipe is manufactured. Exercise care to avoid any damage to the lining. Any point that is damaged can result in the buildup of foreign material over time. This, in turn, destroys the flow characteristics of the piping. The lining will chip out with hammer blows or crack if the pipe is dropped from a truck.

8.8.0 Pipe Joints

Fitting pipe together properly in the trench is the most important aspect of laying pipe. If the joints are not properly fitted, leaks occur. This is not just an economical concern. It is also a safety concern. Leaks can undermine the embedment, causing piping to shift and blow-outs to occur.

Most pipe manufacturers publish fully detailed instructions for the assembly and installation of their pipe. This is the best source of information if there are uncertainties about a procedure.

8.8.1 Mechanical Joint Pipe

Mechanical joint pipe generally comes in 18' lengths. The wall thickness varies for different pipe classes and sizes.

The joint itself consists of a special bell on the pipe, including a flange, with a cast-iron gland or following ring, a rubber gasket, and necessary bolts and nuts (*Figure 29*).

On Site

Crane Safety

Workers in heavy construction routinely use a variety of lifting devices, as shown here. Cranes vary in size and carry different types of loads. They are among the most commonly used pieces of heavy equipment and can also be the most dangerous. You must understand and follow all safety requirements when working around cranes.

106SA04.EPS

Tests have proven that a rubber gasket sealed in a joint lasts as long as the pipe. The gasket does not deteriorate because the rubber is compressed in a relatively cool, air-free space, unexposed to light. Little or none of the gasket is exposed to water passing through the pipe. Follow these steps to assemble a mechanical joint:

Step 1 Dip the gasket in a pail of soapy water. If the temperature is below freezing, use glycerin instead of water.

Step 2 Slip the gland on the pipe, with the lip extension of the gland toward the joint.

Table 5 Approved Piping

Material	Source	Standard
Cast-iron water pipe (cast-in-metal mold)	ANSI	A21.6-197
Cast-iron water pipe (cast-in-sand lined mold)	ANSI	A21.8-1975
Ductile iron pipe (cast-in-metal mold)	ANSI	C150-C151
	AWWA	A21.50, A21.51
Polyvinyl chloride pipe up to 12" (Schedule 40, 80, 120)(Class 100, 150, 200)*	ASTM	D1785-76
	AWWA	C-900

*Polyvinyl chloride pipe 14" to 48" is governed by C-905

106T05.EPS

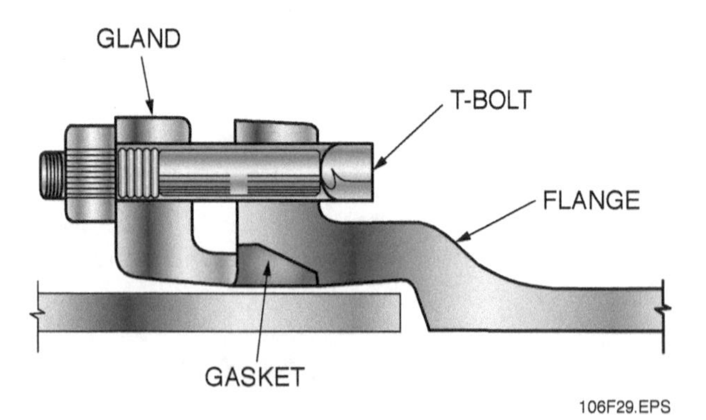

Figure 29 Mechanical joint.

Step 3 Slip the gasket on the pipe with its thick edge towards the gland.

Step 4 Clean the inside of the bell and spigot ends of the pipe and push the spigot end to its seat in the bell.

Step 5 Press the gasket into position with your fingers (or use a hammer), ensuring that it is evenly seated all around the joint.

Step 6 Move the gland into position against the face of the gasket.

Step 7 Insert the bolts and make them finger tight.

Step 8 Tighten the bolts evenly all around using a ratchet wrench (*Figure 30*).

All bolted joint accessories (metal parts) must be cleaned and thoroughly coated with bitumen or other acceptable corrosive-retardant material after installation in accordance with *NFPA 24*.

8.8.2 Slip Joints

Slip joints are basically push-on joints. The inside of the socket, the gasket, and the plain end of the pipe must be kept clean throughout the assembly. Joints are watertight only if they are clean during assembly. Follow these steps to assemble a slip joint:

Step 1 Remove all foreign matter from the socket (such as mud, sand, cinders, gravel, pebbles, trash, or frozen material).

Step 2 Inspect the gasket thoroughly to be certain it is clean.

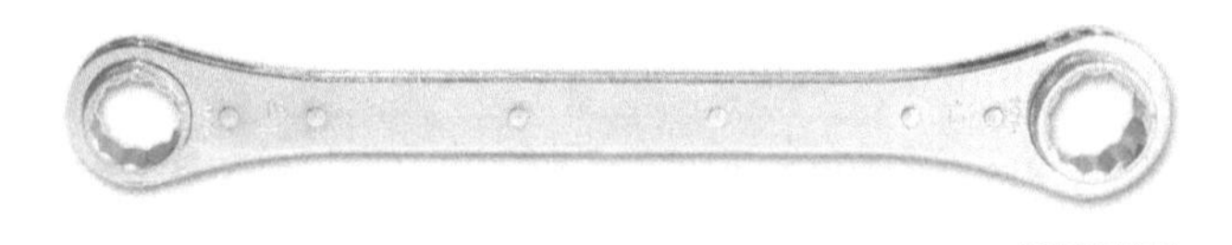

Figure 30 Bolt-thru ratchet wrench.

CAUTION

Foreign matter in the gasket seat may cause a leak. Be sure to wipe the gasket with a clean cloth.

Step 3 Loop the gasket and place it into the socket, with the round bulb end entering first.

Step 4 Inspect to ensure that the gasket heel is evenly seated around the retainer seat.

NOTE

When installing slip joint pipe in subfreezing weather, the gaskets, prior to their use, must be kept at a temperature of at least 40°F. Keep gaskets in a heated area or immerse them in a tank of warm water. Gaskets immersed in water must be dried before being placed in the pipe socket.

Step 5 Apply a thin layer of manufacturer-approved lubricant to the inside surface of the gasket that will come in contact with the plain end of the pipe.

Step 6 Clean the plain end of the pipe of all foreign matter on the outside from the end of the pipe to the stripes.

CAUTION

In cold weather, frozen material may cling to the pipe that must be removed. In some cases it is desirable to apply a thin layer of lubricant to the outside of the pipe about 3" back from the end of the pipe. Do not allow the plain end or trench side to touch the ground after lubricating since foreign material may adhere to the plain end and cause a leak. Lubricants which were not furnished with the pipe must not be used.

Step 7 Align the plain end of the pipe and the socket.

Step 8 Insert the plain end of the pipe into the socket until the plain end comes in contact with the gasket.

Step 9 Force the plain end of the pipe past the gasket until the plain end makes contact with the bottom of the socket. The first painted line on the plain end pipe will have disappeared into the socket and the front edge of the second stripe will be approximately flush with the bell face. If assembly is not accomplished with the application of reasonable force by the methods indicated, the plain end of the pipe must be removed to check for the proper positioning of the gasket, adequate lubrication, and removal of foreign matter in the joint. *Figure 31* shows the steps for assembly.

8.8.3 Joining PVC Pipe

When joining PVC pipe, as with all other pipe, the ring and ring groove, as well as the spigot end and the bell end of the pipe, must be clean. The rubber ring will not seat properly if dirt or foreign material is present. Improper seating of the rubber ring can cause leaks. It is also very important that the ring is not lubricated. In recent years, many manufacturers have begun shipping PVC pipe with the gasket already installed. Follow these steps to join PVC pipe:

Step 1 Clean the ring with a dry cloth.

Step 2 Insert the ring in the groove with the color marking facing the outside of the bell.

Step 3 Clean the pipe end around the entire circumference from the end of the spigot to 1" beyond the reference mark.

Step 4 Lubricate the spigot end of the pipe. Be sure to cover the entire circumference, paying particular attention to the beveled end of the spigot. The coating should be the equivalent of one brush coat of enamel paint. Do not lubricate the ring or the ring groove in the bell. Lubrication in those places can cause the rubber ring to be displaced.

Step 5 After lubricating the spigot end of the pipe, take precautions to keep it clean and free of dirt or other foreign material. If the lubricated end gets dirty, it will have to be cleaned and then re-lubricated.

Step 6 Insert the spigot end of the pipe into the bell end so that it is in contact with the rubber ring.

Step 7 Align the pipe sections and push the spigot end in until the reference mark on the spigot end is flush with the end of the bell.

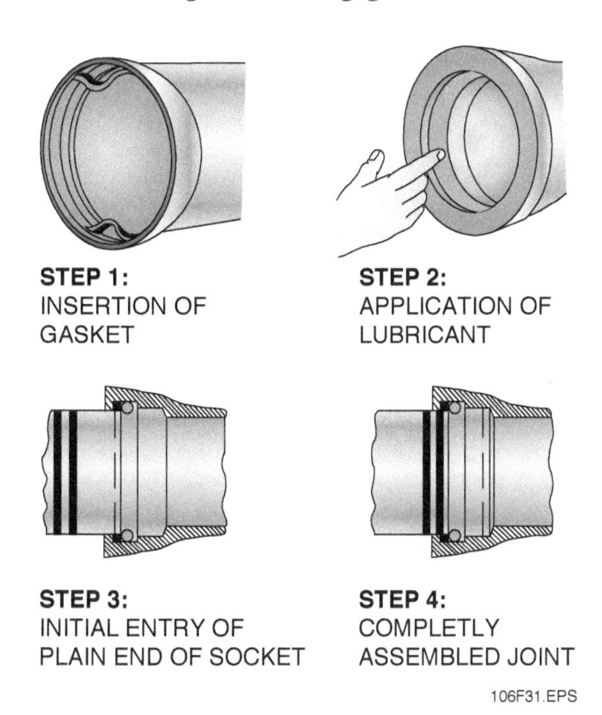

Figure 31 Steps for slip joint pipe assembly.

NOTE

The recommended method for assembly uses a bar and a wood block; however, pullers such as the lever or come-along may be used.

8.9.0 Tools Used for Joining Pipe

Many types of tools are used to join underground pipe. The following sections will explain the basic types of tools used to join pipe.

8.9.1 Bar and Block

This method of joining pipe can be used as long as the ground on which the pipe is being laid is firm. A wooden block is placed against the open end of the pipe, opposite the end that is being joined. Drive the bar into the trench bottom, then apply pressure forward against the wood block to drive the pipe home. The wood block gives a straight center-line push. It also protects the pipe or coupling from damage by the bar. *Figure 32* shows the bar and block method.

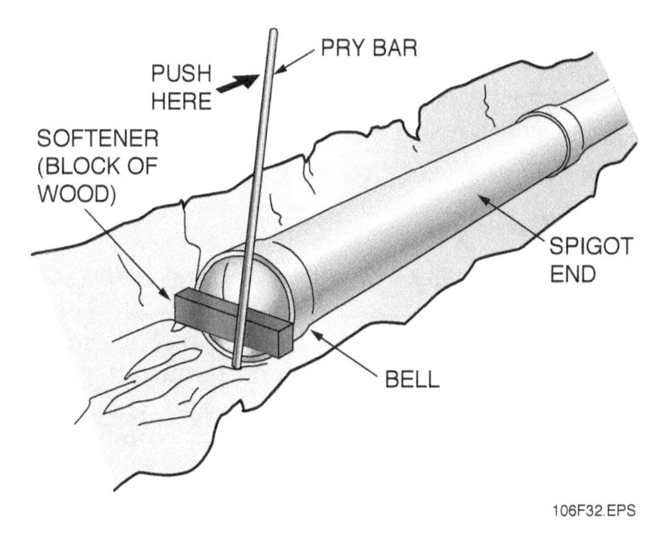

Figure 32 Bar and block method.

8.9.2 Lever Puller

Lever pullers are often used to pull sections of pipe together. The handle is used to pull the chain. One side of the puller is attached to the bell end of the pipe. This is the stationary side. The other end of the chain is placed on the spigot side of the pipe. This is the end where the pulling action takes place. By pumping the handle back and forth, the gears pull the chain through the tool to join the two sections of pipe. *Figure 33* shows a lever puller.

WARNING

Never use the saw without the lower blade guard properly attached. Always wear appropriate personal protective equipment, including eye protection and a hard hat, when using the saw.

8.10.0 Cutting Pipe in the Field

There are occasions when cutting pipe in the field is required. When this is done, the end of the cut pipe must be conditioned to assure a proper joint. There are many tools for cutting pipe, including the abrasive saw.

8.10.1 Ductile Iron Pipe

Use a power rotary cutter to cut ductile iron pipe. Because of its strength, iron pipe cannot be cut with chain cutters. Hinged four-wheel cutters with appropriate cutter wheels work best in close quarters, and for large pipe, a manual rotary cutter can be used. The cut end must be beveled at about ¼" at a 30-degree angle. This can be done with a coarse file or a portable grinder. The key here is to remove any rough, sharp edges that can damage the gasket.

CAUTION

Ensure that backfill material is clean and free of rocks. Rocks in contact with ductile iron pipe can eventually eat a hole through the pipe.

8.10.2 PVC Pipe

PVC pipe is easily cut with a fine-toothed hacksaw, handsaw, or a portable power saw. If a portable power saw is used, use a steel blade or abrasive wheel. It is also very important that the cut be square. This can be better achieved if the pipe is marked around its entire circumference prior to cutting.

As with ductile iron pipe, PVC must also be beveled on the end. This is easily done with beveling tools, which are available through pipe suppliers. A course file or rasp, a portable sander, or an abrasive disc can also be used. The angle and taper can be determined by using an end beveled at the factory as a guide.

8.11.0 Sleeving

The purpose of a sleeve is to remove undesirable physical loads that otherwise would be applied to the pipe. Examples where sleeving of underground pipe is desirable are roadway crossings, railroad crossings, and floor and grade beam penetrations. If the pipe is laid in shifting soil, sleeving will also be required.

In most cases, sleeving should be vertical or horizontal. Sleeving should always be aligned and positioned as called for in the drawings. Improper alignment can cause the pipe to bind when passing through, making the load condition worse rather than better.

Under roadways and railroad crossings, where vehicular load is the problem, the pipe would be expected to lie in the bottom of the sleeve. Where earthquake protection or protection against shifting soil is the objective, the pipe should be centered in the sleeve.

The most commonly used material for sleeving is steel pipe. In soil-covered underground applications, the sleeve material class should be at least equal to that of the pipe being protected. Where the sleeve function is primarily a form for concrete, other materials may be used. These include lightweight pipe, sleeves formed from sheet metal, conduit, and treated fiberboard. Cost, availability, specifications, and local codes will govern this decision.

8.11.1 Size and Location

In *NFPA 24*, the earthquake protection provisions call for at least 1" nominal clearance all-around

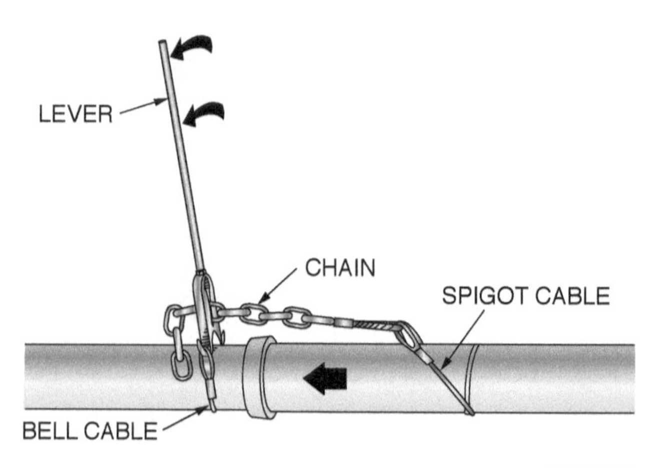

Figure 33 Lever puller.

for pipes 1" through 3½" and 2" all-around for pipes 4" and larger. If the sleeve is made of pipe, a nominal pipe size 2" larger than the pipe being protected may be used for pipe 1' through 3½". For pipes 4" and larger, a nominal pipe size 4" larger than the pipe must be used.

When passing through floor and grade beams, sleeves should always be located from a fixed object such as the center line of a beam, a column, or a fixed point on the face of a wall. The size of the pipe to go through the sleeve, as well as the size and type of sleeve, must be specified.

8.11.2 Caulking

It is necessary to caulk between the pipe passing through the sleeve and the inside of the sleeve. This prevents water from passing through the sleeve from one area to another. With soil cover, soil intrusion must be prevented in a similar manner. This reduces the potential for pipe fractures caused by stresses or minor movement of the soil.

8.12.0 Tapping Lines

Three ways of tapping a water line require evaluation. For a dry tap over 2", the line can be cut and tapped with a mechanical tee, with either a cut-in sleeve or a solid sleeve. These sleeves are manufactured in short and long lengths. The 12" length allows more flexibility in cutting the pipe and is preferable if space permits. For a wet tap over 2", a tapping sleeve and tapping valve should be used. For a wet tap that is 2" or less, a corporation stop (an assembly used to inject chemicals into the center of the stream) is used. A cut-in sleeve is to be used with a mechanical joint tee for a dry tap of an existing underground water main. A solid sleeve can also be used for this purpose. The solid sleeves are less expensive and more readily available. *Figure 34* shows a cut-in sleeve.

> **WARNING**
>
> Before you cut through a wall or partition, determine what is inside the wall, or on the other side of the partition. Avoid hitting water lines or electrical wiring.

The following general procedure applies where the outlet size is greater than 2". Three elements are required: a mechanical or flanged tee, a mechanical joint sleeve (long or short), and a gate valve. Follow these steps to tap a line:

Step 1 Turn the water off at both ends of the section to be tapped.

Step 2 Clean the exterior of the main so that it can accept the joint for both the tee on one end and the sleeve on the other end. All dirt, sand, and clay must be completely removed from the section involved.

Step 3 Cut out a segment of the main, leaving a gap long enough to slip the tee into position.

Step 4 Deburr and bevel the main ends as required to allow the fitting system to slide into position and seal properly.

Step 5 Inspect and clean the inside of the main as necessary to eliminate any foreign material.

Step 6 Slide the gland, gasket, and sleeve (in that order) onto the proper side of the main.

Step 7 Place the gland and gasket onto the main on the other side.

Step 8 Slip the tee into place on the bell side. Support it on the main, and support the other side from the bottom of the ditch.

Step 9 Make up the sleeve to the tee on the other side.

Step 10 Rotate the tee until the side outlet is horizontal and positioned properly.

Step 11 Backfill, thrust block, or support the tee in the proper position, leaving room for observation of the joint under pressure.

Step 12 Make up both joints on the main, when the tee is positioned.

Step 13 Inspect again in the side outlet to identify and remove any foreign material.

Step 14 Install the gate valve on the side outlet.

Step 15 Follow standard procedures and requirements for installing the balance of the pipe, thrust blocking, testing, chlorinating, and inspection.

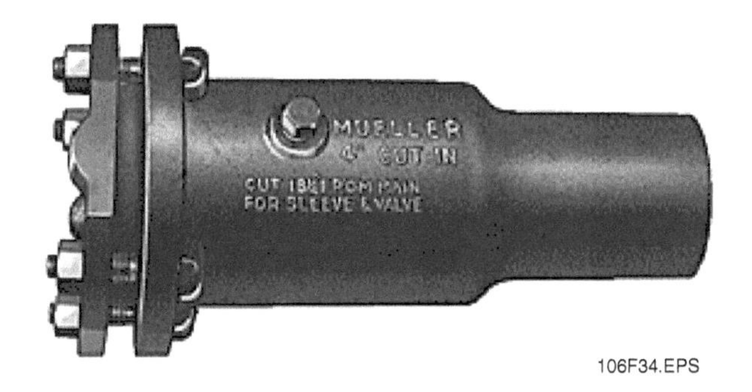

Figure 34 Cut-in sleeve.

8.12.1 Tapping Wet Lines

Figure 35 shows a tapping sleeve and the tapping valve that goes with that sleeve. This is a special gate valve with a larger-than-normal opening that permits the tapping drill to cut a full size hole in the line being tapped.

This arrangement is used for tapping a line without disrupting water service to existing users. The elements required are a tapping sleeve, a tapping valve, and a tapping drill. Follow these steps to make a tap with a tapping sleeve:

Step 1 Measure the outside diameter to determine the size and type of water main. This measurement determines the thickness and the type of sleeve gaskets required. Pit-cast pipe requires ⅝" gaskets. Centrifugally-spun pipe takes ¾" gaskets.

Step 2 When pipe is exposed, carefully clean the entire section where the sleeve will fasten, allowing at least 6" of additional clean surface on each side of the intended mounting place.

Step 3 Install the sleeve on the underground, with the side outlet properly positioned, following the manufacturer's instructions.

Step 4 Attach the tapping valve to the outlet flange.

Step 5 With the valve closed, mount the drill. The drill is in a watertight housing.

Step 6 Open the valve and advance the drill.

Step 7 Cut the appropriate hole in the main.

Step 8 Retract the drill and the coupon (the circular piece of pipe that has been removed by the cutter).

Step 9 Close the valve.

Step 10 Demount the drill assembly and retrieve the coupon.

Step 11 Proceed with the pipe installation.

8.12.2 Tapping with Corporation Stops

Corporation stops are used with outlets of 2" or smaller and are used more often since the advent of residential systems. See *Figure 35(B)*. Corporation stops must be located at least 1' from pipe ends. If two corporation stops are used, one on either side of the main, there must be at least 1' between them. If there are multiple stops on the same side of the main, they also must be separated by at least 1'.

When tapping an iron or steel pipe, a special tapping drill must be used. Some tapping tools screw to the corporation stop, and others are used directly on the main. When tapping through the corporation stop, the corporation stop must be

(A) TAPPING SLEEVE AND VALVE

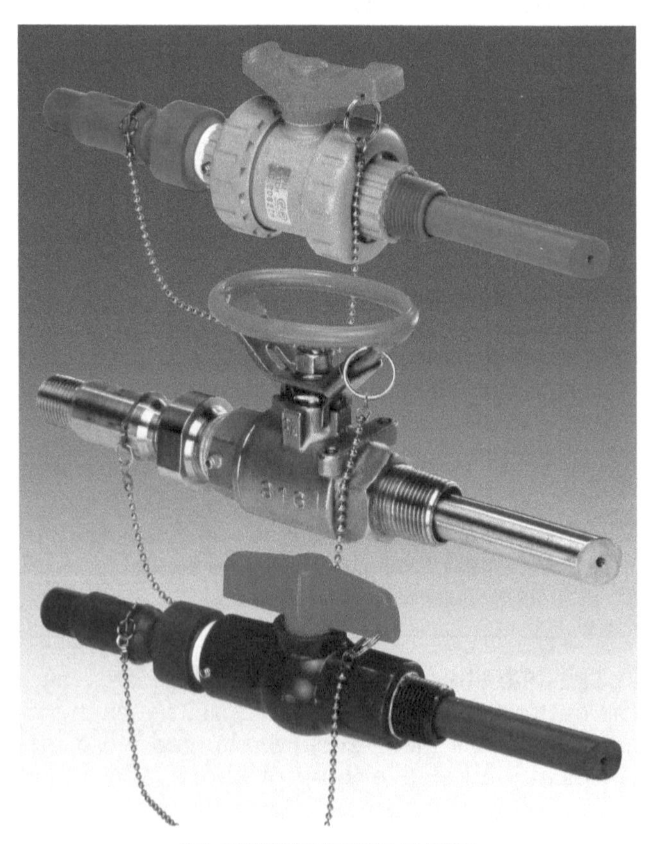

(B) CORPORATION STOPS

106F35.EPS

Figure 35 Tapping sleeve and valve and corporation stops.

open. The tapping tool should have a ratchet-type handle, which is used to turn the drill bit. The use of a light-duty, emulsifiable oil, or a light-duty chemical and synthetic oil as a lubricant is recommended. Special tapping tools have been developed for PVC pipe. Most PVC pipe manufacturers recommend using a shell cutter with internal teeth.

Caution must be used to not drill all the way through to the other side of the main or stop. This can be prevented if the drill bit is turned slowly. Some water leakage may occur. Once the hole has been drilled, the drill bit is backed out of the main, the corporation stop is closed, and the tapping tool is detached. The service line can now be attached and the corporation stop opened.

The threads on the tap must match the threads on the corporation stop. The tool must be sharp; otherwise, a leaking service connection will result. The cutter must be lubricated using a manufacturer-recommended lubricant. This minimizes the effort required to cut and tap, and protects the tool from excessive wear. The threads of the corporation stop should be wrapped with Teflon® tape to make tightening easier. The tapping tool must be clamped on firmly and at an angle of about 45 degrees to the horizontal.

Graphite and oil should be used as a lubricant during drilling and tapping. Powdered graphite and linseed oil are mixed to the consistency of medium weight motor oil. This compound is brushed on to the tool and the stop.

When the stop is fully inserted in the pipe, there should be one to three threads showing. The tool being used for drilling and tapping should also be used for inserting the stop. If the tap was made with a dry-tapping tool, the stop will have to be inserted by hand. Be careful not to overtighten corporation stops.

8.13.0 Service and Saddle Clamps

Service and saddle clamps may also be used for making service connections. Corporation stops screw directly into the saddle. The cutting tool is attached to the corporation stop and is fed through the stop to cut a hole in the pipe. When using this method of connecting service lines, it is important to make the following checks:

- The saddle must have a bearing area of sufficient width along the axis of the pipe. This prevents the pipe from being distorted when the saddle is tightened.
- The saddle must provide full support around the circumference of the pipe.
- The saddle used on PVC pipe must not have lugs that dig into the pipe when the saddle is tightened.
- The surface of the pipe must be clean where the saddle is installed.

8.14.0 Underground Fittings

In underground pipe systems, the elbows are called bends. For instance, a 90-degree elbow is called a quarter-bend. Accordingly, a 45-degree elbow becomes a ⅛ bend; a 22½-degree elbow, a 1/16 bend; and an 11¼-degree elbow, a 1/32 bend. Cross-section drawings of these are shown in *Figure 36* in the mechanical joint configuration.

Tees and crosses are also available. Those shown in *Figure 37* are typical configurations, the particular illustrations being from a selection for AC pipe. Straight and reducing tees are also available from inventory. Straight and reducing crosses are listed in manufacturer offerings, but lead time on other than straight devices could cause completion delays. These fittings must be assembled in accordance with the manufacturer's instructions, and thrust blocking must be provided.

8.15.0 Corrosion Protection of Pipe and Fittings

Buried cast and ductile iron pipe and fittings may require protection from corrosive soil conditions. Two factors have to be considered in the protection process.

A major factor in corrosion is the chemical attack that can take place with highly active soils. These are usually alkaline soils, but this can also occur with highly acidic soils, or soils with a high salt content like those found in coastal areas.

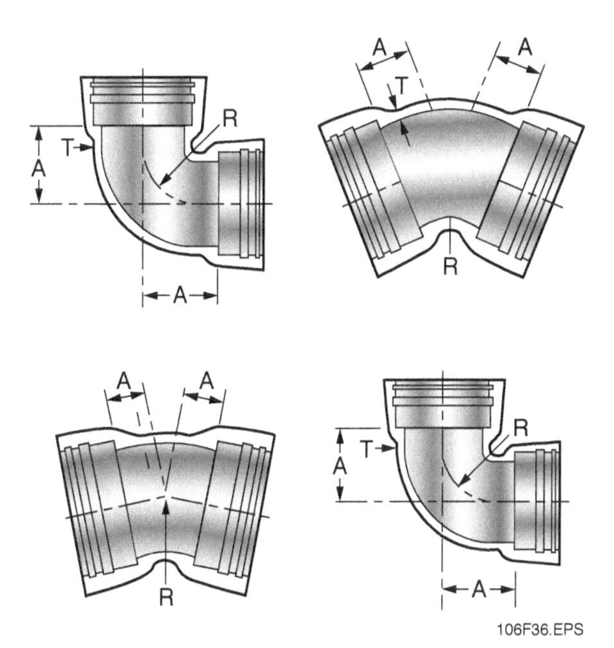

Figure 36 Bends.

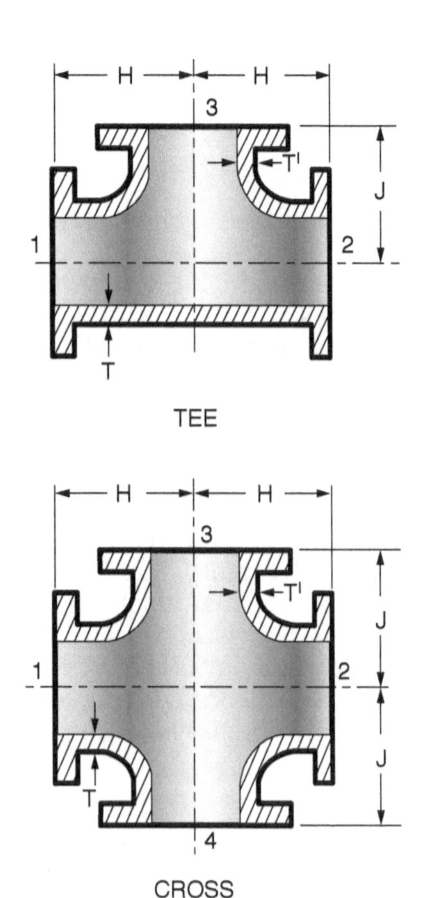

Figure 37 Tees and crosses.

A second factor in corrosion is electrical attack by earth currents (electrolysis). This is a galvanic action similar to electroplating, and is not necessarily related to the chemistry of the soil. In either case, the basic goal is to provide a separation between the metal parts and the soil so that the chemical action is blocked and the electrical currents do not flow.

Cathodic protection is a means of preventing corrosion in underground piping where the earth's electrical currents have an effect in certain soil conditions. This is often done by using metal straps that bridge across gasketed joints and fittings in order to provide continuity along the entire length of the underground piping. An anode is bonded to the pipe in a convenient location. Wiring is run from the anode to an accessible location where the current dissipates.

Protection of bolts, in particular, on fittings requires the use of asphaltic mastic. The bolts and nuts should be completely covered to avoid any moisture contact with those sensitive parts. Failure to provide that kind of coating in highly active soil can result in the bolts and nuts being destroyed in a matter of six months or less.

Another protective procedure for both pipe and fittings uses polyethylene tubing. Encasing the pipe and fittings in polyethylene (*Figure 38*) not only blocks chemical action, but also blocks the stray electrical currents at most levels encountered in the field. Eight mil polyethylene tube is available for this purpose. This is a collapsible tubing, referred to as flat tubing, and comes in oversize dimensions. A complete air- and water-tight enclosure is not necessary. Material and installation procedures are specified in *ANSI/AWWA-C105/A21.5*.

Follow these steps to encase pipe with polyethylene:

Step 1 Lift the pipe with a sling or pipe tongs.

Step 2 Slip a polyethylene tube approximately 2' longer than the pipe over the plain end and leave it bunched up accordion-style (*Figure 39*).

Step 3 Lower the pipe into the trench and make up the joint with the preceding piece of pipe. Shallow bell holes in the bottom of the trench are required to allow the tube to overlap the previous tube at the joints. The sleeve should allow a 1' overlap at each end.

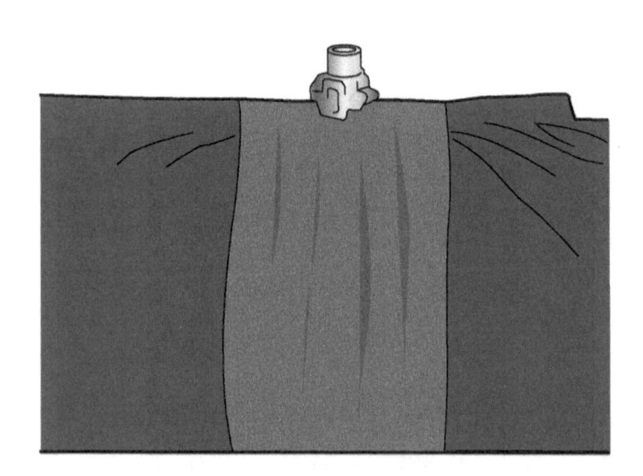

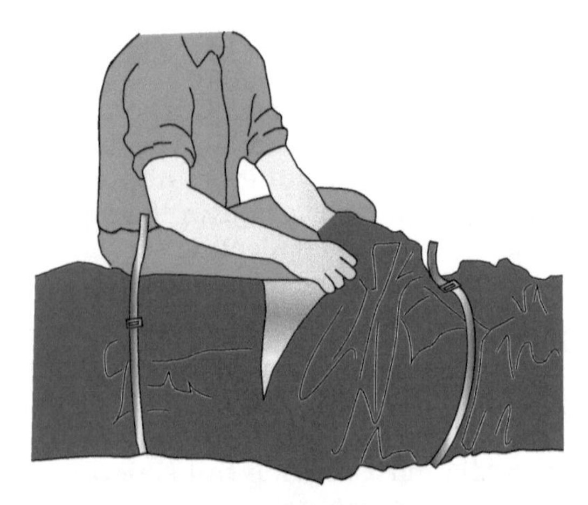

Figure 38 Pipe encased in polyethylene.

Step 4 Remove the sling or tongs from the pipe.

Step 5 Raise the bell a few inches and slip the polyethylene tube along the pipe barrel, leaving approximately 1' of the tube bunched up at each end of the pipe for wrapping at the joints (*Figure 40*).

Step 6 Overlap the made-up joint by pulling one bunched-up tube over the bell and folding it around the adjacent plain end (*Figure 41*).

Step 7 Secure the joint in place with two or three wraps of the polyethylene adhesive tape.

Step 8 Complete the overlap by repeating the same procedure with the bunched-up tube on the adjacent pipe (*Figure 42*).

Step 9 Take up the slack in the tube along the pipe barrel by folding the slack over the top of the pipe, and holding the fold in place with polyethylene adhesive tape (*Figure 43*).

Step 10 Repair any rips, punctures, or other damage to the polyethylene, using the tape, or by cutting open a short length of tube, wrapping it around the pipe, and securing it with tape (*Figure 44*).

Fittings may be protected in a similar manner with the polyethylene. Flat sheets of material may be obtained by slitting the tube open. The flat sheets can then be wrapped around fittings, valves and specialty items, holding them in place with the polyethylene adhesive tape. When joints are made with flat sheets, the two ends should be brought together where possible, folded over in a lock-seam fashion, and then taped in place.

Figure 39 Bunched polyethylene tubing.

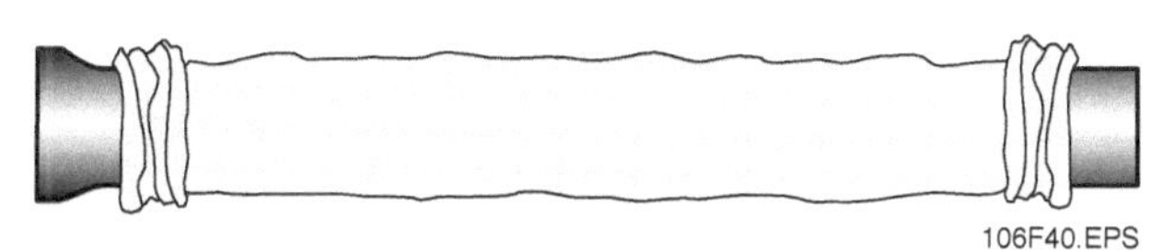

Figure 40 Tube slipped onto pipe.

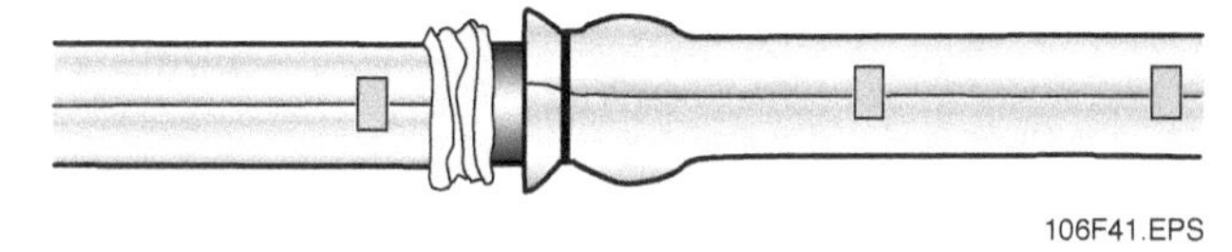

Figure 41 Joint overlapped with one tube.

9.0.0 Thrust Blocks And Restraints

Thrust blocks and restraints are used in underground piping systems to prevent the system components from shifting or moving as a result of internal or external forces. The following sections explain thrust blocks and restraints.

9.1.0 The Purpose of Thrust Blocks

Most underground pipe and fittings have friction connections rather than threaded connections. These connections are the push-on type, which means they push off about as easily as they push on. Normal pressure does tend to force these fittings off, but the major problem is water hammer.

Water hammer is the thrust produced by a flowing stream of water when the direction or volume of flow is changed. When underground joints come apart, tremendous damage results. This kind of blow-out damages things other than the pipe and fittings. The water flow damages streets, parking lots, lawns, buildings and building foundations. Water hammer occurs at every place that a pipeline changes direction; therefore, thrust blocks are placed at all bends, tees, caps, valves, and hydrants.

Thrust blocks are usually made of concrete. Their function is to brace the pipe joint against undisturbed soil. They are wedge-shaped for efficient use of material.

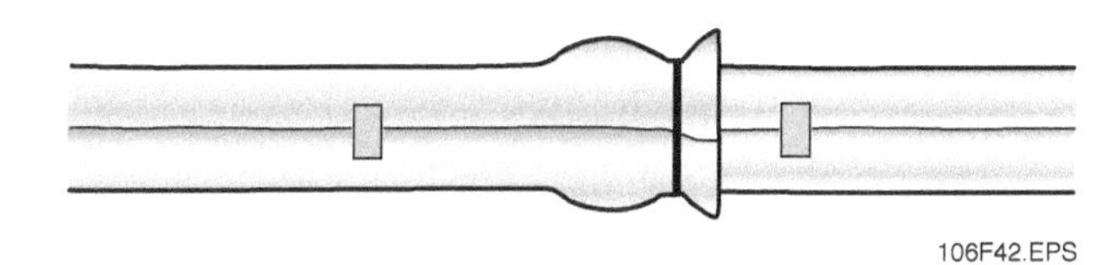

Figure 42 Joint overlapped with both tubes.

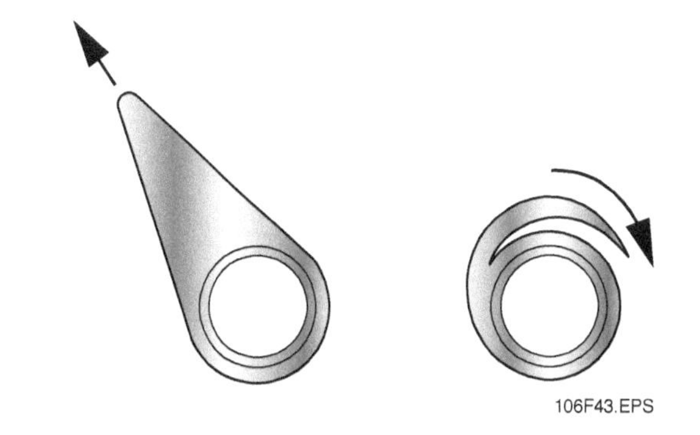

Figure 43 Taking up the slack.

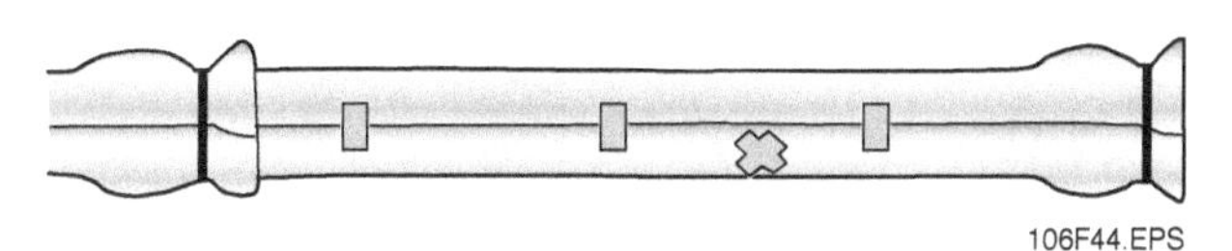

Figure 44 Repaired tubing.

A large bearing area is needed against undisturbed soil. Because concrete has much higher strength than the soil, only a small contact is needed against the pipe. Mass or weight of material is also needed to hold the joint.

On smaller sizes of pipe, the concrete is mixed and placed fairly dry to facilitate easy shaping of the block into a wedge. The widest part of the wedge is against a solid trench wall. For larger pipe sizes (12" or more) and greater pressures, the design of the block should be specified by the engineer to ensure adequate design. *Figure 45* shows a typical thrust block.

9.1.1 Composition of Concrete

The strength of concrete is determined by the proportions of its ingredients: sand, cement, gravel, and water. The time of mixing these can also dramatically affect the strength. The composition of concrete on a particular job is generally specified by the project engineer. Frequently, ready-mix concrete is available. In the event no project engineer is present and a custom mix is needed, a typical mix by volume would consist of the following:

- One part cement
- Two and one half parts sand
- Five parts washed gravel or stone

Notice that the gravel or stone is to be washed. It is very important that no soil or clay be included in the mix. Dirt, soil, and clay will destroy the strength of the concrete.

These ingredients must be thoroughly mixed before being placed in position. The cement is the binder; the water dissolves the cement and allows it to coat the surface of every grain of sand and piece of stone or gravel in the mix. Mix the concrete until everything is a uniform gray color.

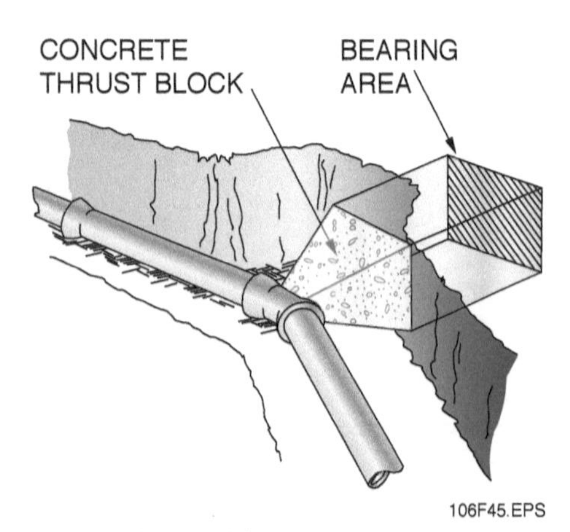

Figure 45 Typical thrust block.

9.2.0 Size of the Thrust Block

The size of the thrust block is determined by four factors:

- *Soil conditions* – The stability of the soil under a load varies with the kind of soil.
- *Pipe size* – The larger the pipe, the more force or thrust the moving water will have.
- *Type of fitting* – The type of fitting determines how much of the thrust will be intercepted, and the direction it will be applied.
- *Maximum pressure* – The magnitude of the thrust is also affected by the pressure in the line.

9.3.0 Soil

Hard soils take higher loads than soft soils. The safe bearing load of a particular soil is usually determined on a local basis by an engineer. There are soil bearing tests that can be used to do this accurately. This information, along with the previously mentioned factors, is used to calculate the size of all thrust blocks. *Table 6* shows the approximate safe bearing loads of different soils.

9.4.0 Fittings

Table 7 lists the approximate amount of thrust at fittings, in pounds. The figures are based on a water pressure of 200 psi.

Using this information, thrust blocks can be calculated using a formula presented in *NFPA 13* and *NFPA 24*. However, that calculation is normally performed by a fire sprinkler system designer or a layout technician.

Table 6 Horizontal Bearing Strengths (Table A.10.8.2(c))

Type of Soil	Bearing Strength, S_b	
	(lb/ft²)	(kN/m²)
Muck	0	0
Soft Clay	1,000	47.9
Silt	1,500	71.8
Sandy Silt	3,000	143.6
Sand	4,000	191.5
Sandy Clay	6,000	287.3
Hard Clay	9,000	430.9

NOTE: Although the bearing strength values in this table have been used successfully in the design of thrust blocks and are considered to be conservative, their accuracy is totally dependent on accurate soil identification and evaluation. The ultimate responsibility for selecting the proper bearing strength of a particular soil type must rest with the design engineer.

106T06.EPS

In general, the following guidelines are specified in *NFPA 13*:

> Thrust blocks shall be considered satisfactory where soil is suitable for their use. Thrust blocks shall be of a concrete mix not leaner than one part cement, two and one-half parts sand, and five parts stone. Thrust blocks shall be placed between undisturbed earth and the fitting to be restrained and shall be capable of such bearing to ensure adequate resistance to the thrust to be encountered. Wherever possible, thrust blocks shall be placed so that the joints are accessible for repair.

Figure 46 shows a comparison of thrust block areas as they relate to pipe size.

Table 7 Approximate Thrust at Fittings at 200 psi

Pipe Size	90° Elbow (¼ Bend)	45° Elbow (⅛ Bend)	Valves/ Tees/Dead Ends
4	5,200	2,840	3,700
6	10,800	5,800	7,600
8	18,600	10,000	13,000
10	27,800	15,100	21,700
12	39,400	21,600	27,800

106T07.EPS

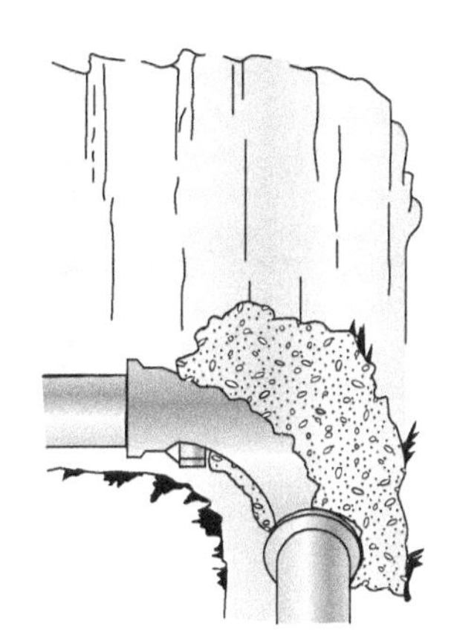

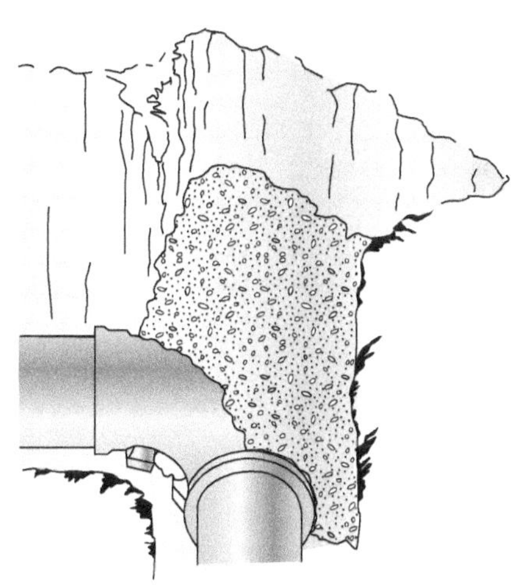

106F46.EPS

Figure 46 Comparison of thrust block areas.

9.5.0 Bearing Areas

Table 8, derived from information in *NFPA 24*, shows approximate horizontal bearing areas against the trench wall for thrust blocks, using a water pressure of 100 psi based on a 90-degree horizontal bend, a soil bearing strength of 1,000 lbs/sq ft, a safety factor of 1.5, and ductile iron pipe outside diameters. *Table 8* shows approximate minimum bearing areas against the trench wall for thrust blocks.

9.6.0 Location

The location of thrust blocks is determined by the direction of thrust and the type of fitting being used. *Figure 47* shows thrust block locations.

9.7.0 Thrust Blocks on Inclines

When pipe must be laid on an incline, it is very important to prevent any downward movement of the pipe. The weight of the pipe can cause the joints to loosen. The pipe must be anchored at the bottom of the incline and then at all changes in direction. Thrust blocks should also be placed every 48' or

Table 8 Required Horizontal Bearing Block Area (Table A.10.8.2(b))

Nominal Pipe Dia. (in)	Bearing Block Area (ft^2)	Nominal Pipe Dia. (in)	Bearing Block Area (ft^2)	Nominal Pipe Dia. (in)	Bearing Block Area (ft^2)
3	2.6	12	29.0	24	110.9
4	3.8	14	39.0	30	170.6
6	7.9	16	50.4	36	244.4
8	13.6	18	63.3	42	329.9
10	20.5	20	77.7	48	430.0

Notes:

(1) Although the bearing strength values in this table have been used successfully in the design of thrust blocks and are considered to be conservative, their accuracy is totally dependent on the accurate soil identification and evaluation. The ultimate responsibility for selecting the proper bearing strength of a particular soil type must rest with the design engineer.

(2) Values listed are based on a 90-degree horizontal bend, an internal pressure of 100 psi, a soil horizontal bearing strength of 1,000 lb/ft^2, a safety factor of 1.5, and ductile-iron pipe outside diameters.

(a) For other horizontal bends, multiply by the following coefficients: 45 degree: 0.541; for 22½ degree: 0.276; for 11¼ degree: 0.139.

(b) For other internal pressures, multiply by ratio to 100 psi.

(c) For other soil horizontal bearing strengths, divide by ratio to 1,000 lb/ft^2.

(d) For other safety factors, multiply by ratio to 1.5.

Example: Using Table A.10.8.2(b), find the horizontal bearing block area for a 6 in. diameter, 45-degree bend with an internal pressure of 150 psi. The soil bearing strength is 3,000 lb/ft^2, and the safety factor is 1.5. From Table A.10.8.2(b), the required bearing block area for a 6 in. diameter, 90-degree bend with an internal pressure of 100 psi and a soil horizontal bearing strength of 1,000 psi is 7.9 ft^2.

For example:

$$Area = \frac{7.9\ ft^2\ (0.541)\frac{150}{100}}{\frac{3,000}{1,000}} = 2.1\ ft^2$$

106T08.EPS

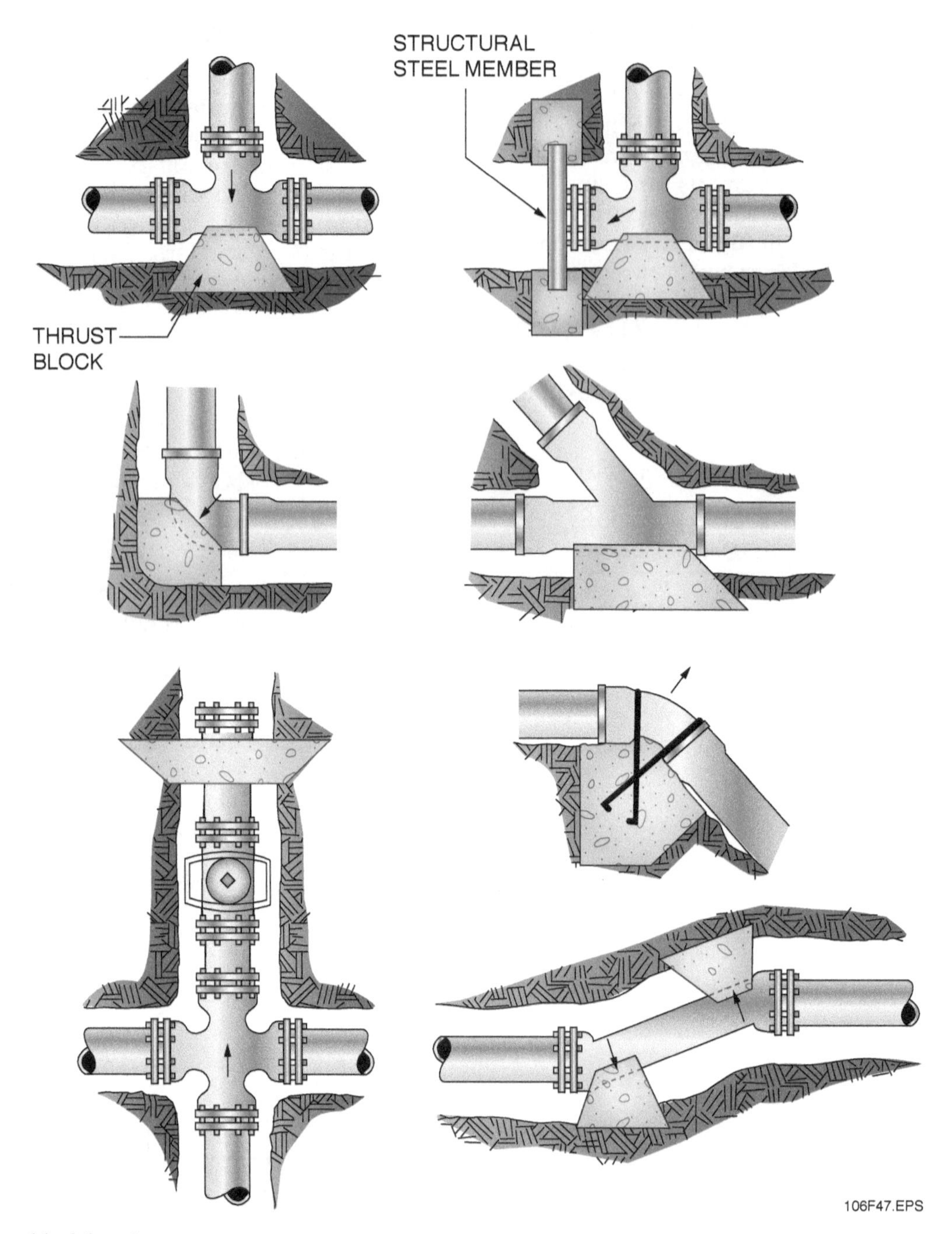

Figure 47 Thrust block locations.

so on the straight run. In addition, pipe going up or down a slope may require a deadman and rodding. The deadman is simply a block, as opposed to a wedge, of concrete in which tie-rods are secured. When installing bell and spigot pipe on an incline, the bell must always face uphill. *Figure 48* shows a concrete block deadman and rodding.

9.8.0 Thrust Blocks at Hydrants

Hydrant run-outs are usually equipped with a valve. There is a distinct advantage to having a gate valve at all hydrant run-outs. This allows the hydrant to be serviced without shutting down the service. The valve can be at the main, at the hydrant, or in between. Regardless of where it is located, each fitting must be separately supported with a thrust block. Never cover up the weep holes. *Figure 49* shows typical thrust blocking at the hydrant.

9.9.0 Thrusts in Soft Soils

In soils that are soft and unstable, such as peat or muck, thrusts are resisted by running corrosion-resistant tie-rods to solid foundations. An alternative to this method is to remove the soft material and replace it with ballast, which is a mass of stone and/or concrete, of sufficient size and weight to resist any thrust developed.

9.10.0 Vertical Bends

Where a fitting is used to make a vertical bend, the fitting is anchored to a thrust block braced against the undisturbed soil. The thrust block should have enough resistance to withstand upward and outward thrusts at the fitting.

9.11.0 Side Thrust

When the pipe line is pressurized, there is side thrust at each deflected coupling. If tamped properly, the soil is usually sufficient to prevent movement from side to side. However, in soft soils it may be necessary to use anchors and restraints to prevent further deflection. Side thrust blocks, or restraints, must be placed on either side of the pipe adjacent to the coupling, not on the coupling.

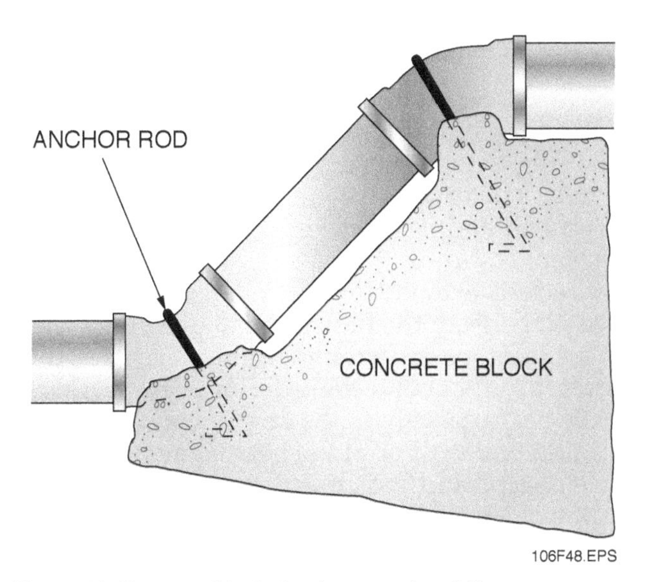

Figure 48 Concrete block deadman and rodding.

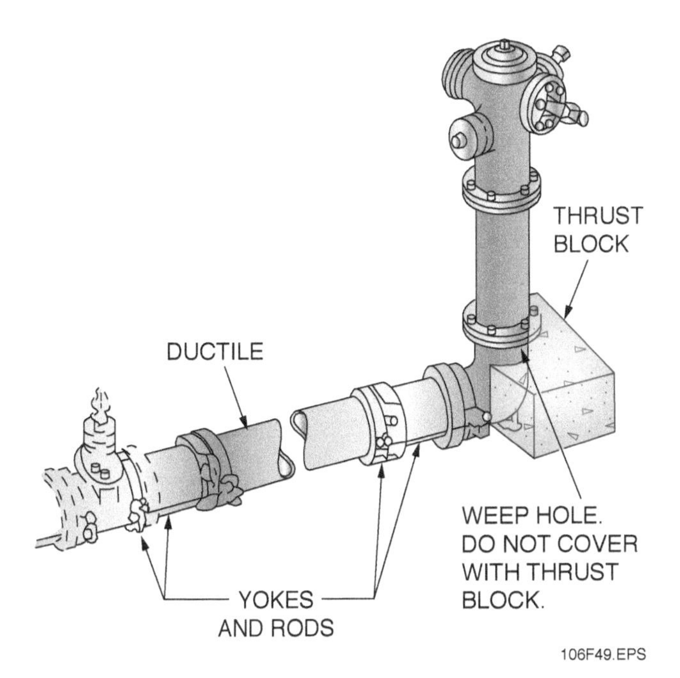

Figure 49 Thrust blocks at hydrant.

9.12.0 Anchors and Restraints

Underground pipe joints are expected to be held in place by the soil in which they are buried. Soft soils are not stable enough to provide this kind of restraint. Restraints, anchors, or tie-rods are used to prevent pipe deflection due to impact from water hammer, load from the backfill, or the vibration from passing vehicles.

One preventive measure is to use thrust blocks along straight runs of pipe, with restraints that physically clamp the pipe to the thrust block. In muck, piles or tie-rods to solid foundations are required. The other option is to excavate the trench below the pipe level, and replace the removed material with more stable material.

NFPA 24 requires that restraining hardware be made of corrosion-resistant material. Usually, hot-dipped galvanized steel material is used in order to meet this requirement. Steel material can be made corrosion-resistant by applying a thick coating of bitumen or other acceptable corrosive-retardant material in accordance with *NFPA 24.* Additional protection from corrosion can be provided by covering the pipe, rod, and clamps with a plastic wrap before backfilling the trench. This has become a common requirement of the inspecting AHJ.

9.12.1 Clamps and Braces

Pipe clamps are used to grip the pipe and provide a tie point for rod connections. *Table 9* shows the minimum dimensions for pipe clamps.

If washers are used, they can be of either steel or iron. They can be either round or square. Iron washers for 4", 6", 8", and 10" pipe must be ⅝" × 3". For 12" pipe they must be ¾" × 3½". Steel washers for 4", 6", 8", and 10" pipe are ½" × 3". For 12" pipe, they are ½" × 3½". *Figure 50* shows different pipe clamping devices.

9.12.2 Tie-Rods

Tie-rods are used with clamps. They allow for balanced force on both sides of the clamp. *NFPA 24* states the minimum rod size allowable is ⅝". Tie-rods must be corrosion resistant and meet minimum size criteria.

NFPA 24 further states that if clamps are used, rods must be used in pairs, two to a clamp. The only exception to this is on assemblies in which an anchor is made by means of two clamps canted on the barrel of the pipe. Provided the AHJ approves, this specific installation may be anchored with one clamp.

Figure 51 shows the use of tie-rods and tie bolts for anchoring a mechanical joint.

Table 9 Minimum Dimensions for Pipe Clamps

Pipe Size	Steel Bar Size	Bolt Size
4	½ × 2	⅝
6	½ × 2	⅝
8	⅝ × 2½	⅝
10	⅝ × 2½	¾
12	⅝ × 3	⅞

NOTE: All dimensions in inches.

106T09.EPS

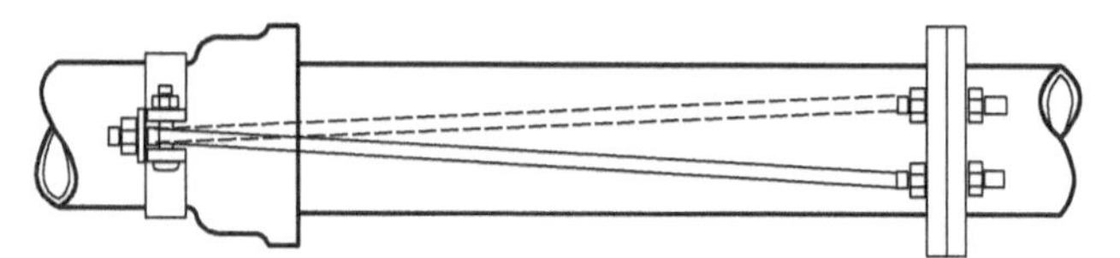

PUSH-ON JOINT

RETAINING GLAND

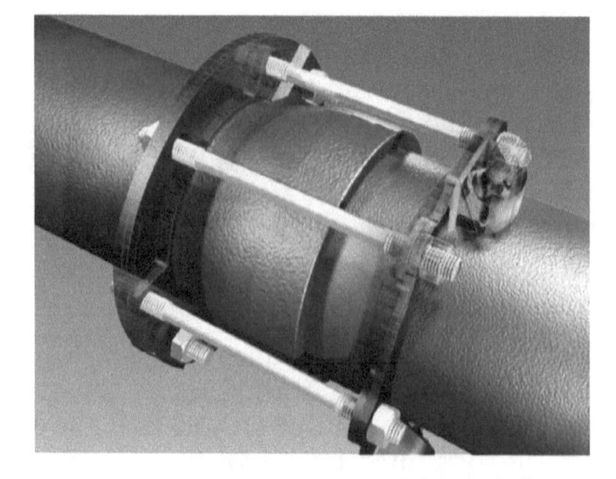

RETAINING GLAND (DUCTILE IRON JOINT)

106F50.EPS

Figure 50 Push-on joint and retaining gland pipe joint restraints.

Table 10 shows the diameter combinations versus numbers of rods to be used, derived from using a pressure of 225 psi and design stress of 25,000 psi.

9.12.3 Retaining-Gland Pipe-Joint Restraint

The retaining gland is a newer type of joint restraint for PVC and ductile iron pipelines. It provides full circle contact and support of the pipe wall. The restraint is accomplished by a series of ring segments mechanically retained inside the gland housing and designed to grip the pipe wall in an even and uniform manner. Refer to *Figure 50* for retaining gland pipe joint restraints.

9.13.0 Flange Spigot Piece Restraint

The mechanical joint pipe-and-fitting fire-line entry into the building should be anchored with restraining rods to prevent movement. The number of rods is determined by the size of the pipe being restrained and its diameter. The use of corrosion-resistant restraining hardware is required.

The rods extend from the interior flange to the base ell beneath the flange spigot piece, and/or to an anchor (one or two underground pipe clamps as required) beyond the exterior of the building grade beam or wall. In the case of horizontal penetration of a wall or grade beam, adequate clearance of pipe and rods must be provided through a pipe sleeve. The void space between the pipe and sleeve must be filled with a waterproofing sealant and mastic caulking material to prevent the flooding of the building through the penetration. Refer to *Table 10* for the number and size of rods to be used with the various pipe sizes.

In the event the rods become loose or broken, a concrete thrust block at the base ell to provide additional restraint may also be required. This will prevent separation of the pipe and fitting. The use of retaining glands with mechanical joint fittings in lieu of rods and clamps may be required by some AHJs.

STAR BOLTS

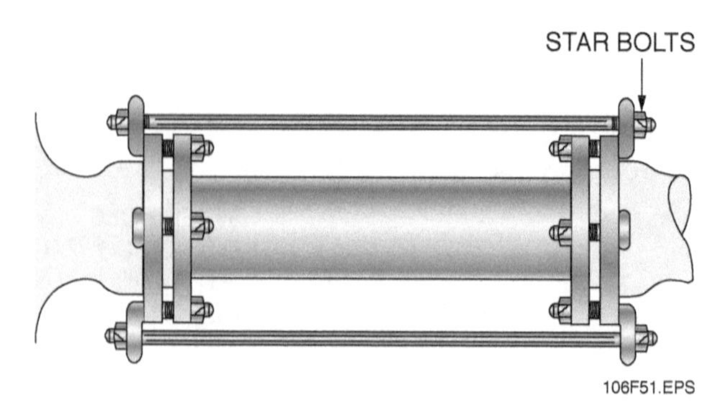

106F51.EPS

Figure 51 Tie-rods as used on a mechanical joint.

Table 10 Number and Size of Rods (Table 10.8.3.1.2.2)

Nominal Pipe Size (in)	Rod Number – Diameter Combination			
	5/8" (15.9 mm)	3/4" (19.1 mm)	7/8" (22.2 mm)	1" (25.4 mm)
4	2	–	–	–
6	2	–	–	–
8	3	2	–	–
10	4	3	2	–
12	6	4	3	2
14	8	5	4	3
16	10	7	5	4

NOTE: This table has been derived using pressure of 225 psi (15.5 bar) and design stress of 25,000 psi (172.4 MPa).

106T10.EPS

10.0.0 In-Building Riser

The in-building riser (*Figure 52*) is used to connect the main fire supply to the building's overhead fire system. The fitting passes under the foundation without joints and extends up through the floor. Provided with installation tabs, the unit has a cast-iron pipe size (CIPS) coupler for easy connection to the underground supply (*AWWA C900* PVC and ductile iron pipe), and an industry standard grooved-end connection to the overhead fire sprinkler system.

The in-building riser is composed of a single extended 90-degree fitting of fabricated 304 stainless steel tubing, with a maximum working pressure of 175 psi. The fitting has a grooved-end connection on the outlet (building) side and a CIPS coupler on the underground (inlet) side.

11.0.0 Backflow Preventers

Piping systems deliver fresh water and take away wastewater. Normally, these two water sources are separated. However, sometimes problems or malfunctions can force water to flow backward through a system. When this happens, wastewater and other liquids can be siphoned into the fresh water supply. This can cause contamination, sickness, and even death. Codes require sprinkler systems to protect against reverse flow. Protection can be in the form of a gap or a barrier between the two sources. Check valves can be effective barriers and can be arranged to provide increasing protection against contamination.

11.1.0 Backflow and Cross-Connections

The reverse flow of nonpotable liquids into the potable water supply is called backflow. Backflow can contaminate a fixture, a building, or a community. There are two types of backflow: back

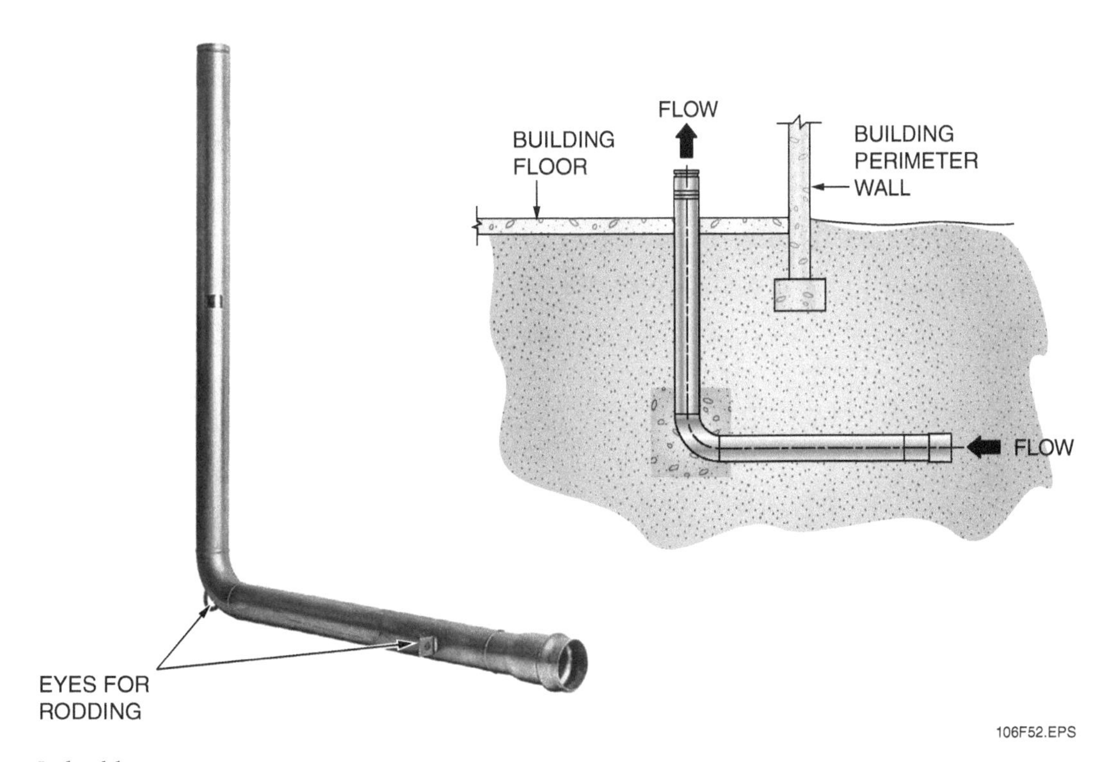

Figure 52 In-building riser.

pressure and back siphonage. Back pressure is a higher than normal downstream pressure in the potable water system. Back siphonage is a lower than normal upstream pressure.

Backflow cannot happen unless there is a direct link between the potable water supply and another source. This condition is called a cross-connection. Cross-connections are not hazards all by themselves (see *Figure 53*). In fact, sometimes they are required. Cross-connections are sometimes hard for the public to spot. Many people might not know that a hose left in a basin of wastewater could cause a major health hazard by creating a backflow.

When something creates a backflow, a cross-connection becomes dangerous. Several things can cause backflow, which, in turn, can force wastes through a cross-connection:

- Cuts or breaks in the water main
- Failure of a pump
- Injection of air into the system
- Accidental connection to a high-pressure source

Sprinkler fitters use backflow preventers to protect against cross-connections. They provide a gap or a barrier to keep backflow from entering the water supply. Many local codes now require backflow preventers for all fixtures in a new structure. Tank trucks filled with sewage, pesticides, or other dangerous substances must also use backflow preventers.

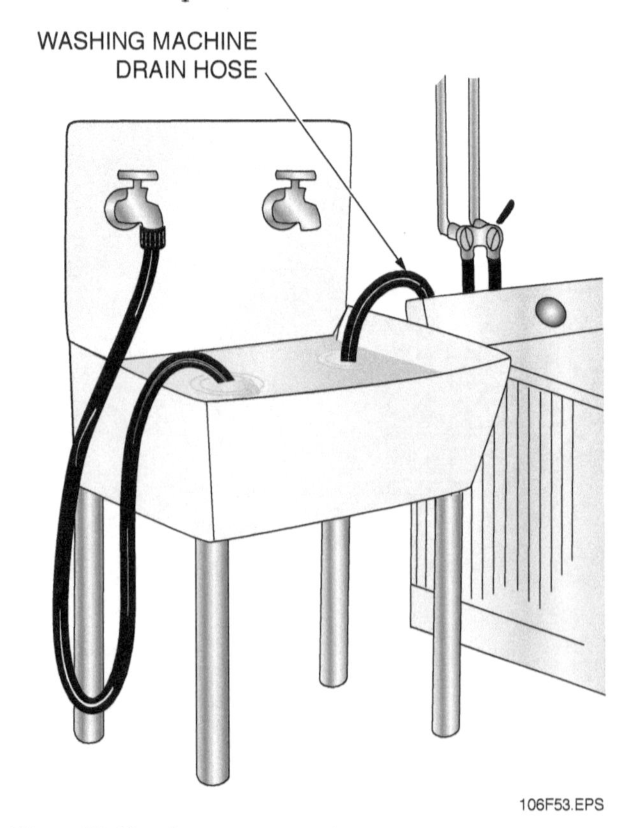

Figure 53 Simple cross-connection.

11.2.0 Types of Backflow Preventers

There are six types of basic backflow preventers, each designed to work under different conditions. Some protect against both back pressure and back siphonage, while others can handle only one type of backflow. Before installing a backflow preventer, always ensure that it is appropriate for the type of installation. Consider several factors:

- Risk of cross-connection
- Type of backflow
- Health risk posed by pollutants or contaminants

Preventers must be installed correctly. Select the proper preventer for the application, and review the manufacturer's instructions before installing a preventer. Never attempt to install a backflow preventer in a system for which it is not designed because it may fail, resulting in backflow. Refer to your local code for specific guidance.

11.2.1 Air Gaps

An air gap is a physical separation between a potable water supply line and the flood-level rim of a fixture (see *Figure 54*). It is the simplest form of backflow preventer. Ensure that the supply pipe terminates above the flood-level rim at a distance that is at least twice the diameter of the pipe. The air gap should never be less than 1 inch.

All faucets must incorporate an air gap. Air gaps can be used to prevent back pressure, back siphonage, or both. Never submerge hoses or other devices connected to the potable water supply in a basin that may contain contaminated liquid. Even seemingly innocent activities such as filling up a wading pool with a garden hose could be a problem if, elsewhere, workers digging with a backhoe accidentally cut through the water main.

11.2.2 Atmospheric Vacuum Breakers

If it is not advisable to install an air gap, consider using an atmospheric vacuum breaker (see *Figure 55*). Atmospheric vacuum breakers (AVBs) use a silicon disc float to control water flow. In normal operation, the float rises to allow potable water to enter. If back siphonage causes low pressure, air enters the breaker from vents and forces the disc into its seat. This shuts down the flow of fresh water (see *Figure 56*).

AVBs are designed to operate at normal air pressure. Do not use them where there is a risk of back pressure backflow. Check the manufacturer's specifications for the proper operating temperatures and pressures. If conditions exceed these limits, consult the manufacturer first.

Do not install an AVB where it will be exposed to higher than normal air pressure. Install an AVB after the last control valve, and ensure that it is at least 6 inches above the fixture's flood-level rim. Install the breaker so that the fresh water intake is at the bottom, and ensure that the water supply is flowing in the same direction as the arrow on the breaker.

11.2.3 Pressure-Type Vacuum Breakers

Install pressure-type vacuum breakers (see *Figure 57*) to prevent back siphonage, back pressure, or both. A pressure-type vacuum breaker (PVB) has a spring-loaded check valve and a spring-loaded air inlet valve. The PVB operates when line pressure drops to atmospheric pressure or less. The float valve opens the vent to the outside air (see *Figure 58*). The check valve closes to prevent back siphonage. PVBs are suitable for continuous pressure applications. PVBs are equipped with test cocks, which are valves that allow testing of the individual pressure zones within the device. They can also be equipped with strainers if required.

11.2.4 Check Valve Backflow Preventers

To keep backflow from reaching the water main, use a dual check valve backflow preventer (DC) (*Figure 59*). These preventers feature two spring-loaded check valves. They work similarly to double-check valve assemblies (DCVs).

DCs protect against back pressure and back siphonage. Because DCs are smaller, they are commonly used in residential installations. These preventers tend to be less reliable than DCVs.

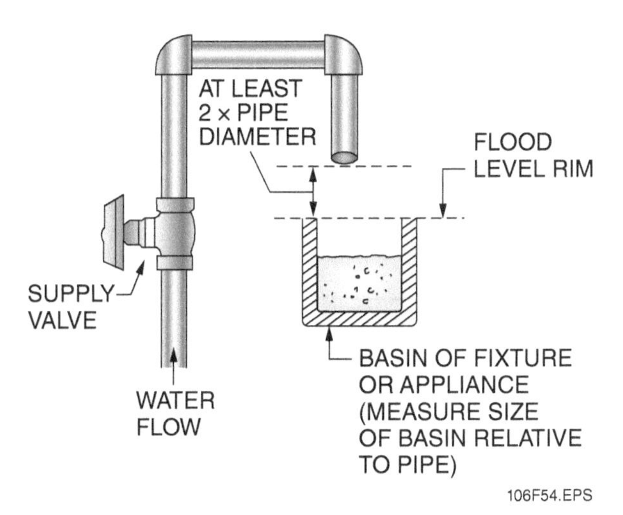

Figure 54 Air gap.

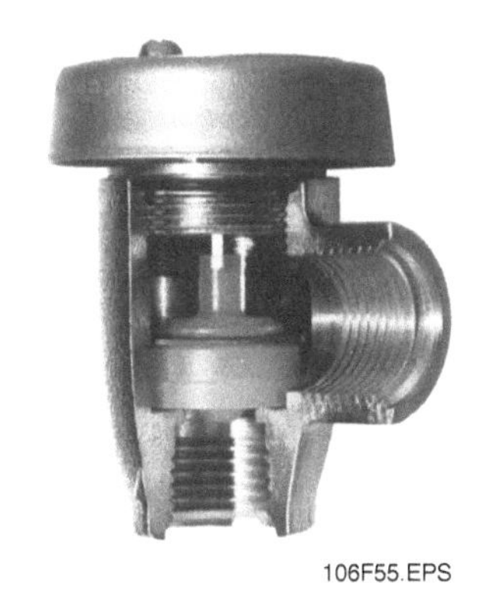

Figure 55 Cut-away of an atmospheric vacuum breaker.

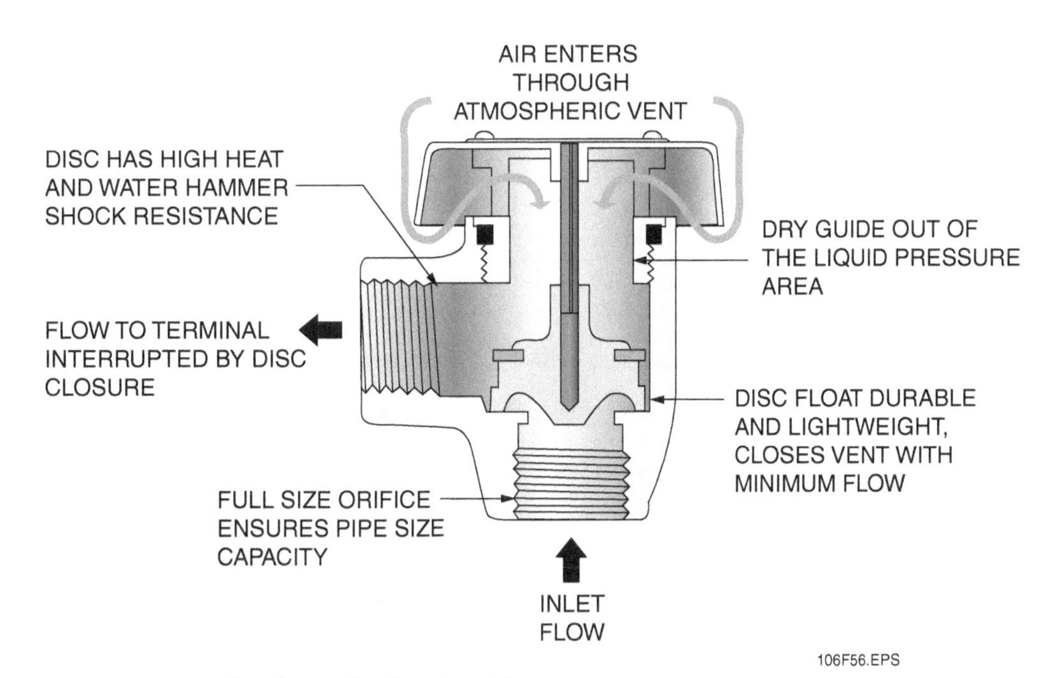

Figure 56 Atmospheric vacuum breaker in the closed position.

Unlike DCVs, they are normally not fitted with shutoff valves or test cocks. This makes it difficult for sprinkler fitters to remove, test, or replace them. Check the local code before installing DCs.

Figure 57 Pressure-type vacuum breaker.

11.2.5 Double-Check Valve Assemblies

Where heavy-duty protection is needed, install a double-check valve assembly to protect against back siphonage, back pressure, or both (see *Figure 60*). This type of backflow preventer has two spring-loaded check valves and includes a test cock. They are designed for installation where both back pressure and back siphonage are threats. During normal operation, both check valves are open; however, backflow causes the valves to seal tightly.

DCVs are available in a wide range of sizes from ¾" up to 10". DCVs have a gate or ball shutoff valve at each end. If required, DCV inlets can be equipped with strainers.

11.2.6 Reduced-Pressure Zone Principle Backflow Preventers

Reduced-pressure zone principle backflow preventers (RPZs) protect the water supply from con-

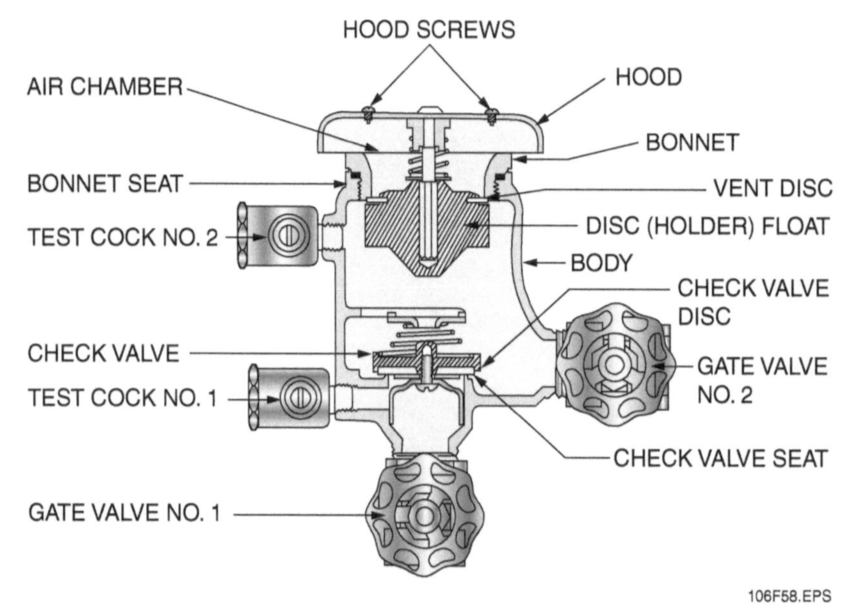

Figure 58 Cutaway illustration of pressure-type vacuum breaker.

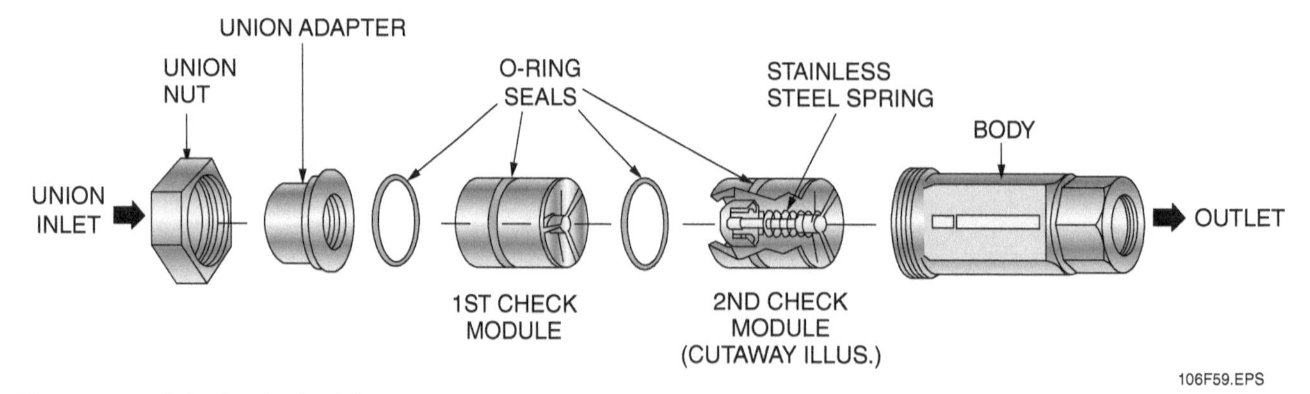

Figure 59 Dual check valve backflow preventer.

tamination by dangerous liquids (see *Figure 61*). RPZs are the most sophisticated type of backflow preventer. They feature two spring-loaded check valves plus a hydraulic, spring-loaded pressure differential relief valve. The relief valve is located underneath the first check valve. The device creates a low pressure between the two check valves. If the pressure differential between the inlet and the space between the check valves drops below a specified level, the relief valve opens. This allows water to drain out of the space between the check valves (see *Figure 62*). The relief valve will also operate if one or both check valves fail. This offers added protection against backflow. RPZs also have shutoff valves at each end.

Use an RPZ if a direct connection is subject to back siphonage, back pressure, or both.

Some codes require RPZs on additional types of installations. RPZs are available for all operating temperatures and pressures. Refer to your local code to see where RPZs are required. Install RPZs at least 12 inches above grade, and provide support for the RPZ (see *Figure 63*). Otherwise, sagging could damage the RPZ. Select a safe location that will help prevent damage and vandalism of the RPZ. Install a manufactured air gap on all RPZ installations (see *Figure 64*). It acts as a drain receptor to prevent siphonage.

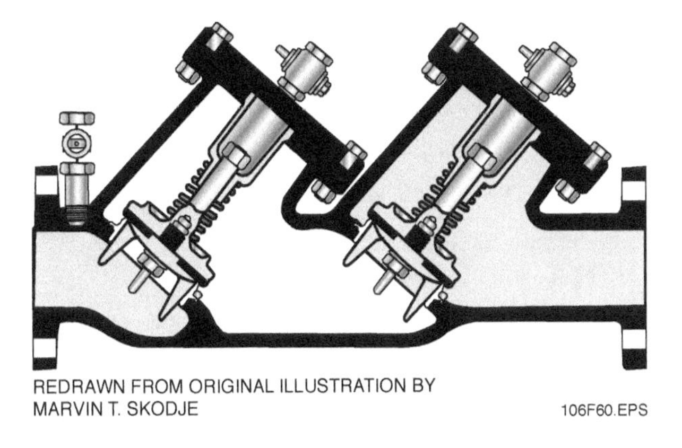

Figure 60 Double-check valve assembly.

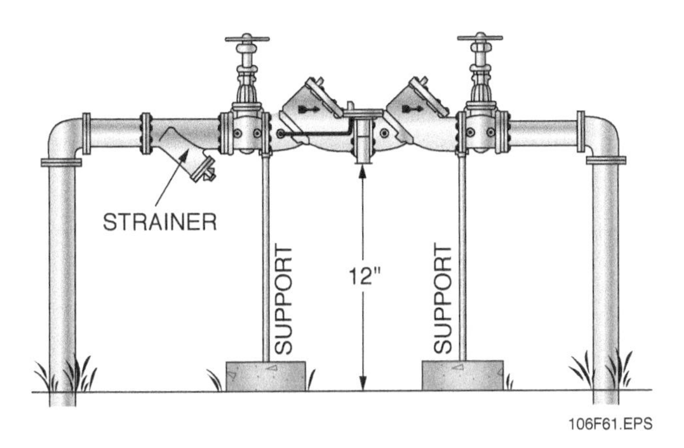

Figure 61 Installation of a reduced-pressure zone principle backflow preventer.

Constant water dripping from an RPZ indicates that the valve disc is not seating properly. Flush the unit, and the problem should correct itself. If it does not, refer to manufacturer's specifications for proper maintenance procedures.

11.2.7 Specialty Backflow Preventers

Often, low-flow and small supply lines require backflow preventers. Use an intermediate atmospheric vent vacuum breaker for this kind of line (see *Figure 65*). The intermediate atmospheric vent is a form of PVB that also protects against back pressure.

12.0.0 Hydrants, Yard Valves, Hydrant Houses, and Related Appurtenances

Sprinkler fitters are required to install several types of fire hydrants and other fire protection equipment. The following sections explain the types of fire protection equipment that the fitter needs to understand.

12.1.0 Fire Hydrants

Fire hydrants are important in the overall fire protection system. The fire department must have quick access to a hydrant even if a building is sprinklered.

CAUTION

To prevent damage to the hydrant, use a hydrant wrench designed for the hydrant you are installing or working on.

The first arriving fireman must make sure the fire sprinkler system is working and has sufficient water pressure. The pumper is connected between a fire hydrant and the sprinkler system to raise pressure and improve water flow in the sprinkler system. This is especially important if the sprinkler system has not controlled or extinguished the fire by the time the fire department arrives.

12.1.1 Hydrant Spacing

The fire department is usually the AHJ over fire hydrants. It determines hydrant locations based on applicable fire codes and ordinances. However, insurance authorities may have more demanding requirements for hydrant placement than the local fire department.

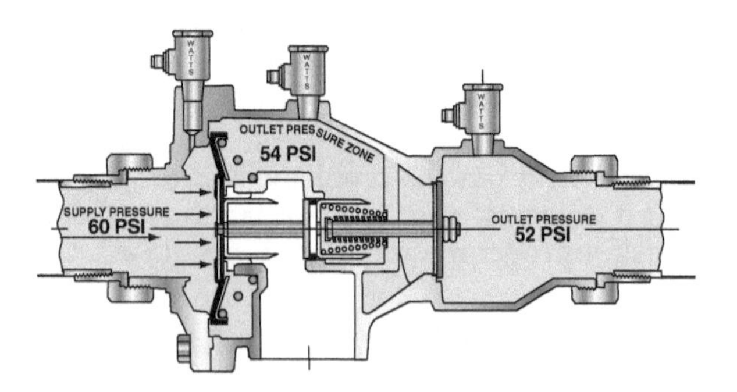

STATIC PRESSURE – 1ST AND 2ND CHECK VALVES TIGHTLY CLOSED. RELIEF VALVE CLOSED.

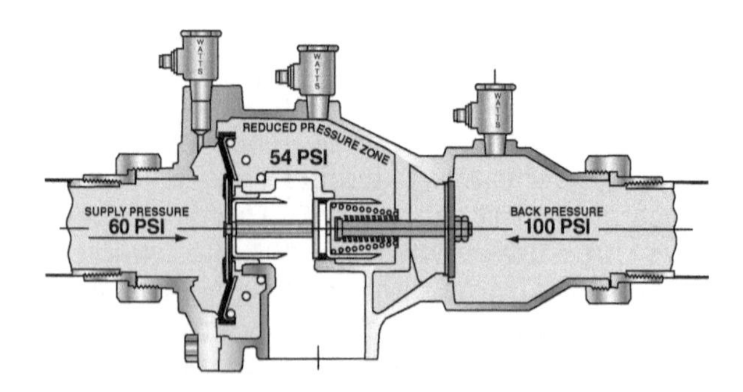

BACK PRESSURE – 1ST AND 2ND CHECK VALVES TIGHTLY CLOSED. RELIEF VALVE CLOSED.

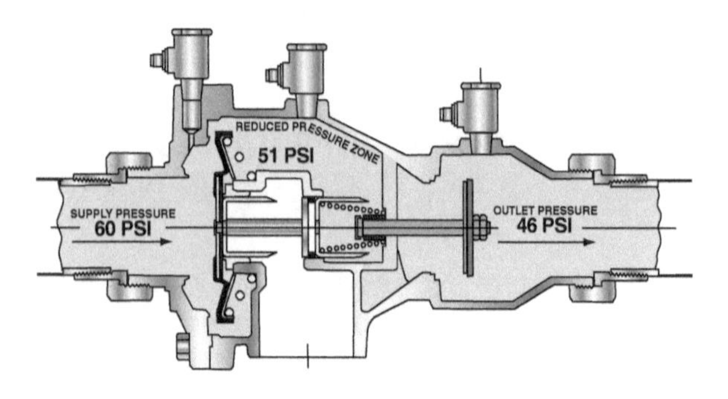

FULL FLOW – 1ST AND 2ND CHECK VALVES FULLY OPEN. RELIEF VALVE CLOSED.

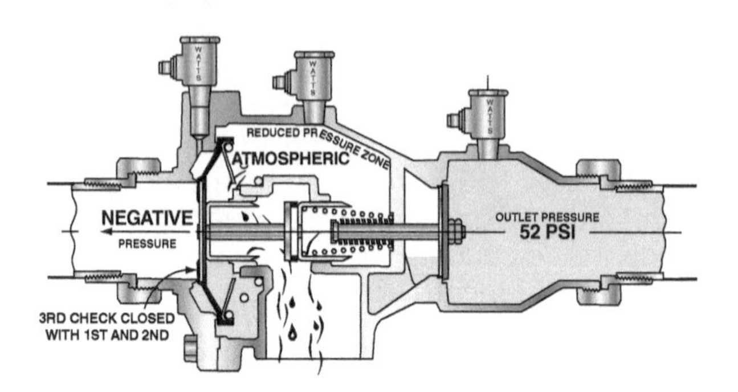

BACK SIPHONAGE – 1ST, 2ND, AND 3RD CHECK VALVES TIGHTLY CLOSED. RELIEF VALVE FULLY OPEN. DISCHARGES WATER IN ZONE.

106F62.EPS

Figure 62 Operation of a reduced pressure zone principle backflow preventer with check valves parallel to flow.

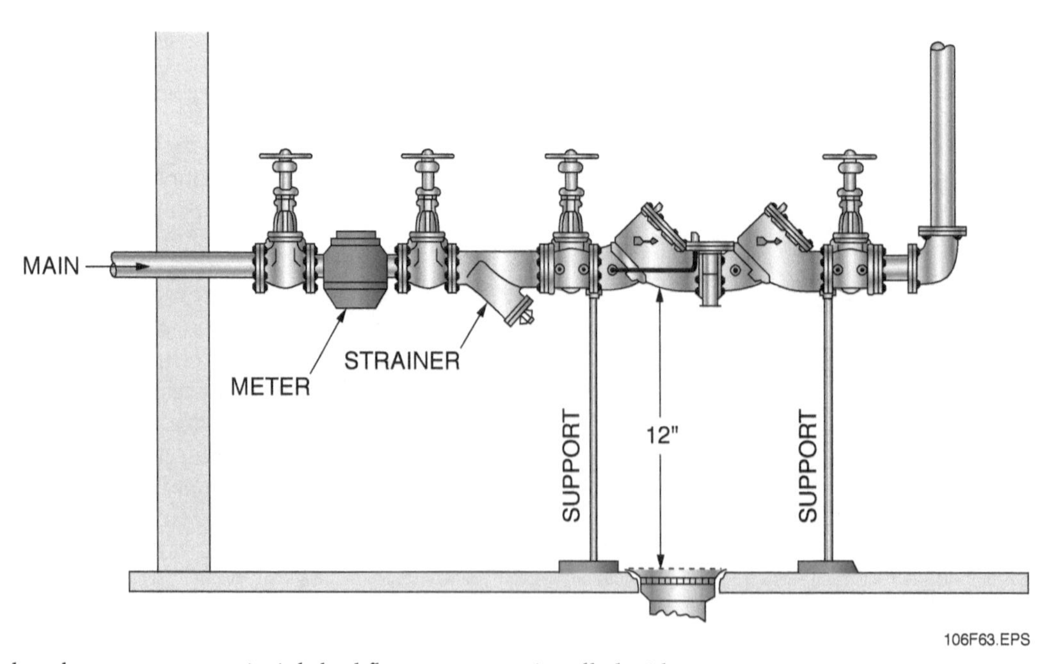

106F63.EPS

Figure 63 Reduced-pressure zone principle backflow preventers installed with supports.

Figure 64 Manufactured air gap.

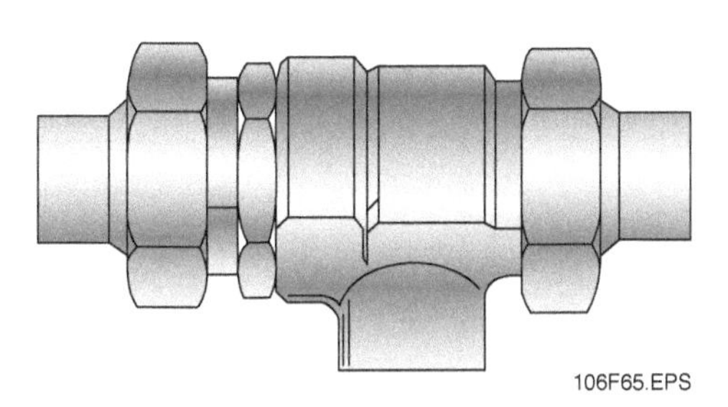

Figure 65 Intermediate atmospheric vent vacuum breaker.

There must be a sufficient number of hydrants located in a manner that will enable the needed flow to be delivered through hose lines to all exterior sides of any important structure. Hydrants must be spaced in accordance with the AHJ.

The spacing between hydrants varies from locality to locality. Spacing can vary from as much as 300' to 800' in industrial areas, and 500' or less in densely built areas.

12.1.2 Separation from Structures

Hydrants must be placed at least 40' from the building being protected. This is a safety precaution to prevent radiant heat or falling debris from making a hydrant unusable. *NFPA 24* makes an exception when local circumstances prevent hydrants from being placed at this distance. They may be located closer, or wall hydrants may be used, provided they are set in locations by blank walls where the possibility of injury by falling walls is small. The requirement for placing hydrants in a location where people are not likely to be driven away by smoke or heat still applies.

12.1.3 Outlets

Standard fire hydrants ordinarily have a minimum of two 2½" connections and a 4½" pumper connection; however, in places such as California and Hawaii, wet barrel street hydrants with only one 2½" outlet are allowed. Make sure you buy hydrants with compatible threads.

NFPA specifies the national standard external threads that can be used in outlets. There are localities with long-standing infrastructures that use street hydrants with threads other than the national standard thread pattern. These hydrants require the use of adapters if standard hoses are used. Before installing a new hydrant, the correct hose thread requirement for that particular locality must be determined.

The center of the hose connection can be no less than 18" above the final grade level. The connection to the main is specified by *NFPA 24* to be at least 6" in diameter.

12.1.4 Hydrant Protection

If the hydrant is placed in a location where it is likely to suffer mechanical damage, precautions must be taken to prevent damage. Guard posts (bollards) made of 3" to 4" pipe, set in and filled with concrete, are a typical protective measure. This protection must be provided in a manner that avoids interference when connecting to or operating the hydrant. A specific example of where protection is important is at the end of a parking lot where vehicles are likely to back into or drive over the hydrant.

12.1.5 Hydrant Valving, Blocking, and Drainage

When hydrant run-outs (piping between the main and the hydrant) are equipped with a valve, the valve can be located at the hydrant, at the main, or anywhere between the main and hydrant. The valve is usually a gate valve in a roadway box. Regardless of the location, the hydrant must be separately supported and thrust-blocked by a concrete slab or flat stones. The hydrant must be braced in place while the pipe connections

Did You Know?

There are people who actually collect fire hydrants. Some people collect only one or two hydrants, while others have begun their own museum of one hundred or more hydrants. One of the primary motivations for collecting hydrants is historic preservation, since old hydrants are typically melted for scrap. Hydrant collections have become important in preserving one aspect of the fire service history. For more information, go to www.firehydrant.org.

are being made and held there until the concrete foundation has been poured and has set. The hydrant must never be supported by the pipe. When pouring the base, take care not to block the drain at the base of the hydrant.

Once the hydrant foundation has set properly (usually 24 hours), it acts as a thrust block and an anchor or hold-down. It also helps prevent washouts caused by the drain of waste water. Frost heave is common in freezing climates, and it must be prevented if at all possible. Freezing of the soil causes it to expand, producing a heaving action. The result is a shifting or raising of the fire hydrant, which creates shearing stress on the pipe. Leaking gaskets and broken pipes can result.

12.1.6 Specifying Hydrants

When hydrants are specified, the following must be included:

- Outlet number, size, and type of threads
- Valve size
- Inlet size
- Barrel type
- Operating nut

There are two basic types of barrels in fire hydrants, wet barrel and dry barrel. Wet barrel hydrants are used in areas where the temperatures remain above freezing. The dry barrel hydrant, which is the more common, is used where there is a possibility of freezing temperatures.

12.1.7 Dry Barrel Hydrants

The base valve on dry barrel hydrants is located below the frost line, keeping water out of the barrel under normal circumstances. Water enters the barrel only when needed and a drain valve (weep hole) at the bottom is open when the main valve is closed. This prevents water from standing in the barrel and minimizes the danger of freezing. As previously noted, never pour concrete over the drain holes (weep holes). Dry barrel hydrants must be inspected semi-annually, preferably in the early spring and the fall. The local AHJ sets those requirements. *Figure 66* shows an exterior view of a typical dry barrel fire hydrant.

> **NOTE**
> Dry barrel hydrants must be open all the way or closed all the way. If not closed all the way, water will constantly drain from the weep holes.

12.1.8 Wet Barrel Hydrants

Wet barrel hydrants are sometimes called California hydrants. They usually have a compression

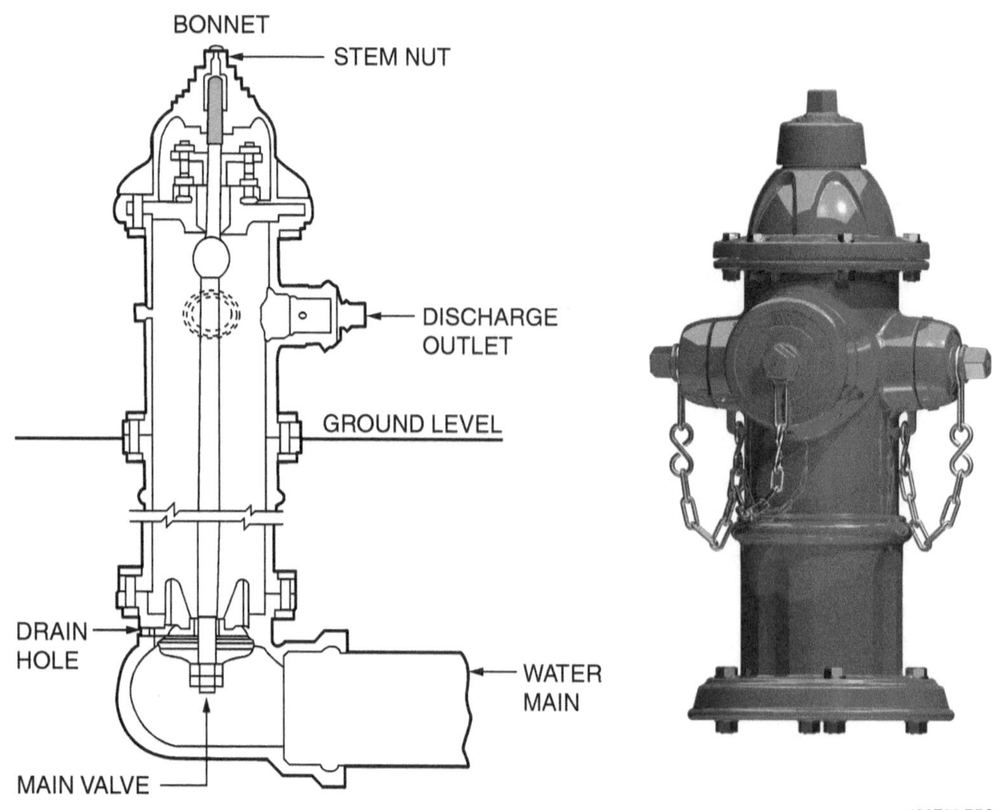

Figure 66 Dry barrel fire hydrant.

valve at each outlet and another valve in the bonnet. If constructed with a valve in the bonnet, that valve controls the flow of water to all outlets. Wet barrel hydrants must be inspected annually to ensure proper functioning. Wet barrel hydrants must never be installed where there is danger of freezing temperatures. *Figure 67* shows an exterior view of a typical wet barrel fire hydrant.

12.1.9 Street Hydrants

The street hydrant most often has a large pumper connection and two 2½" outlets. The actual diameter of the pumper connection varies from locality to locality and must be determined from the local authorities.

12.1.10 Yard Hydrants

Yard hydrants typically have only two 2½" outlets and 4½" pumper connection. Yard hydrants are located on a yard loop. The loop has a fire department connection for pressurizing by a pumper from a street hydrant.

Other sources of pressurization, such as elevated tanks and surface tanks with fire pumps, may also be present. Their principal use is for fighting exposure fires, but they can also be used for interior fire fighting if desired. *Figure 68* shows a yard hydrant.

12.2.0 Yard Valves

Yard valves are valves installed outside a building to control underground water supplies. These valves allow for opening and closing water lines as may be needed for maintenance and testing. *Figure 69* shows a typical yard installation employing a loop.

Most yard valves are iron gate valves designed to be used with indicator posts. As shown in *Figure 70*, the indicator post has a handle arrangement that allows locking the valve either open or shut. A target on the side shows the status of the valve.

Post indicator valves must be set so that the top of the post is 36" above the final grade. They must also be protected from mechanical damage where needed. These have the same exposure to damage as a fire hydrant. The distance from the building, if space permits, at which the valve may be installed is the same as the hydrant, 40', as recommended in *NFPA 24*.

NFPA 24 requires a Listed indicating valve in all water sources to a fire sprinkler system except fire department connections. By exception, a non-indicating valve, such as a gate valve with a roadway box complete with T-wrench, are accepted by some AHJ's. *Figure 71* shows a T-wrench.

Yard valves with indicator posts must be tamper-proof and secure. Some indicating posts are equipped with electric supervisory switching devices that can be used by a proprietary or central station to monitor the position of the posts. Other indicating posts are connected to audible alarms.

106F67.EPS

Figure 67 Wet barrel fire hydrant.

106F68.EPS

Figure 68 Yard hydrant.

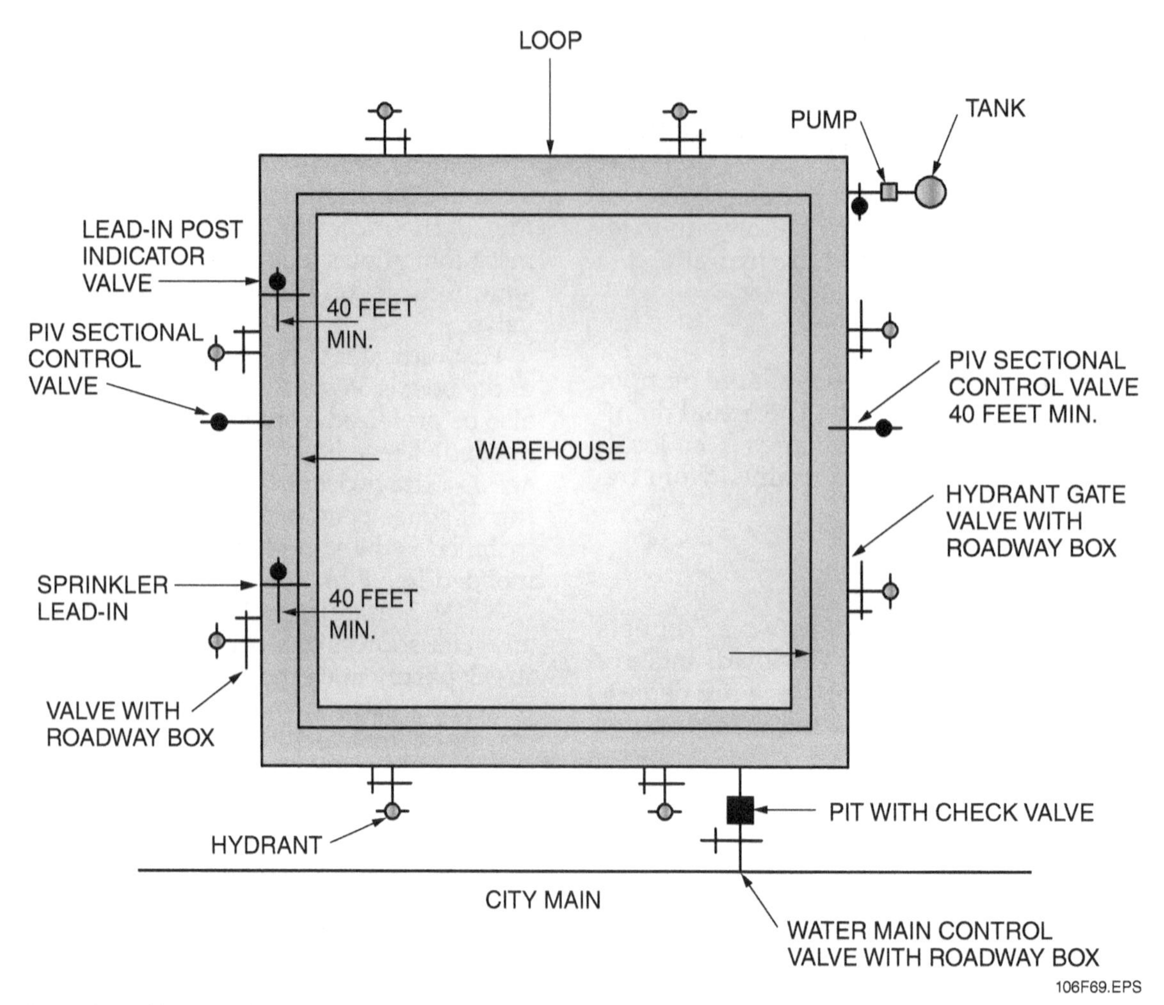

Figure 69 Typical yard loop in plan view.

NFPA 24 provides that if an indicator post is within a fenced enclosure, it can be simply sealed and weekly recorded inspections can be made, provided the enclosure is under the control of the owner.

12.3.0 Hose Houses, Hose Stations, and Equipment

Hose houses are common requirements for industrial plants in rural locations and are used by organized plant fire brigades. Where there are large open spaces, the hose house stores the equipment used to fight exposure fires such as grass fires. The hose house is used to store the fire fighting equipment in the near vicinity of a hydrant. The hose station is a cabinet built with the hydrant inside, the hose connected, and the hose racked for easy pull-out.

12.3.1 Hose Houses

Hose houses must be located near the hydrant to be useful. Hose reels or hose carts must be located so that they can be brought quickly into use at the hydrant. *Figure 72* shows a hose house and a portable hose reel.

NFPA 24 has guidelines and requirements for the construction, location, and equipment for hose houses. These hydrants are intended to be used by plant personnel, a volunteer fire department, or a fire brigade. They must contain an adequate amount of hose and equipment. The quantity and type of hose and equipment is determined by the number and location of hydrants relative to the protected property.

12.3.2 Equipment

Considerations that affect the quantity and type of equipment are: (1) the extent of the fire hazard and (2) the fire fighting capabilities of the equipment users. In all cases the AHJ must be consulted. *NFPA 24* requires hose houses to have hoses and the following equipment in addition to the hose:

- Two approved adjustable-spray solid-stream nozzles equipped with shut-offs for each size of hose provided. *Figure 73* shows an adjustable spray nozzle with shut-off.

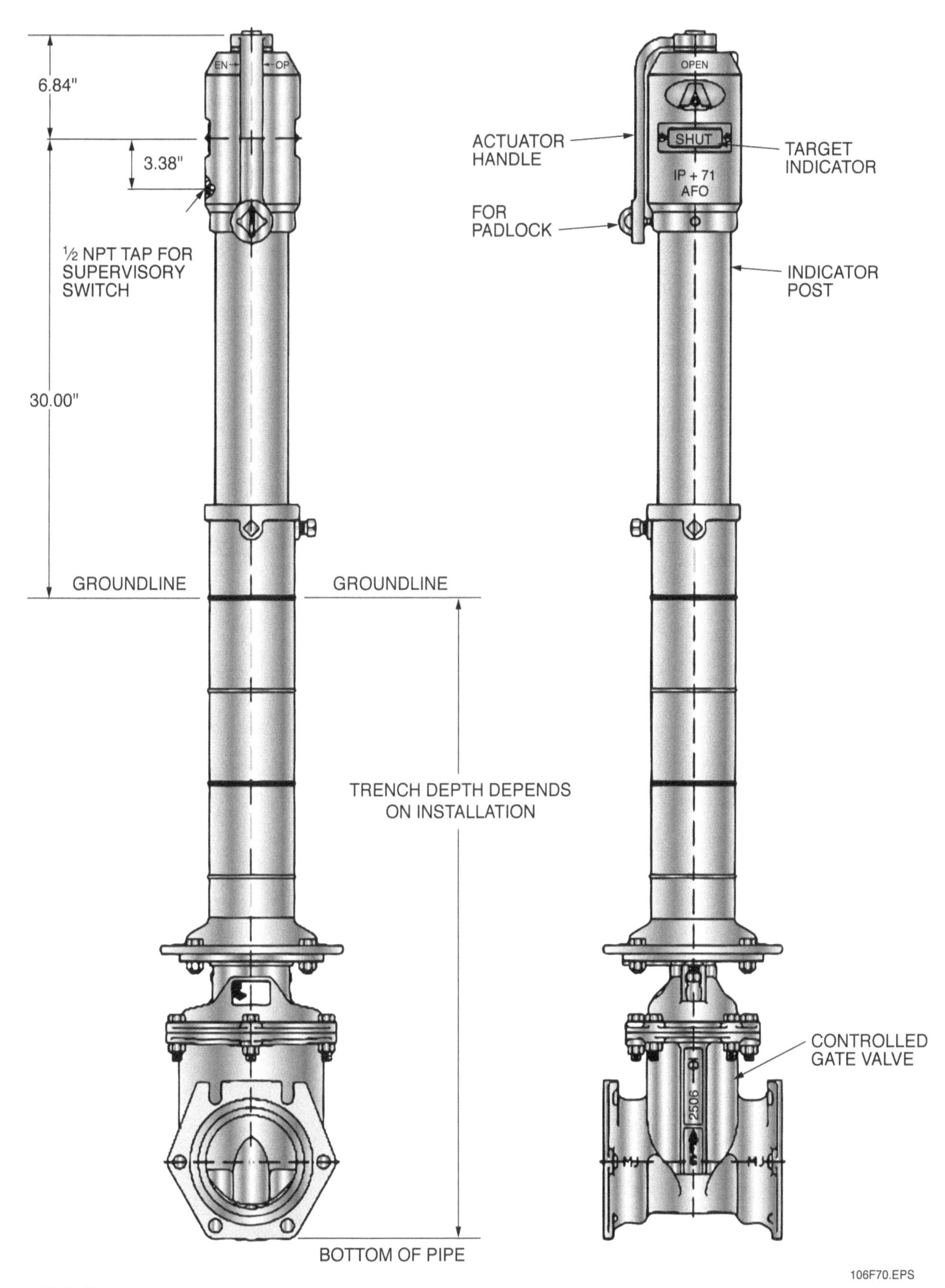

Figure 70 Indicator post.

- One hydrant wrench, in addition to the wrench on the hydrant.
- Four coupling spanners for each size hose provided. *Figure 74* shows a spanner.
- Two hose coupling gaskets for each size hose.

In hose houses that contain more than two sizes of hose and nozzles, the same standard requires that reducers or gated wyes also be included in the hose house equipment. *Figure 75* shows a gated wye.

Other items desirable but optional in a hose house include the following:

- One fire ax with brackets
- One crowbar with brackets
- Two hose and ladder straps
- Two electrical battery hand lights
- Thread adapters

Figure 71 T-wrench.

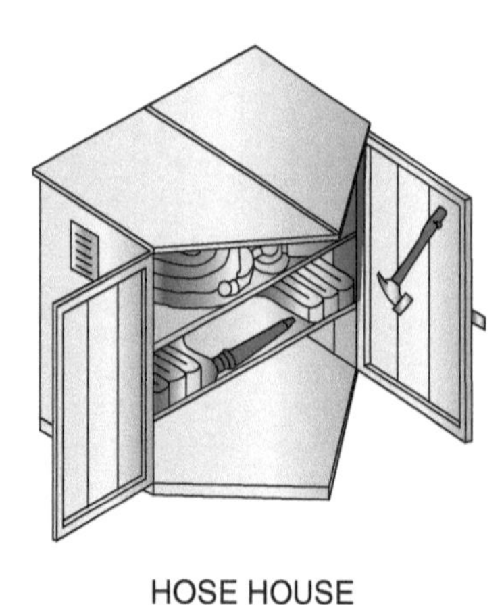

Figure 72 Hose house and hose reel.

12.3.3 Hoses

The hose used must comply with the NFPA standard and be readily accessible. The hose must also be protected from exposure to weather. This can be done by means of hose houses or by placing hose reels or hose carriers in weatherproof enclosures as required by the AHJ.

Figure 73 Adjustable spray nozzle with shut-off.

Figure 74 Common spanner.

The hose couplings must conform to applicable standards. In some localities, the local fire department connections do not conform to NFPA standards. In these cases hydrants and hoses must be special-ordered to meet local standards, and the fire code and/or the fire marshal must be consulted.

The most common term applied to hose is woven jacket, rubber lined. The woven jacket is the outer cover of the hose and can be a single jacket or a double jacket. This is a factor in the durability, flexibility, and weight of the hose. Most jackets are woven polyester or other synthetic material. Rubber lining, usually a synthetic such as Neoprene, adds longer life to the hose because it does not absorb water and facilitates quick drying. The hose that is used for connection with interior emergency standpipes is not recommended for general use. *Figure 76* shows types of hoses.

The old standard for fire hose was linen hose. Unlined linen hose lasts indefinitely when stored in a heated and dry atmosphere, but deteriorates rapidly if allowed to get wet. It also requires the use of a brass, automatic drain valve. Fire department experience is that these hoses degrade after 2 to 3 years, and they no longer use them. Fire hose made of polyester or other synthetics are preferable to linen because they are rot- and mildew-resistant. Unlined linen hose is not permitted in new installations or as replacement hose for use in existing hose houses. When unlined linen hose is replaced, single-jacket lined hose is generally used. The lining can be of a synthetic material other than rubber. Periodic inspection and hydrostatic testing are required for proper hose maintenance; however, unlined linen hose is rarely hydrostatically tested because of the effect of moisture on the material.

106F75.EPS

Figure 75 Gated wye.

12.3.4 Nozzles

The NFPA *Fire Protection Handbook* classifies six general types of spray nozzles for hose streams. They are as follows:

- Open nozzle-fixed spray pattern
- Combination nozzle
- Variable gallonage adjustable stream pattern
- Constant gallonage adjustable stream pattern
- Adjustable gallonage adjustable stream pattern
- Variable gallonage constant pressure-adjustable stream pattern

Nozzles are required for hose houses, hose stations, and hydrant houses. The stream of water directed on a fire plays an important role in reducing the hazard and containing the fire itself. Adjustable spray nozzles can direct a spray of water or can be adjusted to deliver a fog of water. In the case of an electrical fire, a steady stream of water is less effective than a spray of water in containing the fire. A fog or spray of water absorbs heat and can create steam in high heat conditions. The spray created by an adjustable nozzle can also provide protection to the firefighter from radiant and convective heat.

The nozzle must be equipped with a shut-off valve. Shut-off valves can be either a part of the nozzle itself or a separate attachment. They must be kept in good operating order and replaced if they do not function properly. NFPA has established guidelines for the care, use, and maintenance of nozzles and associated accessories.

NITRILE LIGHT DUTY

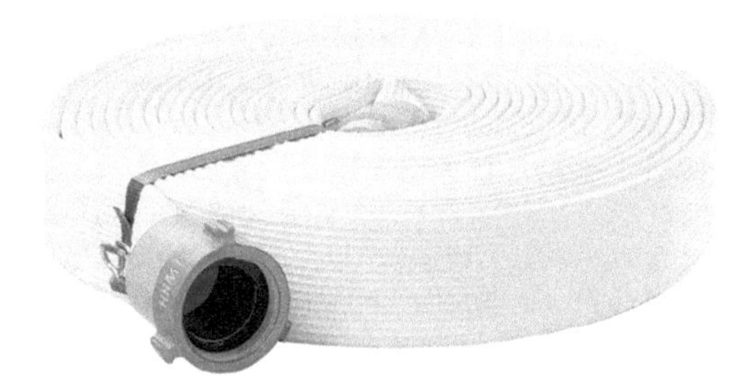

NITRILE HEAVY DUTY

106F76.EPS

Figure 76 Types of hoses.

12.4.0 Hydrant Houses

Hydrant houses are frequently specified for use in rural and industrial areas and in wide-open spaces. *NFPA 24* requires that the hydrant be as close to the front of the hydrant house as possible and still allow enough room in back of the doors for the hose gates and the attached hose. If the hydrant house requires locking, special locks with a brittle shackle must be used. A latch behind a glass plate can also be used. The glass can be broken in case of fire. *Figure 77* shows a hydrant house.

It is common practice to have the hose connected to one of the hydrant outlets at all times. One or more other outlets on the hydrant remain available for additional hoses to be connected if necessary.

The hydrant house must be equipped with a minimum of 100' of lined hose; however, it is preferable to have a 150' hose. To determine the amount of hose needed to protect the property for which the hydrant house has been erected, a loose curving line should be laid from the hydrant house to the building opening, inside the building, and to the farthest point vertically or horizontally the hose might possibly need to reach. The hose should also be long enough to reach all sides of the building while maintaining a safe distance when fighting a fire.

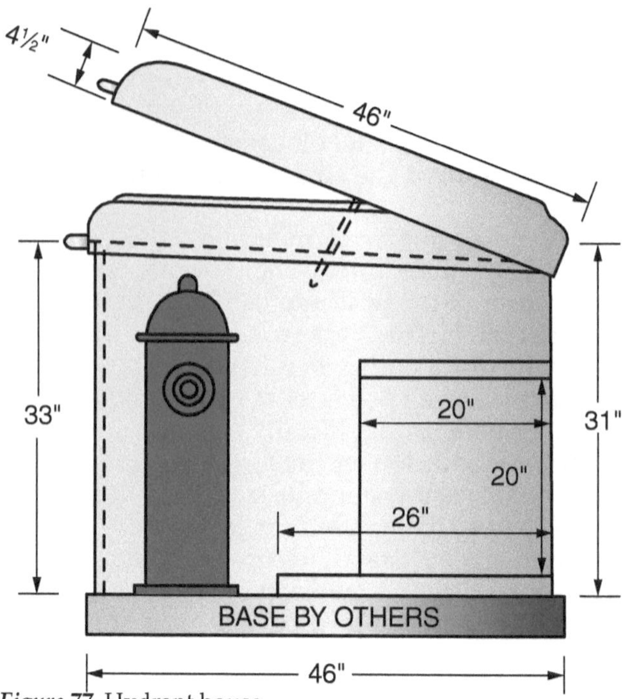

Figure 77 Hydrant house.

13.0.0 Testing, Inspection, Flushing, and Chlorinating

Many installations require the sprinkler fitter to perform testing, inspection, flushing, and chlorinating of underground pipe system installations. The following sections explain these activities.

13.1.0 Testing

Pressure testing of underground water lines as they are being installed is a good practice. It is less expensive to correct errors or defects as the work is being done than to discover them after the entire water supply line has been laid.

For pressure testing purposes, the thrust blocks must be allowed to cure for at least a 24-hour period. Thrust blocks are also tested in a pressure test. Other restraints must be in place as required. A properly conducted pressure test should detect any leaks, loose fittings, or damaged pipe. Pressure tests on water supplies for fire protection must be completed with the pipe joints exposed. *NFPA 24* requires that such tests be applied to water service pipes and rough piping installations prior to covering and concealment. The trench must be backfilled between joints before testing to prevent the pipe from moving during the testing. The joints must remain exposed so leaks can be easily detected.

Testing is done by closing all valves and openings and pressurizing the line with water. Testing equipment must be maintained properly to be reliable and should be drained after each use. Antifreeze is added in the winter to prevent damage from freezing temperatures.

Pressure produces forces that are distributed over every square inch of internal pipe surface. Pressure is generally stated in terms of pounds per square inch, or psi. A pressure of 200 psi means that a force of 200 pounds is applied to each square inch of surface.

Test pressure has to be at least 50 psi greater than the normal maximum supply pressure. The minimum allowable test pressure is 200 psi. Water should be used unless the weather will not allow a water test without freezing.

Additional pressure beyond that available in the local system must be supplied by means of a force pump equipped with a pressure gauge. After pumping to the required test pressure, the pressure gauge must be observed for any drop in pressure. A drop in pressure will indicate a leak in the system. When leaks are observed on the pressure gauge, the first place to look, if the leak location is not obvious, is the force pump itself and its connections.

13.2.0 Inspection

The sprinkler contractor may be required to make formal pressure tests in the presence of the AHJ or the owner's representative. Before these tests are made for official acceptance, tests should be made privately to make sure the installation will pass when the formal test takes place. The contractor will be responsible for completing all required acceptance tests and, subsequently, the test certificate (*Figure 78*) for underground piping. Witnesses for both the contractor and the property owner should sign the completed test certificate.

During the inspection, *NFPA 24* provides limits of allowable leakage. The amount of leakage in piping should be measured at the specified test pressure by pumping from a calibrated container. For new pipe, the amount of leakage at joints shall not exceed two quarts per hour per 100 gaskets or joints, regardless of pipe diameter.

If dry barrel fire hydrants are tested with the main valve open, an additional five ounces per minute is allowable.

13.3.0 Flushing

After the underground pipe has been tested and inspected, the entire system must be thoroughly flushed before being connected to the sprinkler piping. If the system is flushed by others, it is important to obtain documentation. If the proper documentation is not obtained in a timely manner, the flushing will have to be repeated for that purpose.

Flushing is made easy if the inside of the piping is kept clear of foreign matter as the pipe is being laid, and if the open end is kept closed when not attended. Flushing is done under supply pressure through the flanged spigot, hydrants, or other outlets. Care must be taken to section off any connected inside sprinkler or standpipe equipment by valves or blind flanges to avoid lodging foreign material in these systems.

The system is flushed until all foreign material is ejected and the water leaving the system runs clear. Mains supplying sprinkler systems must be flushed at minimum rates as specified by *NFPA 24*, or at the hydraulically calculated water demand rate of the system, whichever is greater. *Table 11* shows the flow required to produce a velocity of 10 feet (3 m) per second in pipes as given in *NFPA 24*. If the flow rate cannot be verified or met, refer to *NFPA 24* for exceptions.

13.4.0 Chlorinating

The water mains of municipalities supply potable water. The term potable water means water that is suitable for drinking, cooking, and domestic purposes. To be potable it must meet the requirements of the water or health authorities.

Local ordinances and/or specifications may require chlorinating of underground systems by the sprinkler contractor. In those instances, the potable water supply lines must be disinfected (sterilized) by chlorinating after passing the pressure test.

Chlorinating is accomplished using either sodium hypochlorite or calcium hypochlorite. This material is available in tablet, granular, or liquid form. Most municipalities use the granular form for convenience and economy. The granular material can be measured and poured like a liquid. If tablets are used, a specific number must be placed in each pipe joint as the pipe is laid. Job specifications specify the number of tablets required in each joint. These must be secured in place by an adhesive such as rubber cement. When tablets are used, the line cannot be flushed prior to chlorinating. The liquid form readily creates fumes, can be toxic, and its strength can vary due to loss of chlorine gas.

The American Water Works Association (AWWA) recommends two methods of chlorinating a water line. The first and most commonly used method is to fill the piping section with a water solution containing 50 parts per million of available chlorine. It is then allowed to stand 24 hours before flushing and placing or returning it to service.

With the second method, the section of piping is filled with a water solution containing 200 parts per million of available chlorine. This concentration is allowed to stand 3 hours before flushing and placing the section in service.

It is not practical to disinfect a potable water storage tank by these two methods. The entire interior of the tank must be swabbed with a water solution containing 200 parts per million of available chlorine and allowed to stand 3 hours before flushing and placing in service. Proper safety precautions must be taken to avoid breathing an excess of chlorine fumes. These fumes are highly toxic.

For potable water filters and similar devices, the quantity used must be approved by the AHJ under the specific prevailing circumstances.

Table 11 Flow Rates Specified by *NFPA 13* (Table 10.10.2.1.3)

Pipe Size (in)	Pipe Size (mm)	Flow Rate (gpm)	Flow Rate (L/min)
2	51	100	379
2½	63	150	568
3	76	220	833
4	102	390	1,476
6	152	880	3,331
8	203	1,560	5,905
10	254	2,440	9,235
12	305	3,520	13,323

106T11.EPS

Contractor's Material and Test Certificate for Underground Piping

PROCEDURE

Upon completion of work, inspection and tests shall be made by the contractor's representative and witnessed by an owner's representative. All defects shall be corrected and system left in service before contractor's personnel finally leave the job.

A certificate shall be filled out and signed by both representatives. Copies shall be prepared for approving authorities, owners, and contractor. It is understood the owner's representative's signature in no way prejudices any claim against contractor for faulty material, poor workmanship, or failure to comply with approving authority's requirements or local ordinances.

Property name	Date
Property address	

Plans	Accepted by approving authorities (names)	
	Address	
	Installation conforms to accepted plans	❑ Yes ❑ No
	Equipment used is approved If no, state deviations	❑ Yes ❑ No
Instructions	Has person in charge of fire equipment been instructed as to location of control valves and care and maintenance of this new equipment? If no, explain	❑ Yes ❑ No
	Have copies of appropriate instructions and care and maintenance charts been left on premises? If no, explain	❑ Yes ❑ No
Location	Supplies buildings	
Underground pipes and joints	Pipe types and class / Type joint	
	Pipe conforms to ______ standard	❑ Yes ❑ No
	Fittings conform to ______ standard If no, explain	❑ Yes ❑ No
	Joints needing anchorage clamped, strapped, or blocked in accordance with ______ standard If no, explain	❑ Yes ❑ No

Test description

Flushing: Flow the required rate until water is clear as indicated by no collection of foreign material in burlap bags at outlets such as hydrants and blow-offs. Flush at flows not less than 390 gpm (1476 L/min) for 4 in. pipe, 880 gpm (3331 L/min) for 6 in. pipe, 1560 gpm (5905 L/min) for 8 in. pipe, 2440 gpm (9235 L/min) for 10 in. pipe, and 3520 gpm (13,323 L/min) for 12 in. pipe. When supply cannot produce stipulated flow rates, obtain maximum available.

Hydrostatic: All piping and attached appurtenances subjected to system working pressure shall be hydrostatically tested at 200 psi (13.8 bar) or 50 psi (3.4 bar) in excess of the system working pressure, whichever is greater, and shall maintain that pressure ± 5 psi for 2 hours.

Hydrostatic Testing Allowance: Where additional water is added to the system to maintain the test pressures required by 10.10.2.2.1, the amount of water shall be measured and shall not exceed the limits of the following equation (For metric equation, see 10.10.2.2.4):

$$L = \frac{SD\sqrt{P}}{148,000}$$

L = testing allowance (makeup water), in gallons per hour
S = length of pipe tested, in feet
D = nominal diameter of the pipe, in inches
P = average test pressure during the hydrostatic test, in pounds per square inch (gauge)

Flushing tests

New underground piping flushed according to ______ standard by (company) If no, explain	❑ Yes ❑ No
How flushing flow was obtained: ❑ Public water ❑ Tank or reservoir ❑ Fire pump	Through what type opening: ❑ Hydrant butt ❑ Open pipe
Lead-ins flushed according to ______ standard by (company) If no, explain	❑ Yes ❑ No
How flushing flow was obtained: ❑ Public water ❑ Tank or reservoir ❑ Fire pump	Through what type opening: ❑ Y connection to flange and spigot ❑ Open pipe

NFPA 13 (p. 1 of 2)

106F78A.EPS

Figure 78 Sample of contractor's material and test certificate for underground piping (1 of 2).

<table>
<tr><td rowspan="1">Hydrostatic test</td><td colspan="3">All new underground piping hydrostatically tested at
______ psi for ______ hours</td><td>Joints covered
❑ Yes ❑ No</td></tr>
<tr><td rowspan="2">Leakage test</td><td colspan="4">Total amount of leakage measured
______ gallons ______ hours</td></tr>
<tr><td colspan="4">Allowable leakage
______ gallons ______ hours</td></tr>
<tr><td>Hydrants</td><td>Number installed</td><td colspan="2">Type and make</td><td>All operate satisfactorily
❑ Yes ❑ No</td></tr>
<tr><td rowspan="2">Control valves</td><td colspan="3">Water control valves left wide open
If no, state reason</td><td>❑ Yes ❑ No</td></tr>
<tr><td colspan="3">Hose threads of fire department connections and hydrants interchangeable with those of fire department answering alarm</td><td>❑ Yes ❑ No</td></tr>
<tr><td rowspan="3">Remarks</td><td colspan="4">Date left in service</td></tr>
<tr><td colspan="4"></td></tr>
<tr><td colspan="4"></td></tr>
<tr><td rowspan="4">Signatures</td><td colspan="4">Name of installing contractor</td></tr>
<tr><td colspan="4">Tests witnessed by</td></tr>
<tr><td>For property owner (signed)</td><td colspan="2">Title</td><td>Date</td></tr>
<tr><td>For installing contractor (signed)</td><td colspan="2">Title</td><td>Date</td></tr>
<tr><td colspan="5">Additional explanation and notes</td></tr>
</table>

NFPA 13 (p. 2 of 2)

106F78B.EPS

Figure 78 Sample of contractor's material and test certificate for underground piping (2 of 2).

SUMMARY

Excavation can be a dangerous procedure if not done properly. Procedures are regulated by OSHA. It can also be dangerous to the pipeline itself, the public, and environment. For this reason, excavation and backfill procedures are monitored and regulated by the DOT, EPA, and other state and local entities.

For the safety of workers, it is important to ensure that trenches are safe places. It is important also to know how to deal with hazardous material when encountered. Using heavy equipment to dig and backfill can be dangerous to workers and to the pipeline itself. Specific procedures are followed to avoid accidents and damage from excavators and backhoes.

Many states are putting the responsibility for a fire sprinkler system's underground pipe on the sprinkler contractor, so it is important that the apprentice fitter be familiar with the proper way to install underground pipe. Installation includes laying, inspecting, testing, flushing and chlorinating. The primary concern of a sprinkler fitter working in an excavation or trench is safety. Always be aware of your surroundings and watch and listen for possible signs of trouble. Whenever unsafe conditions arise, get out of the trench, inform your co-workers, and have them leave the trench. Trench work is extremely dangerous, and every effort must be made by everyone on the job site to ensure that the work progresses safely.

Underground joints do not restrain the pipe in the longitudinal direction. Thrust blocks, restraints, and clamps must be depended upon to secure the connection. As with all other aspects of sprinkler fitting, safety rules and regulations must be followed.

Fire and yard hydrants require proper location and selection to be useful. They must be supported and thrust blocked to avoid blowouts. Threads on outlets must meet local standards.

Hoses are available in several materials. Specifications usually govern the material selection. Wrenches, when supplied, must be selected to fit the local need for reversing caps and connecting hoses, and to fit the hydrant operating nut.

Hydrant houses, hose houses, and hose stations contain equipment other than hoses that must also meet specifications.

Review Questions

1. A cubic yard of soil weighs _____.
   a. 300 pounds
   b. 500 pounds
   c. 3,000 pounds
   d. 5,000 pounds

2. Average soils are usually a mixture of _____.
   a. sand, iron ore, and water
   b. sand, silt, and clay
   c. sand, bark, and lime
   d. clay, silt, and water

3. The factor that determines how a soil will behave during trenching is _____.
   a. its weight
   b. its elevation
   c. the water table
   d. its texture or composition

4. In any trench that is over 4' deep, ladders must be placed every _____.
   a. 15'
   b. 25'
   c. 50'
   d. 75'

5. Increases or decreases in soil moisture content can affect the stability of a trench.
   a. True
   b. False

6. Trench walls are sloped to make it easier to get into and out of the trench.
   a. True
   b. False

7. Which of the following types of shoring is best used near existing structures?
   a. Hydraulic shores and spreaders
   b. Interlocking steel sheeting
   c. Vertical sheeting
   d. Concrete blocks

8. Trenches have a maximum allowable slope of 1.5 horizontal to 1 vertical in _____.
   a. stable rock
   b. Type A soil
   c. Type B soil
   d. Type C soil

9. A stable foundation for pipe in a trench consists of bedding, shoring, and possibly a stabilization layer.
   a. True
   b. False

10. Bedding extends from the top of the stabilization layer, if present, to _____.
    a. the middle of the pipe
    b. the top of the pipe
    c. the bottom of the pipe
    d. a thickness of 10"

11. The spoil from a new ditch *cannot* be used as fill if it is cut through _____.
    a. rock
    b. Type A soil
    c. Type B soil
    d. Type C soil

12. Most authorities allow buried piping to be located at any depth that is convenient so long as there is enough protective covering over the pipe.
    a. True
    b. False

13. Which of the following is *not* a characteristic of PVC pipe?
    a. Ease of transport
    b. Loss of strength due to potable water corrosion
    c. Cost effectiveness
    d. Availability in sizes compatible with ductile iron fittings

14. An 8" ductile iron pipe with a thickness class of 53 will be _____.
    a. 0.32" thick
    b. 0.36" thick
    c. 0.38" thick
    d. 0.39" thick

15. The main problem with pipe joints that are not fitted properly is that _____.
    a. they are difficult to disassemble
    b. the pipeline may be misaligned
    c. they are difficult to assemble
    d. they may leak

16. In *NFPA 24,* earthquake protection provisions call for how many inches of nominal clearance around pipes 4" and larger?
    a. 1"
    b. 2"
    c. 3"
    d. 4"

17. Chemical and electrical sources of pipe corrosion can be reduced by using which of the following thicknesses of polyethylene tubing to encase the pipe and fittings?
    a. 4 mil
    b. 6 mil
    c. 8 mil
    d. 10 mil

18. How many pounds of thrust is developed on an 8" 90-degree elbow?
    a. 10,000 pounds
    b. 10,800 pounds
    c. 13,000 pounds
    d. 18,600 pounds

19. Thrust block calculations are typically done by the _____.
    a. sprinkler fitter
    b. shop foreman
    c. layout technician
    d. AHJ

20. Backflow preventer assemblies are designed to _____.
    a. allow the reverse flow of polluted water to enter the main water supply when required
    b. prevent the reverse flow of polluted water from entering the potable water supply
    c. allow backflow of pollutants that are objectionable but not toxic
    d. allow backflow of pollutants that are objectionable and toxic

Trade Terms Quiz

Fill in the blank with the correct trade term that you learned from your study of this module.

1. A(n) ___________ is a narrow excavation (relative to its length) made below the surface of the ground.
2. Cutting the excavation walls to the angle of repose of the soil being used is called ___________.
3. A(n) ___________ uses two spring-loaded check valves plus a hydraulic spring-loaded pressure differential relief valve.
4. A direct link between the potable water supply and water of unknown or questionable quality is a(n) ___________.
5. A(n) ___________ protects employees from cave-ins, from material that could fall or roll into an excavation, or from collapse of adjacent structures.
6. The capability of soil to hold a heavy weight is termed ___________.
7. A(n) ___________ is a hole that allows 360-degree access to a pipe.
8. Horizontal members of a shoring system placed parallel to the excavation face whose sides bear against the vertical members of the shoring system are referred to as ___________.
9. A structure that provides support to an adjacent structure, underground installation, or the walls of an excavation is a(n) ___________.
10. ___________ is a structure such as a metal, hydraulic, mechanical, or timber system that supports the sides of an excavation and is designed to prevent cave-ins.
11. A valve in the backflow preventer that permits testing individual pressure zones is a(n) ___________.
12. A backflow preventer that does not have a test cock and that uses two spring-loaded check valves to prevent back siphonage and back pressure is a(n) ___________.
13. A(n) ___________ prevents back siphonage by allowing air pressure to force a ceramic disc into its seat and block the flow.
14. ___________ is a chemical process in which water or liquids accumulate and cause a breakdown of surrounding minerals such as dirt.
15. A chemical process in which carbon accumulates and causes a breakdown of surrounding minerals such as dirt is called ___________.
16. The greatest angle above the horizontal plane that a material will rest without sliding is called the ___________.
17. A(n) ___________ is a pre-manufactured steel box that provides a shielding system by allowing workers to lay pipe within the box while the box is inside the excavation.
18. A flat, relatively narrow key that is integral to a shaft, produced by milling a longitudinal groove is called a(n) ___________.
19. A(n) ___________ is a structure that is able to withstand the forces imposed on it by a cave-in and thereby protect employees within the structure.
20. A(n) ___________ is installed on a reduced-pressure zone principle backflow preventer to prevent back pressure and back siphonage.
21. ___________ is an undesirable condition that results when nonpotable liquids enter the potable water supply by reverse flow through a cross-connection.
22. The horizontal members of a shoring system installed perpendicular to the sides of the excavation, the ends of which bear against either uprights or wales are called ___________.

23. ___________ is a method of protecting workers from cave-ins by excavating the sides of a trench to form one or a series of horizontal levels or steps, usually with vertical or near-vertical surfaces between them.

24. ___________ are vertical members of a trench shoring system placed in contact with the earth and usually positioned so that individual members do not contact each other.

25. A rock fragment or mineral in the soil that has a diameter of 0.002 to 0.05 millimeter and is smaller than fine sand and larger than clay is called ___________.

26. ___________ is a specialty backflow preventer used on low-flow and small supply lines to prevent back siphonage and back pressure.

27. A backflow preventer that has a test cock and prevents back pressure and back siphonage through the use of two spring-loaded check valves is a(n) ___________.

28. ___________ is soil that has been previously backfilled or removed from the earth and replaced by any means.

29. The surface or foundation on which a pipe rests in a trench is called ___________.

30. ___________ refers to the use of specially edged timber planks, such as tongue-and-groove planks, that are at least three inches thick.

31. A condition that occurs when individual timber uprights or individual hydraulic shores are not placed in contact with the adjacent member is a(n) ___________.

32. ___________ is a form of backflow caused by lower than normal upstream pressure in the fresh water line.

33. A(n) ___________ is capable of identifying existing and predictable hazards in the area, or pointing out working conditions that are unsanitary, hazardous, or dangerous to employees, and who has authority to take prompt corrective measures.

34. A laboratory test that determines moisture content, cohesion factor, and soil gradation of a particular soil is a(n) ___________.

35. ___________ is a depression in the earth that is caused by unbalanced stresses in the soil surrounding an excavation.

36. Timber or other material used as a temporary prop for excavations and which may be sloping, vertical, or horizontal is a(n) ___________.

37. ___________ are used to prevent groundwater flow along the pipeline, thus preventing erosion to the pipe bedding material.

38. A(n) ___________ is installed in a water line that uses a spring-loaded check valve and a spring-loaded air inlet valve to prevent back siphonage.

39. Any man-made cut, cavity, trench, or depression in an earth surface, formed through earth removal is a(n) ___________.

Trade Terms

Angle of repose
Atmospheric vacuum breaker
Back siphonage
Backflow
Bedding
Bell hole
Benching
Carbonation
Competent person
Compression strength
Cross braces
Cross-connection
Disturbed soil
Ditch breakers
Double-check valve assembly
Dual-check valve backflow preventer
Excavation
Hydration
Intermediate atmospheric vent vacuum breaker
Manufactured air gap
Pressure-type vacuum breaker
Proctor
Protective system
Reduced-pressure zone principle backflow preventer
Shield
Shore
Shoring
Silt
Skeleton
Sloping
Spline
Subsidence
Support system
Test cock
Tight sheeting
Trench
Trench box
Uprights
Wales

Cornerstone of Craftsmanship

Travis Torres

Service Manager
American Automatic Sprinkler
Fort Worth, Texas

Travis Torres graduated from the AFSA apprentice program in 2003. He is now the service manager at American Automatic Sprinkler. The company president says that Torres is definitely one of his future leaders. His story is an example of how hard work and a positive attitude pay dividends in promotions and success.

How did you choose a career in the sprinkler fitting field?
I moved to the Fort Worth area to go to college. Then I got married and needed to go to work to support my wife and child. A friend told me about the apprenticeship program. I bought a house and started working full time.

What types of training have you been through?
I went through the apprenticeship program. Now that things are more stable, I am going back to college to get a degree in business or accounting, something that will help me advance in the business side of the trade.

What kinds of work have you done in your career?
I started as a helper and went from apprentice, to fitter, to foreman. The past year and a half, I was given the opportunity to manage the service division. We service, maintain, and test our existing systems and handle any remodeling that needs to be done.

What do you like about your job?
I like that the industry is always evolving. The code changes, or the way we do things changes. I like a job that challenges my mind. Recently, the state mandated that everyone get their National Institute for Certification in Engineering Technologies (NICET) certification. Soon I will be back in school to get that training.

I also like dealing with people on a personal level. I meet many different people with creative minds, like architects, engineers, and designers. It is not always easy, but I try to find a way to create a win/win result.

What factors have contributed most to your success?
Hard work. Dedication. Persistence. It may sound corny, but in reality, it is what it is. I was young and working around men 10 or more years older than me. It was hard taking criticism and learning to lead. But it was worth the effort.

This company was my starting point. They take care of their people, and I have had some wonderful opportunities.

What advice would you give to those new to the sprinkler fitter field?
The possibilities are endless. Strive for what you want and don't settle. There is an opportunity around the corner for everybody. You just have to be open to see it and take the chance when the door opens. Everyone has potential. You just have to work hard, and it will be there for you.

Trade Terms Introduced in This Module

Angle of repose: The greatest angle above the horizontal plane that a material will rest without sliding.

Atmospheric vacuum breaker: A backflow preventer designed to prevent back siphonage by allowing air pressure to force a ceramic disc into its seat and block the flow.

Back siphonage: A form of backflow caused by lower-than-normal upstream pressure in the fresh water line.

Backflow: An undesirable condition that results when nonpotable liquids enter the potable water supply by reverse flow through a cross-connection.

Bedding: The surface or foundation on which a pipe rests in a trench.

Bell hole: A hole that allows 360-degree access to a pipe. Bell holes are commonly used at a field joint or for field repair.

Benching: A method of protecting workers from cave-ins by excavating the sides of a trench to form one or a series of horizontal levels or steps, usually with vertical or near-vertical surfaces between levels.

Carbonation: A chemical process in which carbon accumulates and causes a breakdown of surrounding minerals, such as dirt.

Competent person: A person who is capable of identifying existing and predictable hazards in the area or working conditions that are unsanitary, hazardous, or dangerous to employees, and who has the authority to take prompt corrective measures to fix the problem.

Compression strength: The ability of soil to hold a heavy weight.

Cross braces: The horizontal members of a shoring system installed perpendicular to the sides of the excavation, the ends of which bear against either uprights or wales.

Cross-connection: A direct link between the potable water supply and water of unknown or questionable quality.

Disturbed soil: Soil that has been previously backfilled or removed from the earth and replaced by any means.

Ditch breakers: Used to prevent groundwater flow along the pipeline which causes erosion of the pipe bedding material. The breakers are constructed of sandbags placed under, beside, and over the pipe in the trench.

Double-check valve assembly (DCV): A backflow preventer that prevents back pressure and back siphonage through the use of two spring-loaded check valves. DCVs are larger than dual-check valve backflow preventers and are used for heavy-duty protection. DCVs have test cocks.

Dual check valve backflow preventer (DC): A backflow preventer that uses two spring-loaded check valves to prevent back siphonage and back pressure. DCs are smaller than double-check valve assemblies and are used in residential installations. DCs do not have test cocks.

Excavation: Any man-made cut, cavity, trench, or depression in an earth surface, formed through earth removal.

Hydration: A chemical process in which water or liquids accumulate and cause a breakdown of surrounding minerals, such as dirt.

Intermediate atmospheric vent vacuum breaker: A specialty backflow preventer used on low-flow and small supply lines to prevent back siphonage and back pressure. It is a form of pressure-type vacuum breaker.

Manufactured air gap: An air gap that can be installed on a reduced-pressure zone principle backflow preventer to prevent back pressure and back siphonage.

Pressure-type vacuum breaker: A backflow preventer installed in a water line that uses a spring-loaded check valve and a spring-loaded air inlet valve to prevent back siphonage.

Proctor: A test performed in a laboratory that determines moisture content, cohesion factor, and soil gradation of a particular soil.

Protective system: A method of protecting employees from cave-ins, from material that could fall or roll from an excavation face or into an excavation, or from the collapse of adjacent structures. Protective systems include support systems, sloping and benching systems, and shielding systems.

Reduced-pressure zone principle backflow preventer (RPZ): A backflow preventer that uses two spring-loaded check valves plus a hydraulic, spring-loaded pressure differential relief valve to prevent back pressure and back siphonage.

Shield: A structure that is able to withstand the forces imposed on it by a cave-in and thereby protect employees within the structure. Shields can be permanent structures or can be designed to be portable and moved along as work progresses. Additionally, shields can be either pre-manufactured or job-built in accordance with *29 CFR 1926.652 (c)(3)* or *(c)(4)*.

Shore: Timber or other material used as a temporary prop for excavations. May be sloping, vertical, or horizontal.

Shoring: A structure such as a metal hydraulic, mechanical, or timber system that supports the sides of an excavation and is designed to prevent cave-ins.

Silt: A rock fragment or mineral in the soil that has a diameter of 0.002 to 0.05 millimeter and is smaller than fine sand and larger than clay.

Skeleton: A condition that occurs when individual timber uprights or individual hydraulic shores are not placed in contact with the adjacent member.

Sloping: A method of protecting employees working in excavations from cave-ins by cutting the excavation walls to the angle of repose of the soil being used.

Spline: A flat, relatively narrow key that is integral to a shaft, produced by milling a longitudinal groove.

Subsidence: A depression in the earth that is caused by unbalanced stresses in the soil surrounding an excavation.

Support system: A structure, such as an underpinning, bracing, or shoring, that provides support to an adjacent structure, underground installation, or the walls of an excavation.

Test cock: A valve in a backflow preventer that permits the testing of individual pressure zones.

Tight sheeting: The use of specially edged timber planks, such as tongue-and-groove planks, that are at least 3" thick. Steel sheet piling or similar construction that resists the lateral pressure of water and prevents loss of backfill is also called tight sheeting.

Trench box: A pre-manufactured steel box that provides a shielding system by allowing workers to lay pipe within the box while it is inside the excavation.

Trench: A narrow excavation (in relation to its length) made below the surface of the ground. In general, the depth is greater than the width, but the width of a trench (measured at the bottom) is not greater than 15' (4.6 m).

Uprights: The vertical members of a trench shoring system placed in contact with the earth and usually positioned so that individual members do not contact each other. Uprights placed so that individual members are closely spaced, in contact with, or interconnected to each other, are often called sheeting.

Wales: Horizontal members of a shoring system placed parallel to the excavation face whose sides bear against the vertical members of the shoring system or the earth.

Additional Resources

This module presents thorough resources for task training. The following resource material is suggested for further study.

Excavations. Washington, DC: OSHA Publications Office.

Multimedia Apprenticeship Training Supplement for Fire Sprinkler Fitters (CD set: Level 1 through Level 4). American Fire Sprinkler Association. www.sprinklernet.org.

NFPA 13, Standard for the Installation of Sprinkler Systems, Latest Edition. Quincy, MA: National Fire Protection Association.

NFPA 24, Standard for the Installation of Private Fire Service Mains and Their Appurtenances, Latest Edition. Quincy, MA: National Fire Protection Association.

Figure Credits

Kennedy Valve, Module opener

Doss Brothers, Incorporated, 106F01

National Aeronautics and Space Administration, 106F04

Trench Shoring Services, Inc., 106F05–106F07, 106F09, www.shoring.com

John Hoerlein, 106SA02, 106F28 (flat plate compactor)

Port Industries, Inc., 106SA03

U.S. Pipe and Foundry Co., LLC, 106F25, 106F31, 106F38 (photo), 106F45, 106F46

Team EJP/New Concept Tools, 106F28 (pipe tamping bar)

Manitowoc Crane Group, 106SA04

Pollardwater.com, 106F30

Mueller Company, 106F34

American Flow Control, 106F35A, 106F36, 106F37, 106F70

Primary Fluid Systems, Inc., 106F35B

National Fire Protection Association, 106T06, 106T08, 106T10, 106T11, 106F78

Reprinted with permission from *NFPA 13-2013, Installation of Sprinkler Systems*, Copyright © 2012, National Fire Protection Association, Quincy, MA 02169. The reprinted material is not the complete and official position of the NFPA on the referenced material, which is represented only by the standard in its entirety.

Los Angeles Department of Water and Power, 106F49

EBAA Iron, Inc., 106F50 (ductile iron joint)

Ford Meter Box Company, Module opener, 106F50 (retaining gland)

Ames Fire and Waterworks, 106F52

Watts Regulator Company, 106F53-106F57, 106F59, 106F61, 106F63–106F65

US EPA Office of Ground Water and Drinking Water, 106F60

Courtesy of American AVK Co., 106F66 (photo), 106F67, Fresno, CA, Minden, NV, www.americanavk.com

Dixon Valve & Coupling Company, 106F72 (photo), 106F76

Wholesale Fire & Rescue, Ltd., 106F73–106F75

MODULE 18106-13 — ANSWERS TO REVIEW QUESTIONS

Answer		Section
1.	c	2.0.0
2.	b	2.1.2
3.	d	2.1.3
4.	b	3.0.0
5.	a	4.0.0
6.	b	5.2.0
7.	c	6.1.2
8.	d	6.4.4
9.	b	7.1.0
10.	c	7.1.2
11.	a	7.2.1
12.	b	7.6.0
13.	b	8.3.0
14.	b	8.6.0
15.	d	8.8.0
16.	b	8.11.1
17.	c	8.15.0
18.	d	9.4.0
19.	c	9.4.0
20.	b	11.1.0

MODULE 18106-13 — ANSWERS TO TRADE TERMS QUIZ

1. Trench
2. Sloping
3. Reduced-pressure zone principle backflow preventer
4. Cross connection
5. Protective system
6. Compression strength
7. Bell hole
8. Wales
9. Support system
10. Shoring
11. Test cock
12. Dual-check valve backflow preventer
13. Atmospheric vacuum breaker
14. Hydration
15. Carbonation
16. Angle of repose
17. Trench box
18. Spline
19. Shield
20. Manufactured air gap
21. Backflow
22. Cross braces
23. Benching
24. Uprights
25. Silt
26. Intermediate atmospheric vent vacuum breaker
27. Double-check valve assembly
28. Disturbed soil
29. Bedding
30. Tight sheeting
31. Skeleton
32. Back siphonage
33. Competent person
34. Proctor
35. Subsidence
36. Shore
37. Ditch breakers
38. Pressure-type vacuum breaker
39. Excavation

NCCER CURRICULA — USER UPDATE

NCCER makes every effort to keep its textbooks up-to-date and free of technical errors. We appreciate your help in this process. If you find an error, a typographical mistake, or an inaccuracy in NCCER's curricula, please fill out this form (or a photocopy), or complete the online form at **www.nccer.org/olf**. Be sure to include the exact module ID number, page number, a detailed description, and your recommended correction. Your input will be brought to the attention of the Authoring Team. Thank you for your assistance.

Instructors – If you have an idea for improving this textbook, or have found that additional materials were necessary to teach this module effectively, please let us know so that we may present your suggestions to the Authoring Team.

NCCER Product Development and Revision

13614 Progress Blvd., Alachua, FL 32615

Email: curriculum@nccer.org
Online: www.nccer.org/olf

❑ Trainee Guide ❑ AIG ❑ Exam ❑ PowerPoints Other ______________________________

Craft / Level: Copyright Date:

Module ID Number / Title:

Section Number(s):

Description:

Recommended Correction:

Your Name:

Address:

Email: Phone:

Glossary

Alloy: Any substance made up of two or more metals.

Ambient temperature: The surrounding air temperature.

American Fire Sprinkler Association (AFSA): A non-profit, international association representing merit shop fire sprinkler contractors, dedicated to the educational advancement of its members and promotion of the use of automatic fire sprinkler systems.

Angle of repose: The greatest angle above the horizontal plane that a material will rest without sliding.

Atmospheric vacuum breaker: A backflow preventer designed to prevent back siphonage by allowing air pressure to force a ceramic disc into its seat and block the flow.

Authority Having Jurisdiction (AHJ): The person or agency that decides the acceptability of the systems and devices being installed in accordance with code requirements.

Back siphonage: A form of backflow caused by lower-than-normal upstream pressure in the fresh water line.

Backflow: An undesirable condition that results when nonpotable liquids enter the potable water supply by reverse flow through a cross-connection.

Bedding: The surface or foundation on which a pipe rests in a trench.

Bell hole: A hole that allows 360-degree access to a pipe. Bell holes are commonly used at a field joint or for field repair.

Benching: A method of protecting workers from cave-ins by excavating the sides of a trench to form one or a series of horizontal levels or steps, usually with vertical or near-vertical surfaces between levels.

Block: Device used to secure pipe stored in tiers.

Borates: Boric acid salts used as a flux.

Brazing: A method of fusing metals with a nonferrous filler metal using heat above 840°F, but below the melting point of the base metals being joined.

Bushing: A pipe fitting that connects a pipe with a fitting of a larger nominal size. A bushing is a hollow plug with internal and external threads to suit the different diameters.

C-factor: Refers to the roughness of the inside of the sprinkler pipe. The higher the C-factor, the smoother the pipe.

Capillary action: The movement of a liquid along the surface of a solid in a kind of spreading action.

Carbonation: A chemical process in which carbon accumulates and causes a breakdown of surrounding minerals, such as dirt.

Cast-iron fitting: A cast-iron fitting is generally used in water applications and has a classification of either 125 or 250.

Chamfering: Beveling and squaring the end of CPVC pipe to make a clean end for a fitting.

Check valves: Allow water to flow in one direction only.

Chlorinated polyvinyl chloride (CPVC): A common thermoplastic pipe used in the fire sprinkler industry.

Close nipple: A nipple that is about twice the length of a standard thread that is threaded from each end with no shoulder.

Companion flange: A flange that is used to connect a fitting or flange directly to threaded pipe.

Competent person: A person who is capable of identifying existing and predictable hazards in the area or working conditions that are unsanitary, hazardous, or dangerous to employees, and who has the authority to take prompt corrective measures to fix the problem.

Compressed gas: Gas stored under pressure in cylinders.

Compression strength: The ability of soil to hold a heavy weight.

Concentric reducing coupling: A reducing coupling that maintains the same center line between the two pipes that it joins.

Control valve: Controls the water supply to a sprinkler system.

CPVC compound: An orange thermoplastic material used to produce fire sprinkler pipe and fittings.

CPVC resin: A white thermoplastic material that accounts for 85 percent of the weight of finished CPVC products.

Cross: A pipe fitting with four branches that are all at right angles to each other.

Cross braces: The horizontal members of a shoring system installed perpendicular to the sides of the excavation, the ends of which bear against either uprights or wales.

Cross-connection: A direct link between the potable water supply and water of unknown or questionable quality.

Cut groove: A groove that is formed around pipe by cutting away the pipe material.

Deluge sprinkler system: A type of sprinkler system that uses open sprinklers rather than closed sprinklers. When a fire triggers the detection system, the deluge valve is released, producing immediate water flow through all sprinklers in a given area.

Dielectric: A nonconductor of electricity.

Dimple: A slight surface depression, or the act of creating one.

Disturbed soil: Soil that has been previously backfilled or removed from the earth and replaced by any means.

Ditch breakers: Used to prevent groundwater flow along the pipeline which causes erosion of the pipe bedding material. The breakers are constructed of sandbags placed under, beside, and over the pipe in the trench.

Double-check valve assembly (DCV): A backflow preventer that prevents back pressure and back siphonage through the use of two spring-loaded check valves. DCVs are larger than dual-check valve backflow preventers and are used for heavy-duty protection. DCVs have test cocks.

Drain valve: Used to drain all or parts of a sprinkler system.

Drop/suspended ceiling: A ceiling that is not part of the structural framework of a building. This ceiling is installed suspended from the floor above, or from the roof, and is commonly used to provide space for services such as cables, recessed lighting, and piping.

Dry pipe sprinkler system: Dry pipe systems are filled with pressurized air or nitrogen. Water flows into the system only when activated by a fire.

Dual check valve backflow preventer (DC): A backflow preventer that uses two spring-loaded check valves to prevent back siphonage and back pressure. DCs are smaller than double-check valve assemblies and are used in residential installations. DCs do not have test cocks.

Ductile: In reference to ductile iron, a type of cast iron in which magnesium is added to the molten gray iron to reduce brittleness.

Eccentric reducing coupling: A reducing coupling that displaces the center line of the smaller of the two joining pipes to one side.

Elbow: A fitting that makes an angle between adjacent pipes. An elbow is always 90 degrees unless another angle is stated on the fitting or in the drawing specifications. Also known as an ell.

Electrolysis: The decomposition of a substance by the passage of electricity through it.

Escutcheon: A protective faceplate for sprinklers.

Excavation: Any man-made cut, cavity, trench, or depression in an earth surface, formed through earth removal.

Extra heavy: When applied to pipe, extra heavy means pipe thicker than standard pipe. When applied to fittings and valves, extra heavy indicates units suitable for higher working pressures.

Flammable: Material that is easily ignited and burns rapidly.

Flange-thread: A fitting that provides for a threaded connection on one end and a flanged connection on the other.

Flexible drop: Flexible fitting developed to span the distance from the branch line to the sprinkler.

Flux: A chemical substance that aids the flow of solder and removes and prevents the formation of oxides on the pieces to be joined by soldering.

FM Global: A testing laboratory and listing agency that certifies new products for service in particular locations and for specific applications.

Galvanic: Refers to direct-current electricity, especially when produced chemically.

Galvanized: Dipped in zinc during the manufacturing process.

Gasket: A flat sheet or ring of rubber or other material used to seal a joint between metal surfaces to prevent gas or liquid from entering or escaping.

Gib: A plate of metal or other material machined to hold other parts in place, to afford a bearing surface, or to provide a means for overcoming looseness.

Hazard: A possible source of danger or potential injury.

Hydration: A chemical process in which water or liquids accumulate and cause a breakdown of surrounding minerals, such as dirt.

Inside diameter (ID): The distance between the inner walls of a pipe. Used as the standard measure for tubing used in heating and plumbing applications.

Intermediate atmospheric vent vacuum breaker: A specialty backflow preventer used on low-flow and small supply lines to prevent back siphonage and back pressure. It is a form of pressure-type vacuum breaker.

Iron pipe size (IPS): The nominal pipe size of iron pipe.

K-factor: A measure of sprinkler orifice size. The larger the orifice, the larger the K-factor.

Labeled: Equipment or material which has attached a label, symbol, or other identifying mark that is acceptable to the Authority Having Jurisdiction.

Liquid Teflon®: A pipe thread compound that is applied to the male threads on a pipe to serve as a lubricant and sealant.

Liquidus: The state of solder when it has melted by heating and is flowing readily.

Listed: Included in a list of products or services published by a recognized listing agency and acceptable to the AHJ for use in fire sprinkler systems for a specified purpose.

Makeup: A term used to describe the amount of pipe that screws into a threaded fitting.

Malleable iron: A metallic fitting material that is generally used for air and water applications and has a classification of either 150 or 300.

Manufactured air gap: An air gap that can be installed on a reduced-pressure zone principle backflow preventer to prevent back pressure and back siphonage.

National Fire Protection Association (NFPA): Organization that publishes codes and standards with the goal of preventing loss of life and property.

National Pipe Thread (NPT): The U.S. standard for pipe threads. NPT has a 1⁄16" taper per inch from back to front.

NCCER Standardized Curricula: Standardized construction education materials produced by NCCER.

***NFPA 13, The Standard for the Installation of Sprinkler Systems*:** The organization (NFPA) that provides the basic reference to the requirements (*NFPA 13*) for installing fire sprinkler systems.

Nipple: A piece of pipe that is threaded on both ends and is less than 12" in length. Any pipe over 12" is referred to as cut pipe.

Office of Apprenticeship (OA): The U.S. Department of Labor office that sets the minimum standards for training programs across the country.

On-the-job-learning (OJL): Job-related learning acquired while working.

Outside diameter (OD): The distance between the outer walls of a pipe. Used as the standard measure for ACR tubing.

Oxidation: The process by which the oxygen in the air combines with metal to produce tarnish and rust

Oxyacetylene: A mixture of oxygen and acetylene that is used to produce an extremely hot flame for cutting, welding, or brazing metal.

Plain-end: Pipe that has no groove on the end and has all imperfections removed.

Potable: Refers to water that is fit for human consumption.

Preaction sprinkler system: Uses automatic sprinklers attached to a piping system. The piping system contains air and may or may not be under pressure. There is a supplemental detection system installed near the sprinklers.

Prepared-end: Pipe whose ends have been prepared by the manufacturer or fabrication shop.

Pressure-type vacuum breaker: A backflow preventer installed in a water line that uses a spring-loaded check valve and a spring-loaded air inlet valve to prevent back siphonage.

Proctor: A test performed in a laboratory that determines moisture content, cohesion factor, and soil gradation of a particular soil.

Protective system: A method of protecting employees from cave-ins, from material that could fall or roll from an excavation face or into an excavation, or from the collapse of adjacent structures. Protective systems include support systems, sloping and benching systems, and shielding systems.

Purge: The release of compressed gas to the atmosphere through some part or parts, such as a hose or pipeline, for the purpose of removing contaminants.

Rack: Device used to support pipe in tiers.

Reduced-pressure zone principle backflow preventer (RPZ): A backflow preventer that uses two spring-loaded check valves plus a hydraulic, spring-loaded pressure differential relief valve to prevent back pressure and back siphonage.

Regulations: Rules and specifications for the sprinkler industry provided by various organizations.

Rolled grooves: Grooves that are formed by pressing the metal around the pipe.

Shield: A structure that is able to withstand the forces imposed on it by a cave-in and thereby protect employees within the structure. Shields can be permanent structures or can be designed to be portable and moved along as work progresses. Additionally, shields can be either pre-manufactured or job-built in accordance with *29 CFR 1926.652 (c)(3)* or *(c)(4)*.

Shore: Timber or other material used as a temporary prop for excavations. May be sloping, vertical, or horizontal.

Shoring: A structure such as a metal hydraulic, mechanical, or timber system that supports the sides of an excavation and is designed to prevent cave-ins.

Shoulder-to-shoulder: A term used to describe the face-to-face dimension of a fitting.

Silt: A rock fragment or mineral in the soil that has a diameter of 0.002 to 0.05 millimeter and is smaller than fine sand and larger than clay.

Skeleton: A condition that occurs when individual timber uprights or individual hydraulic shores are not placed in contact with the adjacent member.

Sloping: A method of protecting employees working in excavations from cave-ins by cutting the excavation walls to the angle of repose of the soil being used.

Solder: An alloy that is melted and used to join metallic surfaces.

Soldering: A method of joining metals with a nonferrous filler metal using heat below 840°F and below the melting point of the base metals being joined.

Solidus: The solid or stable phase of solder.

Spline: A flat, relatively narrow key that is integral to a shaft, produced by milling a longitudinal groove.

Sprinkler: A heat-sensing device used in sprinkler systems to discharge water onto a fire.

SprinklerNet: An online source for current information related to automatic fire sprinklers. www.sprinklernet.org.

Standard dimensional ratio (SDR): The pipe wall thickness in proportion to its outside diameter.

Standpipe: A horizontal or vertical supply main running through buildings and having hose valves on each floor.

Street elbow: An elbow that has one male thread and one female thread.

Subsidence: A depression in the earth that is caused by unbalanced stresses in the soil surrounding an excavation.

Support system: A structure, such as an underpinning, bracing, or shoring, that provides support to an adjacent structure, underground installation, or the walls of an excavation.

Sway bracing: Bracing of sprinkler pipe to minimize pipe movement during an earthquake.

Sweat joint: A soft solder joint, or a joint soldered below 840°F.

Takeout: The distance from one pipe stop in a fitting to the center line of the fitting.

Tee: A fitting that has one side outlet 90 degrees to the run.

Teflon® tape: Tape that is made of Teflon® that is wrapped around the male threads of a pipe before the pipe is screwed into a fitting. Teflon® tape serves as both a lubricant and a sealant.

Test cock: A valve in a backflow preventer that permits the testing of individual pressure zones.

Thermoplastic: Refers to plastic piping that is soft and pliable when heated, but hard when cooled.

Tight sheeting: The use of specially edged timber planks, such as tongue-and-groove planks, that are at least 3" thick. Steel sheet piling or similar construction that resists the lateral pressure of water and prevents loss of backfill is also called tight sheeting.

Transverse: The lateral shifting of pipe.

Trench: A narrow excavation (in relation to its length) made below the surface of the ground. In general, the depth is greater than the width, but the width of a trench (measured at the bottom) is not greater than 15' (4.6 m).

Trench box: A pre-manufactured steel box that provides a shielding system by allowing workers to lay pipe within the box while it is inside the excavation.

Underwriters Laboratories® (UL): A testing laboratory and listing agency that certifies new products for service in particular locations and for specific applications.

Uprights: The vertical members of a trench shoring system placed in contact with the earth and usually positioned so that individual members do not contact each other. Uprights placed so that individual members are closely spaced, in contact with, or interconnected to each other, are often called sheeting.

Wales: Horizontal members of a shoring system placed parallel to the excavation face whose sides bear against the vertical members of the shoring system or the earth.

Water hammer: Shock wave in the piping caused by a sudden change in water flow.

Wet pipe sprinkler system: The simplest, most common type of sprinkler system. Wet pipe systems are filled with water at all times, compared to dry pipe systems which contain no water until the system is activated by a fire.

Wick: To draw solder into a joint.

Wrought fittings: Copper tube fittings that are soldered or brazed as opposed to threaded.